Regional and Local Aspects of Air Quality Management

Advances in Air Pollution

Objectives

The last decade has shown an increase in public and government concern around environmental issues due to air pollution, in particular that generated by man-made processes associated with the lifestyle of modern society. Pollution is widespread throughout the world and elimination of the risks to human health is of paramount importance.

Atmospheric pollution consists of the adverse effects of a variety of contaminants emitted into the atmosphere by natural and anthropogenic processes. These phenomena are complicated because an apparently inert contaminant can be transformed by chemical reaction into an adverse one, during its transport through the atmosphere.

Nowadays, the physical and chemical processes undertaken during an air pollution episode are to a large extent understood. The modelling of such processes has shown a constant and remarkable growth in the last four decades. Also, the development of efficient and economical monitoring devices has been very successful in recent years. Modelling and monitoring studies constitute only an initial activity. The results provide useful information for possible future implementation of emission regulations and control strategies.

The definition of efficient control strategies cannot be achieved without a good and clear knowledge of the complete pollution process, *i.e.* emission, transport and transformation.

Volumes in the series cover areas of current interest or active research in air pollution and include contributions by leaders in the field. Topics for the series consist of: air pollution assessment, management, monitoring and modelling; aerosols and particles; chemical transformation modelling; chemistry of air pollution; comparison of modelling with experiment; data analysis and observation; emission inventories; fluid mechanics for air pollution; global studies; health problems; indoor air pollution; laboratory studies; process studies; regulatory bodies; turbulence modelling at small and meso scales; urban air pollution, and urban and suburban transport emission.

G. Passerini
Universita Politecnica delle Marche
Italy

R. Rosset
Laboratoire d'Aerologie
France

H. Power
University of Nottingham
UK

T. Tirabassi
Institute FISBAT-CNR
Italy

Regional and Local Aspects of Air Quality Management

Editors:

D.M. Elsom & J.W.S. Longhurst

WIT*PRESS* Southampton, Boston

Regional and Local Aspects of Air Quality Management

Series: Advances in Air Pollution

D.M. Elsom
Faculty of Social Sciences and Law, Oxford Brookes University, Oxford, UK.

J.W.S. Longhurst
Faculty of Applied Sciences, University of the West of England, Bristol, UK.

Published by

WIT Press
Ashurst Lodge, Ashurst, Southampton, SO40 7AA, UK
Tel: 44 (0) 238 029 3223; Fax: 44 (0) 238 029 2853
E-Mail: witpress@witpress.com
http://www.witpress.com

For USA, Canada and Mexico

Computational Mechanics Inc
25 Bridge Street, Billerica, MA 01821, USA
Tel: 978 667 5841; Fax: 978 667 7582
E-Mail: infousa@witpress.com
http://www.witpress.com

British Library Cataloguing-in-Publication Data

A Catalogue record for this book is available
from the British Library

ISBN: 1-85312-952-6
ISSN: 1369-5886

Library of Congress Catalog Card Number: 2003116912

The texts of the papers in this volume were set individually by the authors or under their supervision.

Printed in Great Britain by Athenaeum Press Ltd, Gateshead.

Contents

Chapter 9 Air quality management in Greater Tehran Metropolitan area **235**

A. Salahi & Sh. Geranfar

Chapter 10 Regional air quality management in Italy **261**

C. Trozzi

CHAPTER 1

Local and regional aspects of air quality management: an overview

D.M. Elsom[1] & J.W.S. Longhurst[2]
[1]*Faculty of Social Sciences and Law, Oxford Brookes University, Oxford, United Kingdom*
[2]*Air Quality Research Group, Faculty of Applied Sciences, University of the West of England, Bristol, United Kingdom*

1 Introduction

Resolution of local and regional air pollution problems requires the development of an appropriate scientific and decision-making framework within which effective air quality management may be undertaken. The goal of air quality management is to ensure pollution concentrations do not exceed levels (air quality standards) that pose a threat to people or the environment. In areas where such levels are currently exceeded, the aim is to implement a plan that will achieve healthy air quality levels as soon as is practicable and to ensure that healthy air quality is then maintained.

This international edited collection of case studies is drawn from nine countries around the world – Argentina, Australia, Colombia, India, Iran, Italy, Mexico, United Kingdom and United States. They describe the development and implementation of selected aspects of local or regional management frameworks and/or measures adopted in the pursuit of achieving and sustaining acceptable air quality. Identifying those aspects that are particularly successful will help assess the potential for transferability of air quality management approaches between nations. This collection complements an earlier selection of international studies compiled by the editors [1].

2 Air quality management

Air quality management tools and components include health- and environmental-based air quality standards, air quality and meteorological monitoring networks, emissions inventories, numerical prediction models, regulatory instruments, and a range of cost-effective measures to reduce emissions from different types of sources. These constituents form the basic scientific and decision-making framework that is needed for effective air quality management capability in specific localities and regions. The case studies highlight that air quality management demands a substantial financial investment in equipment, facilities and staff training. It needs political commitment from national, regional and local government and support from stakeholders within the community. Formulating and implementing a comprehensive and effective air quality management strategy requires considerable co-operation and co-ordination between national, regional and local institutions representing the various jurisdictions in urban areas such as the environment, transport, planning, energy, industry and health. The business community, academic scientists, labour unions, non-governmental organisation (NGOs) and the public need to be involved in the development and final selection of the mix of policies and measures adopted. Wide consultation is vital to encourage a shared responsibility or partnership approach to improving air quality. Public consultation and education programmes help increase the community's understanding of, and stimulate strong support for, the need for specific air quality management measures. Communicating air quality information to the public is important, especially the provision of short-term air quality alerts to those whose health make them particularly sensitive to high pollution concentrations.

Many of the case studies highlight that motor vehicles are a major cause of the air pollution problems, especially ozone, fine particles and nitrogen dioxide. It is evident that similar broad measures (vehicle-targeted, fuel-targeted and transport management measures) are adopted in the air quality management strategies in each area. They are often supported by a range of economic instruments to increase their effectiveness, shorten the time taken to become effective, or help fund all or part of the cost of the measures. In those case studies where domestic and industrial sources are important for specific pollution problems, the measures implemented to tackle these sources are outlined.

Ideally, when a range of measures to improve air quality are proposed in an air quality management strategy there should be an evaluation of their relative effectiveness for short- and long-term air quality improvement. A good example where this has been attempted is in the Perth air quality management plan described by Leanne Graham and Howard Bridgman. The criteria used to provide a ranking of the relative effectiveness of the measures include the direct and indirect emission reduction potential, the timing of effectiveness (time frames for an action to have an effect and to be fully implemented), cost effectiveness, equity (identification of the stakeholders who would bear the costs of the action and the individuals or groups within the community who may be disadvantaged), technical feasibility, enforceability, public acceptance and public health benefits.

This international collection of case studies provides insights into a wide range of aspects of local and regional air quality management but for convenience the following overview is grouped into sections.

3 Air quality management in Australia and Great Britain

Leanne Graham and Howard Bridgman review the development since 1998 of the National Environment Protection Measures for Ambient Air Quality (Air NEPM) in Australia. Air NEPM set legally binding air quality standards for six key urban pollutants to be achieved by 2008. State and territory governments are required to establish nationally accredited ambient air quality monitoring networks and to report annually on compliance with these standards. Although there has been a steady decrease in the maximum 1-h ozone concentrations since 1979, there is no downward trend in the maximum 4-h ozone value, and the number of days exceeding the Air NEPM ozone standard has increased. Nitrogen dioxide and carbon monoxide concentrations show a decrease. PM_{10} concentrations, monitored since 1995, show a slight downward trend, with occasional years when very high levels may occur. Some high concentration events are due to bush fires or controlled burning of vegetation to reduce the bush fire hazard near urban centres. Higher peaks in autumn and winter indicate the effects of wood smoke. Generally, the pollutants of most concern in urban areas are photochemical smog and fine particles, with motor vehicles being the most significant source. To achieve Air NEPM goals by 2008 in the major urban areas, State Governments have developed action plans, improvement plans and strategies (referred to collectively as air quality management plans). The air quality management plans combine the best available science with consultative strategy development, monitoring and review. They incorporate community involvement in decision making and a partnership approach to implementation.

Graham and Bridgman evaluate the air quality management plans for four major urban areas: Brisbane, Melbourne, Perth and Sydney. Each plan proposes a range of cost-effective actions and targets similar priorities. Reducing transport emissions is the most important strategic aspect of all four plans. The plans aim to integrate air quality goals with urban and transport planning and to improve the emissions performance of new and in-service vehicles (new vehicle emission standards, new fuel quality standards, alternative fuels). The introduction of an inspection programme for in-service vehicles will improve vehicle maintenance and so improve fuel economy and lower vehicle emissions. On a fleet-wide basis, it is estimated that regular servicing of vehicles could achieve exhaust emission reductions of 16% for VOCs, 25% for carbon monoxide and 9% for nitrogen oxides. Servicing would also reduce evaporative VOC emissions significantly, by, for example, ensuring that fuel filler caps are properly sealed. Regulations to lower summertime petrol volatility will help too. Regular servicing has significant fuel savings benefits of about 5% for the most poorly maintained cars. Vehicle owners who provide evidence that their vehicle has been serviced within the past 12 months would be charged less for registration than those who do not. Sydney favours the introduction of 400 public transport buses powered by CNG while Melbourne advocates the introduction of zero-emission vehicles. Shaping urban development into a more sustainable form through growth occurring in strategic regional centres is intended to lower levels of car dependency. Perth's TravelSmart programme encourages the use of alternate transport modes (public transport, cycling, walking) through motivation, information and empowerment. A pilot scheme in south Perth achieved a significant reduction in car use on an average day and a decline in the number of car trips per day. For example, the overall result was a 14% decline in vehicle kilometres travelled and a 25% increase in public transport journeys (March–June 1999 compared to March–June 2000). The air quality management plans target non-transport emission sources including new regulations being proposed for solid fuel burning for domestic heating (with improved design for domestic wood heaters), agriculture, and forest and bush fire management that are all intended to reduce particle pollution. Economic incentives will target reductions in major industrial emissions and there is financial help for small businesses to adopt cleaner production programmes.

Jim Longhurst, Clare Beattie, Tim Chatterton, Derek Elsom, Nurul Leksomo and Nicky Woodfield evaluate the British air quality management framework (Air Quality Strategy) designed to identify and improve areas of poor air quality. Local authorities have the responsibility to review and assess local air quality against human–health–based air quality objectives. Each air quality objective has a date by which the Government, after taking into account costs, benefits and technical feasibility, judge it should be attained. The objectives apply in areas where the public may be exposed for the averaging time of the relevant objective such as building facades, public open spaces, pavements and gardens. This means an objective will not be exceeded in a location where there is no relevant public exposure. The Government supports local authorities in implementing the Air Quality Strategy through the provision of detailed guidance. The Government has also developed high–quality national monitoring networks, compiled emissions inventories at various geographical scales, provided training for local authority personnel and offered additional financial resources to assist local authorities in purchasing equipment and technical resources. A key support element has been the development of web, telephone and e-mail help desks.

An area assessed as likely to experience air quality exceeding an air quality objective has to be declared an Air Quality Management Area (AQMA). These are local areas where national actions on their own will not secure the air quality objectives. By mid-2003, over 30% of local authorities had declared, or intended to declare, an AQMA. Exceedances of nitrogen dioxide (particularly, the annual mean objective) and PM_{10} objectives are the cause of most AQMAs being declared and over 95% of the designated AQMAs are a result of traffic emissions. These are being declared alongside motorways and trunk roads, across road networks, as well as along individual road sections where low traffic speeds, high percentages of heavy goods vehicles and canyon-like street topographies lead to poor air dispersion. An unexpected outcome has been the declaration of AQMAs in smaller towns where traffic flows have been modest but an exposed population has been identified. A small number of AQMA declarations are due to sulfur dioxide emissions from industrial point sources, or combinations of emissions from a number of such point sources, or other sources such as shipping.

For each AQMA a local authority has to develop an Action Plan detailing both the measures to be taken and the time-scale for implementation in order to improve air quality. The original Air Quality Strategy (1997) placed a statutory duty on local authorities to secure each air quality objective. However, a revision of the strategy in 2000 shifted the responsibility for securing air quality objectives to central government and the devolved administrations, whereas local authorities were given a duty to work towards securing the objectives. Local authorities have been granted new powers to assist in delivering improved air quality, for example, vehicle emissions testing. Since air quality is determined by emissions from a variety of sources over varying spatial and temporal scales, an Action Plan needs to integrate solutions into existing policy packages such as transport plans, land use plans, economic development plans and sustainable development plans. Currently, local authorities are formulating and evaluating the likely effectiveness of their Action Plans. Such plans usually involve a range of measures given that it is unlikely that a single measure would be effective. The Air Quality Strategy requires that the proposed measures are evaluated for the air quality improvements likely to be achieved, time taken to be implemented, cost effectiveness, non-air quality impacts (for example, noise and socio-economic impacts), and public and political perceptions of the various measures. Local authorities are also expected to consider how to evaluate the effectiveness of an Action Plan following its implementation. This may involve long-term air quality monitoring, the use of health indicators and, in the case of traffic management measures, monitoring of changes in traffic flows and journey speeds.

The air quality management framework in Great Britain is a continuous process and is subject to a rolling review to reflect developments in European Union legislation, technological and scientific advances, improved air pollution modelling techniques and an increasingly better understanding of the socio-economic issues involved. Following the review and assessments that began in 1999, the first round of the process was completed in 2001 and approximately 130 of 407 local authorities declared, or were expected to declare, one or more AQMAs within their area of jurisdiction. The next round of reviews and assessments are due to be completed by the end of 2003, except in London where 2004 is the deadline.

4 The importance of modelling: case study of Mendoza, Argentina

Enrique Puliafito and C. Puliafito describe air quality management in Mendoza, Argentina, a city of 900,000 inhabitants. In particular, they highlight the importance of developing a GIS-based model that enables alternative air quality improvement measures to be explored so enabling an effective air pollution control strategy to be implemented. Such a model is also employed to study the potential impact on air quality of new industrial sources and the expected increase in number of motor vehicles.

Puliafito and Puliafito explain the development and calibration of the GIS-based numerical dispersion model and the type of input data provided by the air quality monitoring network and emissions inventories. They describe the distinctive seasonal, weekly and diurnal pollution patterns in the city that highlight the influence of local meteorological conditions. In Mendoza, the Zonda winds, similar to the Föhn or Chinook winds, prevail in the higher layers mostly during the winter months (May to October). During strong frosts these warm winds lead to higher pollution concentrations due to the creation of strong thermal inversion layers. Valley-mountain circulation (upslope flow during the day, downslope flow during the night) also affect pollution levels by affecting the mixing heights and inversion layers.

Various air quality management strategies are explored using the model. For example, the city authorities are planning a major improvement to the public transportation system involving new electric trains and trams. This will replace diesel buses that currently provide 35% of daily passenger trips (cars account for another 45% of trips). By making assumptions about the expected population and traffic growth for 2010, the model enables a comparison to be made of the air quality impact with and without the new transport system. Various assumptions about the number of car users who may switch to the new system are also being explored. The estimated impact in 2010 is that the new public transport system would lower carbon monoxide and hydrocarbons emissions significantly whilst maintaining current levels of nitrogen dioxide and total suspended particulate even in the face of increased population and number of motor vehicles. In contrast, a significant increase in emissions would occur if no improvements in the public transport system were implemented.

The value of having such an air pollution model was further demonstrated in relation to a high pollution event lasting several hours that occurred on 18 May 1995 in the city and during which hundreds of people, especially children, were hospitalised. Since no company acknowledged responsibility the model was employed to investigate possible causes and sources. An epidemiological survey established that the symptoms of the affected persons (respiratory problems, dyspnea, headache, migraine, nausea, vomiting), together with the strong smell described by many people, indicated the cause was a combination of sulfur oxides and hydrogen sulfide in the presence of particulate matter. The ambient concentrations of the gases were estimated to be between 0.05 and 1.0 ppm (unfortunately there were no sulfur dioxide or hydrogen sulfide measurements available that day). Meteorological conditions and the intensity of the health effects indicated that only a few industrial sources could have been responsible for the emission. Simulations were explored that pointed to the release of a sulfur dioxide and hydrogen sulfur pulse of at least 100 ppm during 4 hours from the refinery located in the south of the city. Several years later, during the trial process, personnel of the refinery admitted that gases had been vented through a secondary torch tower without having been burned and that the amount of gases and the timing matched the results of the model. As a direct consequence of this event, improved pollution control equipment was installed in the refinery.

5 Tackling motor vehicle emissions: the USA experience

Richard Baldauf *et al.* review the management of motor vehicle emissions in the United States and give particular attention to the engineering control methods used to reduce emissions from current technology gasoline and diesel-powered vehicles. Since the introduction and widespread use of electronic engine management systems, switching oxygen sensors, and three-way catalyst systems during the mid-1980s, control of emissions from spark-ignition, gasoline-powered engines has focused primarily on creating the conditions of temperature and chemistry in the exhaust that are most conducive to the catalytic reactivity of the exhaust emission control system. Conversely, diesel emission control has focused almost entirely on control of combustion and other engine design parameters. These two differing approaches to emission control are beginning to converge as post-combustion catalytic exhaust control systems are developed for diesel-emission control in order to meet US regulations.

Baldauf *et al.* summarise the adverse effects of vehicle exhaust and evaporative emission compounds (criteria pollutants and air toxics) on humans and/or laboratory animals as well as ecological effects (global warming and atmospheric deposition). The role of the federal, state and local agencies in managing the impacts on human health and the environment is explored in the context of the major legislative actions, notably the Clean Air Act, and the Transportation Equity Act for the 21st Century. Voluntary motor vehicle emission reduction programmes involving small and large businesses, trade associations and the public are outlined. These are technology based (retrofitting vehicles with control technologies, purchasing alternatively fuel vehicles, changing to low-rolling resistance tyres) or activity based (scrapping vehicles, Ozone Action Days, ride sharing, reduced vehicle idling).

It is important that other countries take note of the lessons learned from the decades of US attempts to reduce motor vehicle emissions in the face of rapidly rising number and use of vehicles. Baldauf *et al.* highlight that the US tries to encourage the development of partnerships with developing countries to assist in adoption of best practices for transportation planning. It also encourages technology transfer such as the adaptation and demonstration of existing and proven technologies as, for example, in the clean up of two-stroke motorcycles and mopeds, and the support of urban bus retrofit strategies and alternative fuel conversions. The clear benefits arising from the phasing out of leaded gasoline in order to facilitate exhaust emission control is another transferable lesson.

6 The challenges facing Bogota, Delhi, Mexico City and Tehran

Alexander Valencia and Jose Huertas review the air quality management programmes implemented by the authorities in Bogota, the capital of Colombia. Ozone and PM_{10} are the main problems in this city of 7.5 million inhabitants mainly due to motor vehicles and industrial emissions (food production). A network of 14 automatic monitoring stations ensures that the city has the technological infrastructure to assess progress towards compliance with air quality standards, to activate pollution alert procedures, and to develop and validate numerical dispersion models used in the planning, development and evaluation of a range of control strategies. Detailed and regularly updated emissions inventories for the city's stationary and mobile sources provide essential air quality management tools too. A range of measures have been implemented in recent years to improve air quality but as yet there is no clear trend in annual sulfur dioxide, nitrogen dioxide and PM_{10} concentrations since 1998. This is mainly because of the relatively short period for the measures to have been effective and because of annual variability in atmospheric conditions and the effects of economic recession.

Measures intended to reduce motor vehicle emissions include the introduction of vehicle emission standards and the requirement that owners must submit their vehicle to annual certification. More than 650,000 vehicles are registered in the city, of which 79,000 are used for public transportation. In addition, 230,000 vehicles registered outside the city circulate daily. Private cars are the predominant source of emissions in the city although buses contribute significant amounts too. The use of natural gas for vehicles is considered to have considerable potential to improve air quality in the city. Natural gas was introduced only recently for vehicles although it has been widely adopted for residential use and increasingly for some industrial sources. Its penetration into the vehicle fuel market has been limited due to the effects of the city's high altitude (2,650 metres) on engine performance, the lack of fuel stations (only 8 stations across the entire city) and the small financial benefits. Only 12,000 vehicles, mainly taxis, small size buses and fleets of vehicles used to distribute goods in the city, have been converted to natural gas to date. Nevertheless, the authorities are introducing measures to stimulate its use in public service vehicles by designating exclusive routes in the city centre for such vehicles and by forcing public service fleet owners (passengers transportation, garbage collection) to incorporate natural gas vehicles. Government vehicles will also be converted. Offering reduced taxes when purchasing natural gas vehicles is planned. Congestion increases vehicle emissions so the authorities have restricted the circulation of cars during four peak hours a day, two days a week, and public transport vehicles during 16 hours a day, once a week, according to the last digit of the license plate. Cars using natural gas may be exempted from this restriction in the future. In the absence of a metro or subway, an extensive transportation system (Transmilenio) has been developed recently (42 km by 2003) using exclusive bus lanes for large vehicles (160 passengers) with feeder routes that take passengers from different places to the main lines. As each new bus is introduced, three or more older vehicles used for public transportation are scrapped. Also, 270 km of exclusive bicycle routes have been constructed.

Maximum permissible emissions levels of pollutants are imposed on Bogota's 4,500 industrial sources according to the type of fuel used. Generally, regulations aim to encourage companies to use cleaner fuels instead of investing in technology to control emissions although Valencia and Huertas argue that this strategy should be more aggressive. The predominant industrial pollution sources in Bogota are associated with the manufacture of food as well as service industries. Although 58% of industrial sources use natural gas and LPG, it is the heavy liquid fuels, such as fuel oil, and coal that give rise to the greatest emissions of sulfur oxides, nitrogen oxides and particulate matter. It is likely this situation will continue until there are greater economic incentives or more stringent environmental legislation to use cleaner fuel oils or natural gas. The maximum sulfur content for fuel oil has been successively lowered but it is the recent initiatives involving natural gas that may offer the greatest potential to reduce emissions from industrial sources. These measures include exempting stationary sources that use natural gas from having to get emission permits, maintaining the fuel price policy that guarantees the use of natural gas over other fuels, and a proposal to introduce a fuel tax to penalise polluting companies and to use this income to subsidise the conversion to natural gas of those industries that use coal in their process. Generally, it is programmes such as fuel substitution by less polluting fuels and the reorganisation of public transportation that may help achieve and sustain major improvements in air quality in the city.

Mukesh Khare and Arun Kansal outline the air quality problems and management strategies in Delhi, India, which is amongst the most densely populated and polluted cities in the world. Its current population of 14 million inhabitants is growing rapidly due to large and uncontrolled in-migration from the surrounding states. The number of industrial units in the city has grown from 26,000 in 1971 to 129,000 in 1998. The number of registered vehicles increased from 0.2 million in 1970 through 1.8 million in 1990 to 2.8 million by 2000, with two-wheeled vehicles (scooters, motorcycles, mopeds) comprising 67% of the total. Generally, vehicle emissions are made worse by two-stroke engines, old vehicles, congested traffic, poor quality roads and outdated traffic management systems. Transport accounts for 67% of total emissions, manufacturing industries and thermal power plants for 25%, and domestic fuel combustion for 8%. The daily air pollution load in Delhi increased from around 15,000 tonnes in 1991 to 30,000 tonnes in 1996.

A network of daily and hourly monitors provide air quality information for industrial, residential and traffic intersection locations. Despite a range of measures to curb emissions having been implemented in recent years, air quality has shown little improvement or has even worsened because of the rising population and associated increase of emissions from the growing number of emission sources. Nitrogen dioxide concentrations increased during the period from 1986 to 1999. Suspended particulate matter (SPM) and carbon monoxide levels remain high. High levels of PAH may be due to the high numbers of two- and three-wheeled vehicles and the significant use of coal and kerosene as domestic fuels. Annual lead concentrations have shown a decrease in recent years due to the introduction of unleaded petrol for the use of vehicles with catalytic converters in new four-wheeled vehicles (only unleaded petrol has been available in Delhi since 1998) and because of vehicles being converted to run on CNG. However, benzene levels increased at the same time as unleaded petrol was introduced. No clear trend is shown in sulfur dioxide levels for the period 1986–99 reflecting that the effects of various control measures (specifying industrial emission standards and vehicle fuels with a lower sulfur content) have been offset by the growth in number of emission sources. There is also the possibility that high sulfur dioxide conversion to sulfates is taking place in the presence of high levels of soot particles.

Measures to curb vehicle emissions in recent years include all 7,400 buses and over 4,000 minibuses being powered by CNG engines and the conversion of the entire fleets of taxis (15,000) and three-wheelers (45,000) to CNG–powered engines. To reduce reliance on road transport, the Delhi Metro Rail project has been implemented. It comprises a 200–km network of underground, elevated and surface corridors, with phase 1 (62.5 km) expected to be completed by 2005. Although India is committed to introducing Euro IV standards for all new vehicles by 2005 the greatest challenge is to control emissions from old vehicles. Replacement of older vehicles with newer ones will need fiscal incentives and more rigorous enforcement of bans on vehicles of a specified age. In addition, travel demand management is needed to complement vehicle- and fuel-targeted (technological) measures. Khare and Kansal point out that travel demand management measures will require a policy framework encompassing regulatory, pricing and taxation mechanisms, and need to be supported by effective enforcement so as to encourage the use of clean vehicles and fuels and to modify travel behaviour and transport demand.

Since the 1980s, cleaner fuels and improved cooking appliances have been promoted throughout India to reduce domestic emissions but have been met with mixed success, often due to the lack of involvement of local people, especially women. In Delhi, the use of LPG has increased greatly at the expense of kerosene. Such a shift benefits not only ambient air quality but indoor pollution exposure too. Increased subsidies will need to be introduced to accelerate such air quality improvements.

The city has grown haphazardly with residential and commercial zones mixed together. Despite the Supreme Court of India in 1997 directing a number of polluting industries in Delhi to close down or relocate to designated zones, a survey in December 2000 showed that 71,000 small-scale manufacturing units producing food, metal, electroplating, textile, paper, paper products and repair services were still operating in residential areas. Such units are of low combustion energy efficiency and use low-grade fuel, such as high sulfur content furnace oil. Although emissions data are available for large industrial sources, such as the seven coal- and gas-fired power plants in and around the city, the lack of information about the number and type of small-scale industrial and household units limits the accuracy of the emissions inventories. This, in turn, hinders attempts to assess the likely effectiveness of specific control policies. Delhi faces difficult challenges to ensure its air quality management strategies will be successful in the near future.

Ernesto Jauregui outlines the air quality management challenges facing Mexico City. An extensive automatic monitoring network, emissions inventories and numerical dispersion models have provided the tools for the development of the current air quality management plan (2002–12) whose primary objective is protection of public health. Eighty-nine measures are being implemented to achieve significant reductions in levels of ozone and PM_{10} and to ensure concentrations of carbon monoxide, sulfur dioxide, lead and nitrogen dioxide meet health-based standards. Emissions from motor vehicles, providing transport for the 17.7 million population of the metropolis, are the major cause of the pollution problems. Emissions from old and poorly maintained microbuses, cars and lorries in and around congested central districts are a particular problem. Microbuses provide over half of the trips made by the public in the metropolis. Around 36% of the 22,834 microbuses circulating in the Federal District alone are pre-1992 and half of the 103,694 taxis are older than 10 years. These older taxis ought to have been scrapped as part of a current renewal programme for taxis. The programme also promotes organised taxi stands to replace the current situation whereby taxis cruise the streets looking for passengers and generating additional pollution. The problems caused by vehicle and industrial emissions are exacerbated by the geographical setting of the metropolis in an elevated inland valley (lower oxygen levels lead to incomplete fuel combustion) that is characterised by poor atmospheric ventilation and high insolation.

A series of air quality management programmes, begun in 1979, have made some progress in improving air quality with dramatic improvements in lead, sulfur dioxide and carbon monoxide since the early 1990s. Although ozone and PM_{10} remain severe problems, there have been some improvements. The number of days of exceedances of the daily maximum ozone standard of 0.11 ppm has declined from 319 days in 1995 to 273 days in 2001 – though this still represents 75% of all days in the year. The number of days of ozone alerts (phase 1), initiated when 1-h concentration reaches 0.257 ppm, fell dramatically from 113 days in 1991 to only 2 in 2001. The number of days when the 24-h PM_{10} standard of 150 $\mu g/m^3$ was exceeded fell from 262 days in 1990 to only 30 days in 2001 although wind–blown dust and forest fires cause some annual variability in this downward trend. Generally, the progress achieved in Mexico City suggests that even the largest cities in the developing world can take some significant steps towards the long-term goal of achieving healthy air quality for its population.

Ali Salahi and Shirin Geranfar highlight that the poor air quality experienced in Tehran, Iran, is mainly due to the 1.4 million motor vehicles, including half a million motorcycles, operating on congested roads. Iran's low fuel prices offer little incentive to introduce fuel-efficient technologies. Many vehicles have older engine designs (many are more than 15 years old), consuming 50 per cent more fuel than more up-to-date technologies, and they are not well maintained. Public transportation in this city of 10 million inhabitants is poor, increasing reliance on cars. Airborne particles count for 75% of Tehran's pollution, with passenger cars responsible for a daily total of 10,200 tonnes and public transportation vehicles another 6,100 tonnes. During winter months the presence of a temperature inversion in this high-altitude city, hemmed by mountains to the north, leads to pollution alerts (e.g. December 1999 when children and elderly with heart or lung disease were advised to stay indoors, and primary schools were closed). Acid deposition is also considered a major problem and Salahi and Geranfar provide details of recent wet deposition measurements in the city.

A joint project funded by the Tehran Municipality and the World Bank in 1993–97 provided an opportunity to explore measures that would reduce vehicle emissions to not only improve air quality in the city but reduce greenhouse gases that contribute to global warming. Subsequently, a 15-year air quality management programme began in 2000. Measures to reduce emissions include improved public transport (electric buses, subway and metro), introduction of unleaded petrol, traffic restrictions in the city centre (weekday restrictions for cars based on special passes; restrictions during pollution alerts based on odd and even license plate numbers), relocation of industrial sources out of the city, use of solar energy, and greater use of natural gas in vehicles and homes (the current use of open fires for indoor cooking and heating exposes householders to high levels of indoor suspended particles).

7 Regional air quality management: Italy and United Kingdom

Carlo Trozzi reviews the development phase of Regional Air Quality Management Plans, first introduced in Italy in 1997. Each regional plan is intended to identify those areas that should be given priority for the introduction of remedial action to improve air quality. The plan also assesses the likely impact of a range of alternative measures to enable the most effective measures to be selected for implementation. Plans require the compilation of an emissions inventory, collection of air quality and meteorological data, classification of territory, mortality data and a model to simulate the impacts with and without various measures to reduce emissions. Trozzi explains in detail how each of the components of the plan are developed and compiled. Particular attention is given to the compilation of emissions inventories for point, area, road transport (road, waterborne, air), forest and forest fires. He also explains the spatial disaggregation of the data (from the municipal level to the model's square grid) and its temporal disaggregation (from annual to hourly, daily and monthly values). Projections of emissions of air pollutants are carried out for 5 to 15 years ahead. This enables the simulation model to compare the likely effects of alternative technological, economic and energy scenarios with the 'do nothing' scenario.

One of the most important problems in air quality management is to locate the areas where remedial action should be made a priority. Trozzi describes an innovative methodology, based on multivariate statistical analysis techniques and particularly cluster analysis, to define territories where remedial action should be a priority. The classification is based on environmental sensibility (vulnerability) of areas and pollutant loads. Classification based on vulnerability uses indicators of resident population, forests and other vegetation areas, beds available for tourists, and biotypes (conservation status). A combination of classification based on environmental vulnerability and pollutant loads produces a regional classification for priority of remediation actions. Currently, the development phase for the system of regional air quality planning in Italy has now been completed and the plans now enter the implementation phase.

Andrew Deacon describes the operation and work of the regional air quality management group in Sussex, United Kingdom. This group comprises 12 local authorities, two County Councils (East and West Sussex), the Environment Agency, two Health Authorities and two universities. The Environment Act 1995 introduced the legislative framework for local air quality management in the United Kingdom. An Air Quality Strategy was formulated that requires local authorities to review and assess air quality and to take action where the assessments show that prescribed objectives for specified pollutants would not be met by target dates set by the Government. Such areas have to be declared Air Quality Management Areas (AQMAs). The Sussex Air Quality Steering Group (SAQSG) supported local authorities in their review and assessments by preparing a GIS-based atmospheric emissions inventory for the region and by undertaking air quality modelling work. This highlighted the impact of the major transport links (e.g. the M23 with up to 68,000 vehicles per day) and international Gatwick Airport on air quality in the region. Local authorities found that no AQMAs needed to be declared in Sussex. However, there were a number of areas close to exceeding air quality objectives for nitrogen dioxide and PM_{10} and it was deemed important to ensure they do not become exceedance areas in the future. Also there were concerns about pollutants, notably ozone, dioxins and heavy metals, not covered by the air quality management regime. Consequently, the SAQSG decided to develop a non-statutory air quality strategy. This strategy would integrate air quality considerations into a number of local authority decision-making processes such as land use planning and traffic management. Apart from its role within Sussex, the SAQSG simplifies communication with similar collaborative groups in southeast England (Kent, Hampshire, Surrey) and the central coastal belt of France. Such communication will become increasingly important in dealing with transboundary air pollution issues, especially ozone and PM_{10}.

8 Conclusion

Valuable insights can be gained by examining the air quality management experiences of localities and regions throughout the world. This is particularly important when it is recognised that there is no universal air quality management strategy or single set of measures to improve air quality that will guarantee the delivery and maintenance of healthy air quality. Each locality and region is unique in terms of its air pollution problems, spatial and temporal patterns of emission sources, and economic, social, cultural and physical (topography, climate) characteristics. Nevertheless, this international collection of case studies helps inform the reader of the range of air pollution problems encountered and the more successful air quality management approaches adopted around the world. This should help in the formulation, evaluation, implementation and fine-tuning of the most appropriate framework, strategy and mix of measures for other localities and regions.

Reference

[1] Longhurst, J.W.S., Elsom, D.M. & Power, H. (eds.), *Air Quality Management*, WIT Press: Southampton and Boston, 2000.

CHAPTER 2

Air quality management in Australia

L. Graham & H. Bridgman
School of Environmental and Life Sciences, University of Newcastle, Australia

Abstract

In Australia, the air pollutants of most concern in major urban areas are photochemical smog and fine particles. The major urban airsheds, located on coastal plains backed by low mountain ranges, re-circulate air pollutants within local and regional air flows. The most significant source of air pollution identified by air emissions inventories is motor vehicle emissions. Oxides of nitrogen and volatile organic compounds, from motor vehicle exhaust, combine with sunlight, high temperatures, clear skies and light winds to create high ozone levels and photochemical smog, particularly in summer. Health studies link photochemical smog and particle pollution events with increased mortality in the major urban centers.

National environment protection measures (NEPMs), set in partnership by Commonwealth, State and Territory, and local governments, provide national goals and protocols for monitoring, managing and reporting on air quality. In 1998, following a national inquiry into air pollution problems, the first NEPM established a national pollution inventory. This pollutant Internet database combines annual reporting by industries and governments with community access to current information on the types and amounts of chemicals emitted into the environment. NEPMs set national goals, to be achieved by 2008, for major urban air pollutants: photochemical oxidants as ozone, nitrogen dioxide, sulfur dioxide, carbon monoxide and particles including lead and fine particles. A NEMP for air toxics is currently in development.

New air quality management plans for the major urban centers merge the best available science with consultative strategy development, monitoring and review. These plans incorporate community involvement in decision making and a partnership approach to implementation. A range of cost-effective actions address common priorities. Integration of urban and transport planning, new vehicle emission standards, new fuel quality standards, and alternative fuels, target the reduction of transport-related emissions. New regulations on solid-fuel burning for domestic heating, agriculture, and forest and bushfire management aim to reduce particle pollution. Economic incentives aim to reduce major industrial emissions.

1 Introduction

Australian urban air quality management for the 21st century has moved towards nationally consistent objectives employing a wide range of actions to improve air quality. Major components include legislation, policies, management plans, and risk assessment with the explicit aim of protecting human health and well-being from the harmful effects of air pollution. The Australian Commonwealth, State and Territory Governments co-ordinate a range of networks for integrating leading edge research and science with policy development, implementation, reporting and reassessment to improve air quality.

This review follows the steps leading to air quality management presented in Table 1, as defined in Bridgman [1]. The discussion is illustrated by examples from four major urban areas: Sydney, Brisbane, Melbourne, and Perth (Figure 1). Section 1 outlines the steps toward nationally consistent goals for air quality in Australia. Section 2 discusses the aspects of air pollution science that assist in defining the overall problem and clarifying the issues to be managed: sources and impacts of the major air pollutants; quantification of emissions; meteorological influences and computer modelling of pollution dispersion. Section 3 focuses on urban air quality management plans for Australia's major urban centres.

2 Steps toward a nationally consistent approach to air quality management in Australia

In 1990, an intergovernmental agreement between the Australian Commonwealth, State, and local governments set the scene for developing a national framework for air pollution management (CoA [2]). Following this agreement, in 1994, the National Environment Protection Council (NEPC) was established to set statutory frameworks of agreed national objectives for environmental protection from air, water or soil pollution. These frameworks are set as National Environment Protection Measures (NEPMs) specifying goals and protocols for monitoring and reporting. Implementation of NEPMs is the responsibility of all levels of government (NEPC Act [3]). During 1995–98, the Australian Commonwealth Government continued the development of a nationally consistent approach to air quality management, commissioning an independent inquiry into urban air pollution, a review of emission standards, and the creation of national goals for ambient air quality.

In 1997, the independent inquiry, *Urban Air Pollution in Australia* (AATSE [4]), examined present and likely future trends, sources and emissions and management options for seven major urban air pollutants: airborne lead (Pb), oxides of sulfur (SO_x), oxides of nitrogen (NO_x), carbon monoxide (CO), particle matter less than 10 microns (μm) in diameter (PM_{10}) and photochemical oxidants, such as ozone (O_3), including ozone precursors (volatile organic hydrocarbons, VOCs).

Table 1: Framework leading to active strategies for air pollution management (after Bridgman [1])

a. Air Pollution Science			
Pollutants *(CO, NOx, VOC, SO_2, PM10, 03, standards)*	Meteorology *(wind, stability, temperature, moisture)*	Sources *(industry, traffic, residential, off-road, natural)*	Emissions *(amounts, dispersion, cumulative impacts)*
Collation of Background Material; Computer Modelling (if needed)			
Identification of Potential Impacts			
Health *(respiratory, asthma, illness, acidity impacts)*	Social *(wellbeing, cleanliness, psychological)*	Economic *(crop damage, loss of work days, poor image)*	Environmental *(flora and fauna, diversity, aesthetics)*
Define the Overall Problem			
b. Air Pollution Management			
Develop Possible Management Options			
Tools *(communicate, compromise, technology, education, economic, etc.)*	Policy *(goals, options, consultation, integration)*	Level *(local, regional, state, national, global)*	Framework *(statutory, power, leadership, resources)*
Construct Airshed Management Plan			
Incorporate Airshed Action Strategies			
Implement *(political, procedural, cooperative)*	Monitor *(measurement, representative, research)*	Evaluate *(data analysis, interpretation, comparison, effectiveness)*	Feedback *(community, industrial, responsible agencies)*
Assess Level of Success and Revisit the Structure on a Regular Basis			

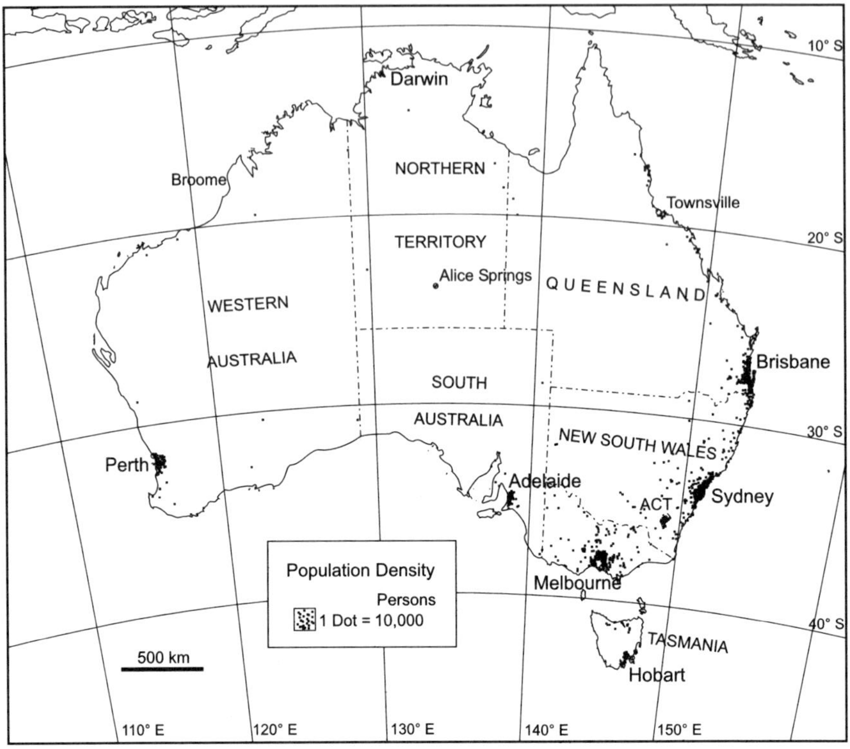

Figure 1: Population distribution in Australia.

The report of the independent inquiry was based on submissions, and a number of previous air quality studies completed in the early to mid-1990s (Bridgman *et al.* [5]; Boyle *et al.* [6]; MAQS [7]). The inquiry identified two major considerations for policy development. Firstly, to achieve clean air requires integration of transport planning, infrastructure development and environmental management, in combination with informed choices and responsible actions of the whole the community – "you and a million others can make a difference" (AATSE [4]). Secondly, reducing emissions to economically and behaviourally practicable levels will continue to result in occasional exceedences of existing and proposed standards in Australian urban centres, due to the effects of coastal conditions coupled with intense sunlight. Based on these considerations the Inquiry recommended a national goal to minimise exceedences of standards in the intermediate and longer term. This recommendation was fulfilled by the 1998 NEPM for Ambient Air Quality (NEPMa [8]), outlined in Section 3.2 (referred to as the Air NEPM).

Table 2: Major air pollutant sources and effects (after NSWSOEa [9])

Pollutant	Sources	Human health and environmental effects
O_3	Secondary pollutant indicating the presence of photochemical smog, formed from reaction of NO, and VOCs in presence of sunlight.	Irritation of eyes and air passages, decreased lung function, lung inflammation, aggravation of asthmatic conditions, increased susceptibility to infection. Indicator of photochemical smog. Damage to vegetation, reduced plant growth and crop yields, increased susceptibility to pests. Damage to fabrics, rubber and construction materials. Long-term low-level exposure is of concern
Particles - Coarse	Grinding, quarrying, mining, diesel fuel vehicles, sea salt, soil, pollens, spores, biomass burning.	Nuisance fallout on materials and vegetation.
- Fine	Motor vehicles, domestic wood combustion, industry and power	Increased rates of respiratory illnesses and symptoms, decreased lung function, excess mortality from heart and lung disease, aggravation of asthmatic conditions.
NO_x	Approximately 10% emitted as NO_2 in the presence of sunlight. Motor vehicles, power stations, bushfires, lightning, biological activity in soil.	Decreased lung function, lung inflammation, aggravation of asthmatic conditions, increased susceptibility to respiratory infections. Leads to the formation of photochemical smog pollutants, i.e. ozone and NO_2. Effects on vegetation, damage to leaf surface, reduced plant growth. Limited research has been done on Australian native vegetation.
VOCs	Motor vehicles and combustion processes.	Eye irritant, coughing & wheezing. Precursor to the formation of ozone in the presence of NOx and sunlight.
CO	Motor vehicles, domestic wood burning, power generation.	Reduces the capacity of blood to carry oxygen; decreased exercise capacity and aggravation of cardiovascular disease.
SO_2	Power generation and metal smelting. Bacteria, volcanic activity.	Decreased lung function; lung inflammation, wheezing, aggravated asthma. Plant damage and growth inhibition, acidification of soils and lakes changes the availability of minerals to plants. Corrosion of construction materials.
Pb	Lead in motor vehicle fuel, metal smelting, and incineration. Erosion of lead ore bodies	Particularly affects children. Altered neurobiological function in children i.e. impaired intellectual development, increased human blood pressure and impaired renal function. Accumulates in soil and food chain.

3 Air pollution science

Air pollution science investigates the interaction between air pollutants and meteorology, inquiring into the characteristics, sources and impacts of air pollutants, air quality standards, quantification of air emissions, meteorological influences and computer modelling of air pollution dispersion. Table 2 (after NSWSOEa [9]) summarises the sources and effects of the major air pollutants.

3.1 The major air pollutants

The inquiry, *Urban Air Pollution in Australia* (AATSE [4]), found that Pb and sulfur dioxide (SO_2) no longer appeared to be major issues within coastal city airsheds. CO emissions generally have decreased and are mainly a local problem in heavily trafficked areas. Initiatives contributing to these emission reductions include the introduction of unleaded petrol and catalytic converters to minimise vehicle emissions; the banning of backyard burning; and the location of major pollutant sources, such as power stations, away from capital city airsheds. Despite reductions in Pb and CO vehicle emissions, the Inquiry identified Australia's vehicle fleet as the greatest single source of air pollutants, with substantial concentrations of ozone and photochemical smog on days with meteorological conditions conducive to smog formation.

Ozone comprises about 85% of the total photochemical smog that occurs in Australian cities and is one of the pollutants of major concern. There is evidence that exposure to O_3 is associated with a small increase in mortality and hospital admissions, primarily for cardiovascular diseases, and mainly among those aged 65 and older. Exercise appears to exacerbate the effects of O_3 on lung function (NEPMb [10] in CASOE [11]). Some plant species, including agricultural crops, demonstrate a reduction in growth and visible injury when exposed to prolonged O_3 concentrations at levels lower than those that cause adverse effects in humans (WPC and DEP [12] and in DEP [13]).

Airborne particles may be solid or liquid and are produced by a wide range of natural and human activities. Inhaleable particles, especially PM_{10}, are associated with respiratory illnesses including increases in the daily prevalence of respiratory symptoms, hospital admissions and mortality. Increasing evidence suggests that $PM_{2.5}$ (particles less that 2.5 μm diameter) may be the major influence on the health effects associated with PM_{10}, since $PM_{2.5}$ may be inhaled more deeply into the lungs (NEPMb [10] in CASOE [11]).

3.2 National Environmental Protection Measure (NEPM) for Ambient Air Quality (Air NEPM)

In 1998, the Air NEPM set the goal to achieve national air quality standards (allowing for exceedences) at performance monitoring stations by 2008. Table 3 summarises the Air NEPM standards (NEPMa [8]), as well as additional standards adopted by Australian states for air pollutants of local concern (EPANAA [14], EPAV [15], EPAQ [16] and PAQMP [17]).

Table 3: Ambient air quality goals. Sources: NEPMa [8], EPANAA [14], EPAV [15], EPAQ [16], PAQMP [17].

Pollutant	Averaging period	NEPM	EPAN	EPAV	EPAQ	EPAWA	WHO[d]
Ozone (O_3) ppm	1 hour	0.10[a]	0.08	0.12	0.098		0.08
	4 hours	0.08	0.06		0.079		0.06
	8 hours			0.05			
	24 hours				0.03		
	100 pdgs				0.03		
Nitrogen dioxide (NO_2) ppm	1 hour	0.12[a]	0.105	0.15	0.16		0.11
	4 hours				0.046		
	24 hours			0.06			
	1 year	0.3			0.01		0.021
Fine particles							
- as TSP μg/m³	1 year		30		90		
- as PM_{10}	24 hours	50[b]			150		
- as PM_{10}	1 year				50		
- as $PM_{2.5}$		Forthcoming					
- visibility rducing km local visual distance	1 hour			20	20	20	
Sulfur dioxide (SO_2) ppm	10 min				0.25		0.175
	1 hour	0.20[a]		0.17	0.20		
	24 hours	0.08[a]		0.06	0.04		0.044
	1 year	0.02			0.02		0.02
Carbon monoxide (CO) ppm	15 min						90
	30 min						50
	1 hour			30			
	8 hours	8		10	8	9	10
Lead μg/m³	3 months			1.5	1.5		
	1 year	0.5					0.5

Notes: ppm = parts per million; dpgs = days per growing season; Allowable exceedences: a = 1 day/year; b = 5 days/year; c = 1 day/month, d = World Health Organisation

For example, Victoria, Queensland and Western Australia include a visibility standard. This measure is equivalent to approximately 20 km local visual distance. An additional 8-hour ozone standard protects sensitive vegetation in Victoria (CASOE [11]).

A NEPM standard for $PM_{2.5}$ is forthcoming. The NEPC requires state environmental agencies to continue or to commence studies into $PM_{2.5}$ and $PM_{1.0}$ to ensure adequate data is available during the development of the standard (NEPMc [18]). In Australia, for comparison purposes, the USEPA standard for $PM_{2.5}$ of 65 $\mu g/m^3$, 24-hour average, is used currently (CASOE [11]).

3.2.1 Air quality standards and human health

The Air NEPM standards are designed to protect human health and well-being, for the majority of the population, and are legally binding on each level of government. The main health risk for Australians, in terms of mortality, arises from particulate matter and from hydrocarbons (Beer [19] in CASOE [11]). The results of recent studies on the relationship between air quality and mortality in Sydney (Morgan [20]), Brisbane (Simpson *et al.* [21]) and Melbourne (EPAVM [22]) indicate that in Brisbane and Sydney, short-term mortality (respiratory deaths not due to cancer) is related to increases in concentrations of particulate matter and ozone. In contrast, statistics for Melbourne indicate that short-term mortality is related to nitrogen dioxide and ozone. In addition, hydrocarbons have long-term health effects for Melbourne (Hearn [23] in CASOE [11]). Extrapolating results to all of Australia, gives estimates of about 1,250 to 1,785 deaths per year as a result of hydrocarbons, excluding deaths ascribed to the particulate matter in the hydrocarbons.

Estimated health costs for the major pollutants, provided by the NEPMa [8], based on the USEPA [24], recognise the substantial controversies and uncertainties that pervade attempts to characterise adverse human health in monetary terms. Extrapolated estimates for Australia are (CASOE [11]): -

- CO A$6m loss from one day of earning for 50,000 people,
- NO_2 A$5m cost of respiratory symptoms by 10 to 15 % of population,
- O_3 A$810m cost of up to 10 deaths per year in Australia,
- PM_{10} A$17.2b cost of 2,400 deaths per year in Australia.

Extrapolated estimates of the community costs of the impact of particles and ozone on the health of residents in the City of Brisbane were approximately A$255 to A$462 million each year (Simpson and London [25] in EPAQ [16]). Extrapolated estimates across the southeast Queensland region, based on a population of 2,223,330 and assuming equivalent health effects, were approximately a total health cost of A$754 to A$1,368 million (EPAQ [16]), that is, approximately A$349 to A$615 per capita

3.3 Ambient air quality in Australia's four major urban centres

The Air NEPM requires state and territory governments to establish nationally accredited ambient air quality monitoring networks and to report annually on compliance with the national standards (NEPMa [8]). Figure 2 shows the trend in maximum concentrations of O_3, NO_2, CO, and PM_{10} in Australia's four major urban centres.

Although there has been a steady decline in the maximum hourly O_3 concentrations (Figure 2a), there is no downward trend in the maximum value of O_3 when averaged over four hours. The number of days that exceed the Air NEPM standards has increased. The slight downward trend in O_3 levels during the 1980s reflects tighter vehicle emissions controls on NO_x and VOCs introduced in the late 1970s. However, emission limits set in the mid 1980s were not sufficiently stringent to enable compliance with the 1998 Air NEPM standard (CASOE [11]).

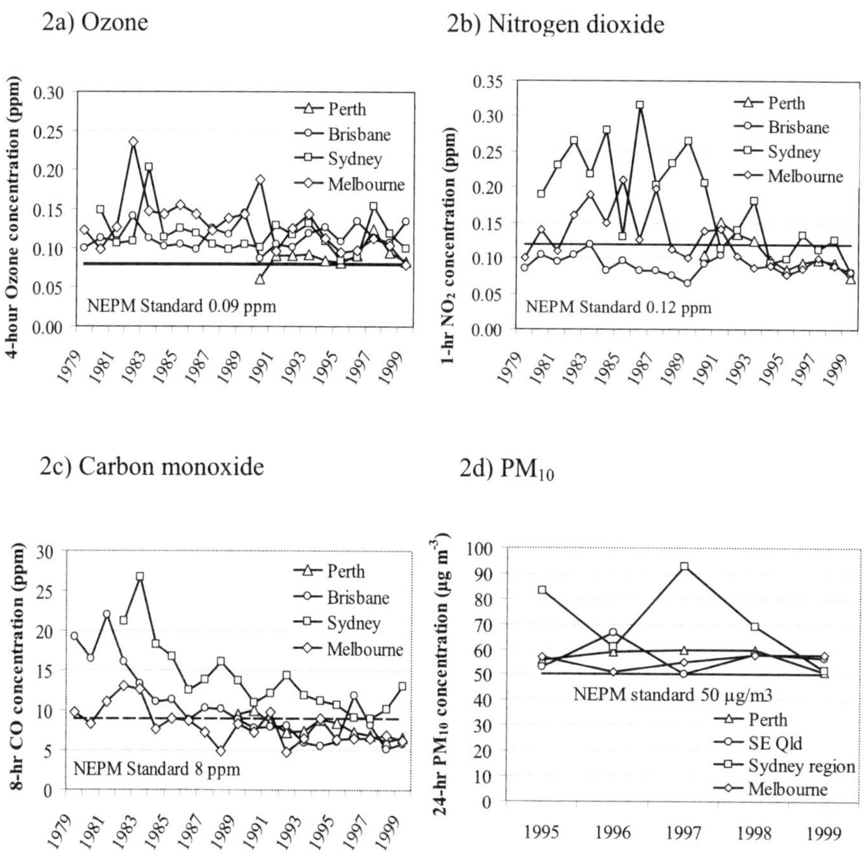

Figure 2: Major air pollutant maximum concentrations in Australia's four major urban centres (CASOE [11], primary data from State environmental authorities NSWSOEb [34] EPAVEI [35], QSOE [36], DEPSK [37]).

NO_2 concentrations show a downward trend, with exceedences of the Air NEPM standard occurring only in Sydney and Brisbane since the early 1990s (Figure 2b). The exceedences may result from the drift of industrial emissions into the urban areas, in combination with vehicle emissions.

The slight downward trend in urban CO concentrations (Figure 2c) reflects successive introduction of new emission standards (AASTE [4], CASOE [11])). Exceedences of the 8-hour Air NEPM standard occur in areas of high traffic density and low traffic flow in the centres of urban areas (CASOE [11]).

PM_{10} concentrations (Figure 2d) show a slight downward trend although there are occasional years when very high levels may occur. Although maximum 24-hour concentrations exceed the Air NEPM standards, median values (50th percentile) and 90th percentile values are well below the Air NEPM standard (CASOE [11]), suggesting that exceedences are infrequent events. Higher concentrations reflect bushfires or controlled burning of vegetation to reduce bushfire hazard near urban centres. The composition of PM_{10} concentrations in Sydney, Melbourne and Brisbane shows particles with two distinct distribution peaks, indicating the source of the particles to be from both combustion and mechanical processes (Keywood *et al.* [26] in CASOE [11]). Higher peaks in autumn and winter indicate the effects of wood smoke.

3.4 Air emissions inventories

An air emission inventory gives a breakdown, by significant sources, of the estimated total mass of specific air pollutants emitted in a study area for a given year. It also includes the spatial and temporal distribution of emissions. Inventories typically group significant sources of air pollution into four general categories: major industrial point sources; motor vehicles; area-based (smaller commercial and domestic aggregated) sources; and biogenic and natural sources. An inventory aids the development and assessment of air quality management options by identifying emission sources that are the major contributors to the inventory, as well as sources where emission reductions readily may be achieved. Emission estimates provide input data for computer modelling to predict air pollution dispersion and photochemical smog formation.

3.4.1 The Australian National Pollutant Inventory (NPI)

The NPI [27], established in 1998, is an Internet database designed to provide the community, industry and government with information on the types and amounts of specific substances emitted into the environment. Australia's State and Territory environment authorities collect annual emissions data from industry, and estimate emissions from non-industry sources and facilities that use less than the specified amount of substances. The national environment authority makes this information available on the NPI database at http://www.npi.gov.au, as well as emission estimation manuals. A new level of accountability by industries is required via annual reporting of emissions estimates and emissions reduction programs. Many larger companies already were reducing their emissions before the advent of the NPI and reporting obligations have caused them to improve their efforts (Rae [28]).

3.4.2 Industrial sources

Emissions from industrial sources include combustion products generated from equipment such as furnaces and ovens, boilers, generators, and compressors used in electricity, gas and water supply as well as metal refining and manufacturing facilities. Emissions from industrial point sources generally are determined using monitoring data supplied by the facilities, and annually updated as part of NPI reporting. Fugitive emissions are not well quantified.

3.4.3 Motor vehicle sources

This emissions source category includes emissions from 'on-road' vehicles, such as passenger vehicles, light commercial vehicles, motor cycles, light trucks and heavy-duty vehicles. Atmospheric emissions from the combustion and evaporation of fuels in on-road vehicles are calculated by combining data on the vehicle fleet, vehicle activity and emission factors. Vehicle activity data, expressed as vehicle kilometers traveled (VKT), is derived using data from traffic flows and commuter behaviour. Vehicle emission factors are in units of g/km, and vary depending on road conditions, the vehicle type, and age and fuel type.

3.4.4 Area-based sources (light industrial, commercial, domestic)

Area-based sources include smaller industrial and commercial facilities and domestic activities and off-road vehicles. These emission sources, although individually minor, are numerous in urban areas, creating a significant cumulative contribution. This emission category includes fuel service station vapour losses, domestic solid fuel combustion, motor vehicle refinishing and marine craft, as well as open burning. The techniques and emission factors commonly used are developed from research trials in Australian urban centers (Carnovale *et al.* [29], Boyle *et al.* [6]).

3.4.5 Natural and biogenic sources

Biogenic emissions emanate from natural processes. Emissions of isoprene and monoterpenes, VOCs emitted by vegetation, take part in the formation of photochemical smog. Biogenic emissions are strongly dependent on temperature and solar radiation, and primarily released during the daytime. Another biogenic emission is NO_x from soil, released during the natural nitrification/denitrification process. The emission factors are dependent on the intensity of solar radiation and the ambient air temperature. The emission rates from both sources are estimated for various daily scenarios, such as: a high oxidant day (>35°C), a summer day (20°C to 35°C) and a winter day (<20°C), (DEPEI [30]).

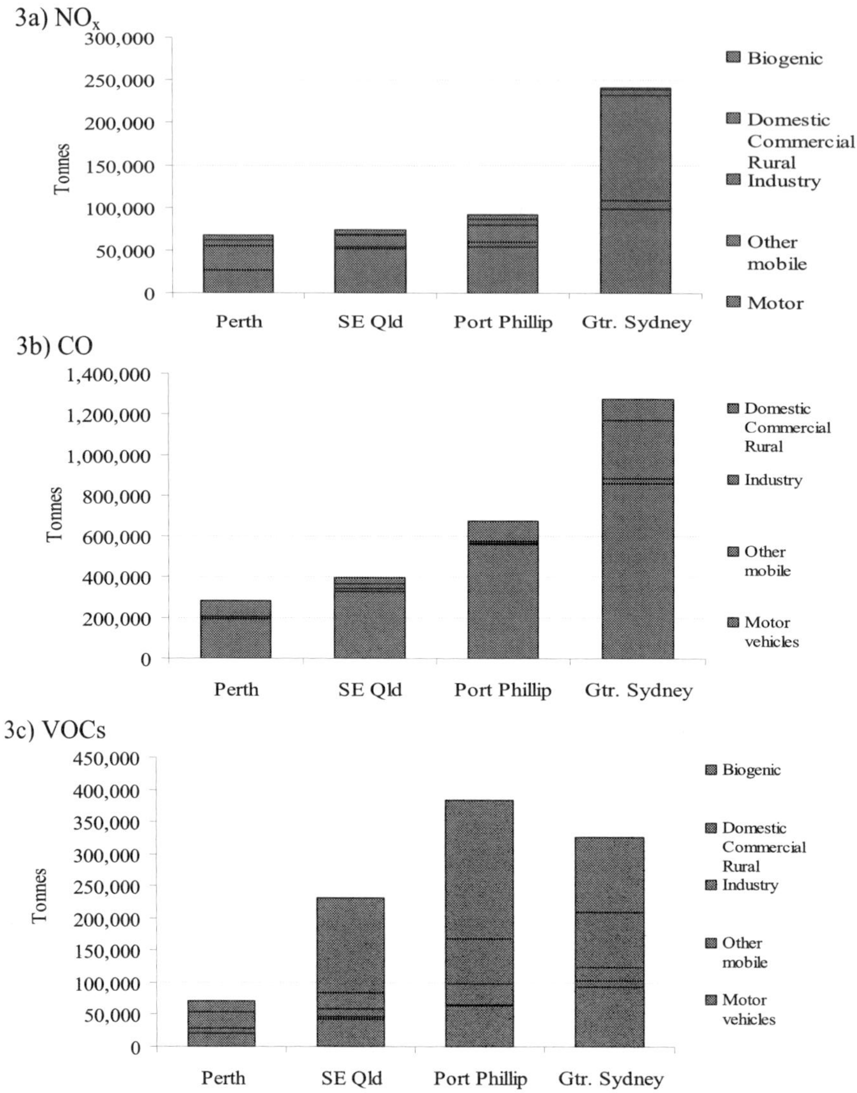

Figure 3: Australian urban emissions of NO_x, CO, and VOCs for Perth, Brisbane (Southeast Queensland), Melbourne (Port Phillip) and Greater Sydney (CASOE [11], primary data from State environmental authorities NSWSOEb [34] EPAVEI [35], QSOE [36], DEPSK [37]).

3.4.6 Air emissions in Australia's four major urban centres

Figure 3 illustrates the broadly similar trends in air emissions estimates for NO_x, CO and VOCs in the 4 major urban areas. Motor vehicles are the major emitters of air pollutants in the four major urban areas. Transport emissions contribute more than 75% of CO emissions, most of the NOx, and a major portion of the VOCs. Emissions inventories for the major urban areas reflect some individual differences relating to local variations in climate, modes for home heating and cooling as well as industry mix.

For example, the Port Phillip air emissions inventory (Greater Melbourne) indicated considerable seasonal variation in emissions from domestic wood heating, contributing 40% of PM_{10} and 46% of $PM_{2.5}$ emissions. In winter, this source contributes about 60% of PM_{10} and 68% of $PM_{2.5}$ to annual emissions. While domestic, commercial and rural sources are the biggest contributors to annual particle emissions, motor vehicles are also a significant contributor to PM_{10} and $PM_{2.5}$. Industry is a significant contributor to total annual emissions of NOx, PM_{10}, VOCs and SO_2. Other mobile sources, particularly large ships are significant contributors to SO_2 emissions (EPAVTS [31]; EPAV [15]).

The Perth inventory (DEPEI [30]) confirms the significant contribution by motor vehicles to emissions of CO, NOx, and total VOCs. Total VOCs and particulate emissions from area-based sources account for a large proportion of overall emissions. Domestic solid fuel (wood) combustion contributes 34% of the area-based total VOCs and 51% of the area-based particulate emissions. The most significant source of SO_2 is from industry. More than 99% of the industrial SO_2 contribution is from the supply of electricity, gas and water.

3.5 Meteorology

Australia's major urban airsheds are located on coastal strips with mountain ranges nearby, as illustrated in Figures 1 and 4. Local meteorology is strongly influenced by thermal contrasts between land and water, and moderately complex topography. Offshore down slope airflows (katabatic flows) develop in the evening and early morning, followed by onshore bay breezes and sea breezes (anabatic flows) in the afternoon. Figure 4 illustrates the generic circulation patterns that enhance air pollution episodes in the four major urban centres.

Sydney's urban area extends across a sedimentary basin surrounding the harbour, backed by low ranges (Figure 4a). Elevated air pollution concentrations in Sydney normally are associated with a high-pressure system located in the Tasman Sea (Hyde *et al.* [32]). Under these conditions, light to moderate local and regional winds predominate. Air pollutants including ozone may be retained, transported and re-circulated in the airshed (EPANMP [33]).

Melbourne is located on the shores of Port Phillip Bay, an almost land-locked water body. Ranges rise to 700 metres (m) within 30 kilometres (km) east of Melbourne. Night-time cooler katabatic flows move pollutants offshore across Port Phillip Bay, with restricted vertical dispersion trapped by a layer of warmer air. During summer and autumn, afternoon sea breezes re-circulate photochemically generated ozone onshore across Melbourne suburbs. The re-circulation, known as the Spillane or Melbourne Eddy, is a major mechanism for ozone pollution events (Figure 4b). During winter and late autumn, slow-moving anticyclones, typified by light winds and low mixing heights, create conditions for poor dispersion, a major mechanism for PM_{10} pollution events (EPAVMP [38]).

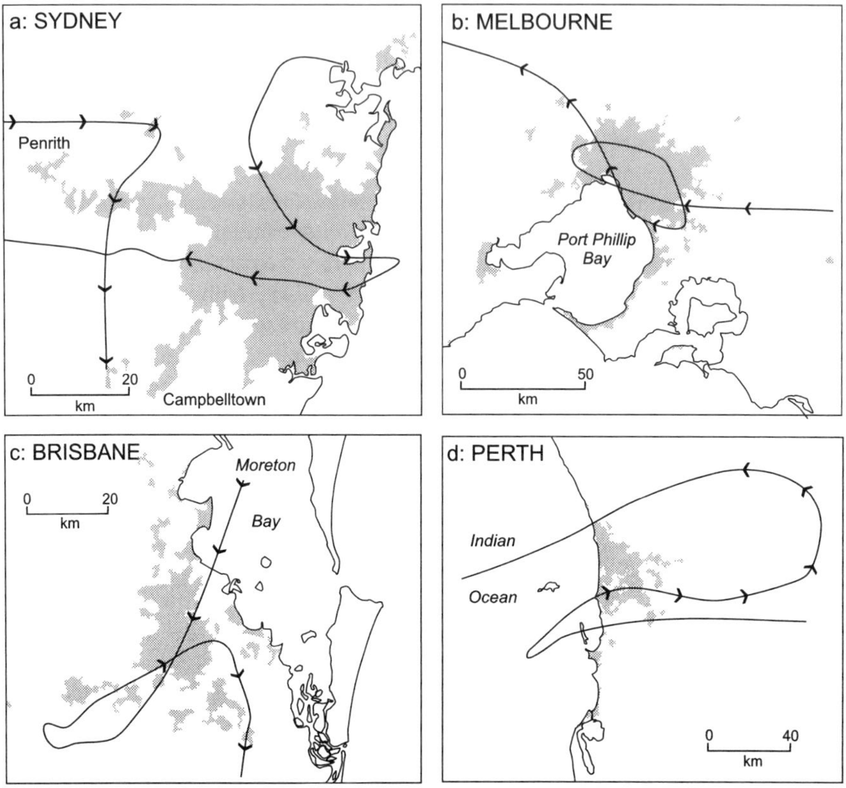

Figure 4: Air pollution re-circulation over major cities (CASOE [11]).

Brisbane is situated on flat to undulating land, backed by river valleys and well-vegetated mountain ranges rising to over 700m within 40km of the city. The topography of the region generates local winds, which play an important role in the dispersal of air pollutants (Figure 4c). Calm and stable meteorological conditions are associated with high levels of ozone and photochemical smog. Two main types of conditions are identified: near-calm, hot conditions under the influence of light north-westerly synoptic winds in summer; and very light north-north westerly synoptic winds in late autumn to early spring (QDE [39]). Concentrations of pollutants increase significantly when these re-circulation patterns are repeated over several consecutive days (EPAQMP [40]).

Perth lies on a gently undulating coastal plain about 25km wide and rising to about 40m above sea level. The coastal plain is bounded to the east by a north-south oriented escarpment, which rises to about 300m. During the hot dry summer, pollutants blown offshore within morning easterly winds are frequently re-circulated onshore in the southwesterly sea breeze, known as the Fremantle Doctor, a regular feature of Perth's summer meteorology (Figure 6d). The worst smog days occur when an offshore low-pressure trough, another common feature of summer meteorology, directs westerly winds onshore (WCP and DEP [12] in PAQMP [17]). On many winter nights, very stable conditions, virtually calm under cloudless skies, trap high concentrations of wood smoke and motor vehicles emissions near ground level.

Progress has been made in estimating the frequency of meteorological conditions that exacerbate air pollution in the major cities of Australia. Photochemical smog occurs when there are high temperatures, clear skies and light winds. Analyses of archived long-term records of the Australian Bureau of Meteorology have produced an indicative measure of smog pollution propensity (Table 4). Values are based on the both the mean number of days with a maximum temperature greater than 30°C and the mean number of clear days. A combination of low rainfall and strong winds can lead to high air pollution events from wind-blown dust. In high temperatures, bushfires elevate concentrations of particles, nitrogen oxides, and ozone.

Table 4: Smog pollution propensities across Australian capital cities (Newton 1997 [41] in CASOE [11]).

	J	F	M	A	M	J	J	A	S	O	N	D
Sydney	9	8	9	9	9	9	11	12	9	8	8	9
Melbourne	12	12	9	4	2	2	3	2	2	3	5	9
Brisbane	13	10	7	4	3	4	5	5	5	6	8	13
Perth	31	29	24	12	7	5	5	7	8	10	14	22

3.6 Air pollution modelling

The increasing application of computer modelling to air pollution research has been accelerated by improvements in the capacity of personal computers and the availability of increasingly reliable air emissions inventories. Computer modelling enables simulation, comparison and validation of ambient air quality concentrations predicted to result from various management options.

As an example, The Air Pollution Model (TAPM, Hurley [42]), released in 2000 by the Australian Commonwealth Scientific and Industrial Research Organisation (CSIRO), has performed well in verification studies to predict meteorology and pollution at the local and regional scales, particularly in locations with complex geography such as coastal areas. TAPM eliminates the need for site-specific meteorological data by using synoptic-scale meteorology and a fluid-dynamics approach to predict terrain-induced airflows and sea breezes. TAPM includes algorithms for chemical reactions and the creation of secondary pollutants such as nitrate and nitric acid aerosols and ozone.

Predictions of population exposure to specific pollutant concentrations are possible. TAPM is being applied to assess the need for monitoring of O_3 and NO_2 in small to medium sized non-industrial cities (CASOE [11]). The research will assist the NEPC and state governments in the implementation and review of the Air NEPM Monitoring Plans.

Another example, the Australian Air Quality Forecasting System (AAQFS), developed by the Australian Commonwealth Government, in partnership with CSIRO, the Bureau of Meteorology, and State environment protection authorities, predicts daily levels of photochemical smog, atmospheric particles and 22 other pollutants. The AAQFS currently provides daily air quality forecasts in less than 20 minutes per region, predicting how air quality will vary, hour-by-hour, suburb-by-suburb, over the next 36 hours, for Sydney and Melbourne. Forecasts are available at http://www.dar.csiro.au/information/aaqfs/.

3.7 Collation of background material to define the overall problem

Table 1 notes the need to collate background material to define the overall problem, prior to developing management options. Sources to be researched include parliamentary inquires, health and population exposure studies, estimates of the costs of air pollution, reviews of monitoring data and meteorology and modelling studies.

This process is well illustrated by the experience of Perth. The Parliamentary Select Committee on Perth's Air Quality was established in May 1997. It was the first step in the State Government's four-year program to develop a basis to improve Perth's air quality, through the development of an air quality management plan. The 4-year program involved collation of background material including a review of ambient air quality trends for five Air NEPM pollutants, as well as TSP, $PM_{2.5}$ and visibility reducing particles (DEPEI [30]). the review identified frequent exceedences of Air NEPM standards for O_3 and PM_{10}, each year, as well as relationships with meteorology. Earlier research measured the magnitude and distribution of photochemical smog concentrations (WPC and DEP [12]) and investigated the nature and sources of fine particles (DEPH [43]). A similar process, with varying levels of detail, was followed by the other major urban centres.

4 Air pollution management

To achieve Air NEPM goals by 2008 in the major urban areas, State Governments have developed action plans, improvement plans and strategies (referred to as air quality management plans (AQMPs) in this discussion). The following sections summarise the key issues in the AQMPs prepared for the greater metropolitan regions of Sydney, Melbourne, Brisbane and Perth.

4.1 Sydney

The New South Wales (NSW) Government's 25-year Air Quality Management Plan *Action for Air* (EPANAA [14]) focuses on the Greater Sydney region, which includes about 70% of the state's population, as well as the major industrial areas. The plan is the outcome of a comprehensive process that commenced with the *Metropolitan Air Quality Study* (MAQS [7]). MAQS involved a detailed air emissions inventory, air quality modelling and health study, coordinated by the NSW Environment Protection Authority (EPAN). Extensive public consultation followed MAQS, through a series of Government Green Papers on developing an air quality management plan, and a smog action plan (EPANAA [14]).

Action for Air integrates urban planning with emission reduction strategies to tackle both photochemical smog and brown haze from fine particle pollution. The actions are grouped, generally, into the following strategic areas:

- Integrated air quality goals and urban and transport planning,
- Transport alternatives,
- Cleaner industries, businesses and homes,
- Managing the impact of open burning, and
- Monitoring and reporting of air pollution levels and progress in air quality management.

Actions for reducing vehicle usage target zero-growth in VKT per capita by 2011 and zero growth in total VKT by 2021. Public transport systems plan to link rail infrastructure with pedestrian and cycle ways, providing alternative transport choices for the journey to work at key centres, such as Parramatta in the city's outer west. Increased freight transport by urban rail and waterways is proposed as an alternative to road transport, reducing the impact of heavy diesel vehicles.

An inspection program has commenced for in-service motor vehicles to improve maintenance and achieve targeted reductions of NO_x, VOCs and CO emissions, as well as improve average fuel economy by 2.5%. Penalty notices are issued to offending smoky vehicles in the inner city. Testing of passenger and light commercial vehicles is planned to extend to outer centres of Newcastle and Wollongong by 2004 (EPANAA [14]). A new website is planned for online reporting of smoky vehicles. The introduction of 400 public transport buses powered by compressed natural gas was announced in 2001. Agreement between the NSW Government and petrol producers and suppliers for reduced petrol volatility will lower summertime VOCs emissions from motor vehicles, lawn mowers and fuel storage facilities and service stations (NSWSOEa [9]).

Action for Air promotes cleaner industries by combining stronger regulation with economic incentives for emission reduction. Load-based licensing of industrial premises, introduced in 1991 and strengthened in 1998, has set limits on the amount of pollutants emitted. Higher fees are payable for targeted polluters in sensitive areas, such as NO_x emitters in western Sydney. Pollutants listed for each industrial activity must be reported in an annual return. The scheme has relatively high threshold levels below which load fees are not payable. While the scheme's full potential to encourage significant emission reductions may not yet be achieved (CASANZ [44]), businesses were estimated to invest $416 million in pollution reduction programs in 2000–2001 (NSWSOEa [9]).

Strategies for cleaner small businesses include codes of best management practice to reduce VOCs emissions in auto-repair shops and surface-coating premises. Measures include the use of high-volume low-pressure spray guns, avoiding spillage, and re-use and re-cycling of materials. An additional $5 million for the cleaner production programs was announced in 2001 (NSWSOEa [9]), to build upon the success of the Solutions to Pollution programs.

Action to reduce fine particle emissions from domestic fuel heaters includes a consultation process to develop an industry code of practice accessible to local councils, domestic solid fuel heater retailers, consumer bodies and relevant licensing and training authorities. A new national design standard for domestic wood heaters was released in 2001. Further action includes guidelines for wood suppliers; regulatory action by local councils on smoky chimneys and incentives for owners of old wood heaters to upgrade to new or alternative fuel heaters (EPANAA [14]).

The major smoke management issue for Sydney air quality relates to planned burning for bushfire hazard reduction and forestry management, since open burning in domestic backyards was prohibited in 1995. *Action for Air* sets the requirement for smoke management guidelines to be developed in consultation with government agencies responsible for rural fire services, forestry and national park management as well as the Bureau of Meteorology.

Daily reporting of air quality data by the NSW EPA, assessed against the Air NEPM standards, began in October 2000. The NSW EPA has been monitoring measurements of $PM_{2.5}$ since 1996 and has provided data for the review of the air quality particles standard, in preparation for a $PM_{2.5}$ national standard by 2003. NSW EPA has continued to collect 2-minute averages for SO_2 in anticipation of a review of the practicality of developing a 10-minute SO_2 national standard to by 2003 (NEPC [45]). Strategies for monitoring and reporting on air pollution levels include support for the national Airwatch schools education program (http://www.airwatch.gov.au), and the community friendly air pollution index, summarised in Figure 5.

Local councils in the Sydney metropolitan area also have developed air quality management plans in partnership with the NSW EPA, integrating landuse and transport planning (SSROC [46]). Local planning controls allow a mix of landuses to minimise distances between residential, employment and service areas, reducing the frequency and length of motor vehicle trips. Traffic calming measures and idling bans reduce vehicle emissions. Policies to reduce emissions from other specific sources involve industrial auditing and dust suppression at construction sites, stockpiles and extractive industries. Community education programs promote informed travel behaviour and product use. Newcastle, a major industrial centre in the north of the greater metropolitan area, has a comprehensive local air quality management and monitoring program (Bridgman

[1]). The NSW Government recently announced a 3-year $6 million fund to tackle local air pollution hotspots (NSWSOEa [9]).

Air pollution control authorities have found it useful, when presenting air quality information to the public, to use an air quality index (or an air pollution index) as a means of combining information about a range of the pollutants. Expressing air quality as an index makes it easier to compare pollutant levels and air quality at different monitoring stations at a glance.

The New South Wales EPA calculates a regional pollution index (RPI) based on measured concentrations of visibility, fine particles, ozone and nitrogen dioxide in the lower atmosphere. Each variable is compared to the Air NEPM Standard, and a value equal to the NEPM standard is taken as 50. Values above 50 are considered highly polluted. Values above 25 indicate medium-level pollution.

The Sydney afternoon air pollution index for January 2000 indicates summer high pollution events.

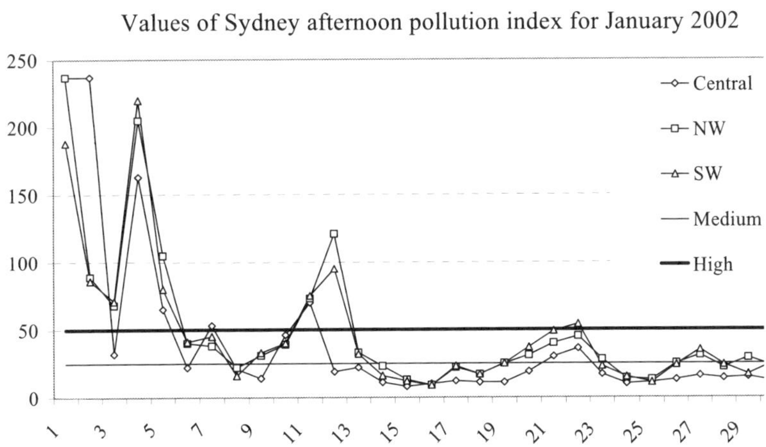

Figure 5: Values of the Sydney afternoon air pollution index for January 2000 (CASOE [11] primary data at RPI [49]) http://www2.epa.nsw.gov.au/airqual/aqupd.asp.

4.2 Melbourne

The goal of the Melbourne *Draft Air Quality Improvement Plan* (*AQIP)* is a safe and ecologically sustainable air environment for a large area defined as the Port

Phillip Region (EPAV [15]). The region includes most of Victoria's population and industry, although the major power generating industries are located further east of the region. The Melbourne *Draft AQIP* builds on a comprehensive inventory of air emissions (EPAVEI [35]), and a series of projections of air quality to the year 2020 (EPAV [15]). Following community response to the draft, the most cost-effective methods for achieving the air quality goals will be determined, using techniques of atmospheric modelling and economic analysis, including risk assessment. The Melbourne *Draft AQIP* will function as one of the management tools to achieve the Air NEPM requirements, in combination with the Air NEPM monitoring plan for Victoria (EPAVMP [38]) and the state environmental planning policies (SEPPs) for air quality (EPAVAQM [47], EPAVAAQ [48]).

The Melbourne *Draft AQIP*, in common with Sydney's *Action for Air*, prioritises actions to reduce pollution from motor vehicles, industry, households and a range of miscellaneous sources. The actions generally are grouped into the following strategic areas:

- Motor vehicle emissions, travel demand and driver behaviour,
- Eco-efficiency and product stewardship in industry,
- Domestic wood combustion and other emission sources, and
- Knowledge as a basis for action.

Proposals to reduce transport emissions focus on two general approaches: improving emissions from the motor vehicle fleet; and influencing travel demand behaviour. Proposals to reduce vehicle emissions, in common with Sydney's *Action for Air*, support national action for more stringent new vehicle emission standards for petrol and diesel engines; an increased level of vehicle servicing; and low sulfur fuels for the in-service vehicle fleet. Actions to influence travel demand propose the integration of landuse and transport planning; promoting alternatives to the car; and driver education to minimise the environmental impacts of cars. In addition, the Melbourne *Draft AQIP* advocates new zero-emission electric and hybrid fuel engine technology.

Emissions reduction in major industries is proposed through enforcement of new regulations, as well as extensive consultation to raise awareness of the ecological impact of industrial production. An audit task force is proposed to ensure that new industrial premises adopt new standards, the latest technologies for cleaner production, and address the management of fugitive emissions. Stronger financial incentives are proposed to address anomalies in the load-based licensing system. A new SEPP for air quality management has introduced the principles of eco-efficiency and product stewardship in industry (EPAVAQM [48]). These principles require both producers and consumers to share responsibility for minimising consumption and waste production at each stage of the product life cycle, including environmentally acceptable disposal of any residual wastes. Supporting actions include a technology information network; open review of design criteria; and environmental accounting and reporting by companies. Emissions monitoring is proposed, with inclusion of data on emission levels per unit of production. Voluntary targets are proposed for reducing the VOC content of various products, and the development of alternatives to VOCs.

Proposals for action to reduce emissions from domestic wood heaters include new emission standards; codes of practice relating to flue heights and correct installation; replacement of fireplaces and older heaters; fuel quality standards; smog alert days and community education.

Other actions propose to reduce emissions from ships and aircraft; minimise emissions from waste burning; and reduce particle emissions from diffuse sources such unpaved roads, construction and demolition sites.

The Melbourne *Draft AQIP* will improve community awareness of air quality issues and develop a greater role for air quality science in the following areas:

- Air pollutants and their impacts,
- Air quality monitoring,
- Maintaining and improving air emissions inventories,
- Airshed modelling, and
- Communicating air quality issues.

To extend monitoring, a mobile air quality monitoring facility will monitor air quality at hotspot locations, such as roadsides and near specific industrial sources remote from fixed-site monitoring stations. Data from all fixed ambient air monitoring stations in the Port Philip region will be relayed to computers to be set up in local council offices for public viewing.

Regular updating of the Port Phillip emissions inventory will be co-ordinated with NPI reporting, reflecting new estimation methods, as well as changes in emission levels over time due to changes in population and the structure of the economy. This will ensure the best available inputs for modelling.

Increased airshed modelling capabilities proposed in the *Draft AQIP* will allow more effective prediction of the consequences of alternative emission scenarios, and improve understanding of community exposure to pollutants in the Port Phillip Region. Further research is proposed to assess the relative importance of NO_x and VOCs in O_3 formation. Modelling of secondary particle formation and nitrogen deposition will support analysis of nutrient load runoff strategies for the Port Phillip Bay catchment.

Effective communication is acknowledged by the *Draft AQIP* as a prerequisite for better understanding of air quality issues by the whole community. A two-way flow of information is emphasised for stimulating debate and obtaining feedback on actions. Strengthened commitment to the AAQFS, providing the community with short-term air quality predictions, will alert those with health conditions sensitive to air-pollution events. The State Environmental Planning Policy for Air Quality Management (SEPPAQM) addresses community concerns about the local impacts of combined pollution sources from industry, local traffic, and domestic wood smoke. Neighbourhood Environment Improvement Plans provide local air quality monitoring and assessment against health-based intervention levels. Actions to reduce emissions from local sources are cooperatively managed by all stakeholders (PVAE [50]).

4.3 Brisbane

The *South-East Queensland Regional Air Quality Strategy* (*SEQRAQS)* plans to protect the health and well-being of present and future residents, and to preserve the region's ecological integrity and amenity (EPAQ [16]).

Table 5: Performance Targets for *SEQRAQS* (EPAQ [16]).

Outcome	2011 Target
Achievement of targets related to passenger vehicle use, occupancy and average trip time, set by Inter-Regional Transport Plan for SE Queensland.	Inter Regional Transport Plan for SE Queensland targets to be achieved.
Harmonisation with world standards for passenger vehicles.	100%
Increased share of land freight task by low-polluting vehicles, e.g. Rail, or vehicles using low-emission fuels such as CNG or LPG.	25% of land freight task
Demonstrable reduction in emissions of VOCs, NO_x and particles from industrial and commercial sources.	15% reduction of mass weighted
Demonstrable reduction in non-transport derived emissions of priority pollutants.	15% reduction in per capita emissions
Despite population growth, no increase in emissions of VOCs, NO_x and particles from domestic sources.	No change relative to 1998
Expansion in regional airshed monitoring and modelling.	75% increase in annual monitoring measurements recorded
Demonstrated reduction of VOCs, NO_x and particles in emissions from the transport sector.	18% reduction of mass emissions.
Reduction in the number of complaints about air quality.	50% reduction in the number of incidents reported to Local Government or EPAQ compared with 1999 (or other approved base year)

The region covers metropolitan Brisbane and surrounding areas including about two-thirds of the state's population. Significant heavy industries include petrochemical refineries and an intermediate peak-load power station. Strategy development followed air quality studies, including an air emissions inventory (DEEI [51]) and wind-field modelling to develop a better understanding of the regional airshed (DEMA [52]). These studies were conducted within a regional planning framework, involving the Queensland Government, the Commonwealth Government and local governments, to develop a coordinated approach to managing the region's dynamic population and economic growth (DLGP [53]).

Table 6: Signature actions identified by Brisbane's *SEQRAQS* (in the numerical sequence of Strategy rather than in priority EPAQ [16]).

	Signature Action
1	Random testing of on-road vehicle emissions to test each year at least 15% of petrol engine light vehicles in the region that are more than 5 years old.
2	Investigate a cost effective truck and bus emissions testing program and develop supporting test facilities, as part of the certificate of inspection.
3	Public education to raise awareness of health and environmental impacts, and practical steps to avoid and reduce vehicle emissions, and the importance of individual effort to achieve large cumulative effects
4	Develop best practice environmental management guidelines for each major industry sector, commencing with identified priority industries.
5	Work with major companies and relevant industry associations to develop specific measures for VOC emission reductions.
6	Review the impact of power generation on the airshed. Set emission reduction targets, with consideration to market factors, technology, energy conservation, and low-emission fuels.
7	Prepare a model local government air quality management plan and manual, with consideration to land use planning.
8	Develop sub-regional or local air quality plans within 3 years, to implement through cooperative government land use planning.
9	Revise the fire-permitting system to incorporate prevailing air quality.
10	Fire management for native forests, with consideration to potential impacts on health, community safety and air quality
11	Native forest management alternatives to burning
12	Expand capacity to monitoring $PM_{2.5}$ across the whole monitoring network.
13	Community education about health effects and community costs of air pollution and steps to reduce domestic air emissions
14	Establish an system for annual analysis of the nature, number, cause and trends in air pollution complaints.

The pollutants of immediate concern in the region are photochemical smog, and visibility-reducing, inhaleable particles. The actions generally are grouped into the following strategic areas:

- Transport emissions and landuse planning,
- Industrial emissions,

- Domestic activities and controlled burning,
- Monitoring, modelling and research of regional airshed issues, and
- Raising awareness and meeting community expectations.

To assess the progress of the Strategy, targets for emission reductions have been set (Table 5), representing goals rather than prescriptive limits.

Brisbane's *SEQRAQS* provides details for strategy implementation as well as implementation of individual actions, in contrast to Sydney's *Action for Air* and Melbourne's *Draft AQIP*. The involvement of stakeholders in implementation is an important factor in ensuring that the overall plan and individual actions meet their objectives (Macintosh, *et al.* [54]). An implementation unit, lead by the Queensland Environment Protection Agency (EPAQ) with community and stakeholder representation, is responsible for coordinating and monitoring implementation of actions, and technical work associated with reviewing and updating of the Strategy. Procedural timeframes have been identified for the 180 separate actions, grouped into 42 sub-strategies. A nominated lead agency has the overall responsibility for the particular action, including progress reporting to the implementation unit. Collaborating agencies are required to assist the lead agencies in their particular areas of interest, responsibility or expertise as required. The timing of actions will be continually revised based on current priorities and resource availability, subject to normal budgetary processes. Signature actions, considered essential precursors to measurable improvements in air quality, provide the community with milestones in the Strategy's implementation, to be delivered by state and local governments within 3 years (Table 6).

4.4 Perth

The *Perth Air Quality Management Plan* (PAQMP [17]) was developed to achieve and maintain clean air throughout the Perth metropolitan region over the next 30 years. The *PAQMP* is based on solid research and planning. Scientific studies of Perth's air quality during the 1980s and 1990s investigated photochemical smog and brown haze (DCE [55], WPC and DEP [12], DEPH [43]). In 1998, a parliamentary committee commenced a four-year program to develop a long-term management strategy, the *PAQMP*, based on local studies as well as national and international air quality reviews (AASTE [4]) and health studies (i.e. Dockery *et al.* [56, 57] PAQMP [17[). The parliamentary review (LA [58]) identified the need for an effective public transport system, fully integrated with its urban design. An imaginative blend of urban residential densities was recommended to create a community that uses the fullest range of transport methods, and is not car dependent. Industrial activity providing employment and economic wealth was to be achieved without detracting from the city's attractive natural and social environments.

The *PAQMP* is a combination of state and local government regulation and planning measures, education, and voluntary actions. A staged implementation plans to deliver 126 actions over 30 years. The actions are grouped into eight areas:

- Health effects research,
- Monitoring, modelling and research,
- Land use and transport planning,
- Vehicle emissions management,
- Domestic activities emissions,
- Burning emissions management,
- Industry emissions management, and
- Community information and education.

As examples of actions in the Perth *AQMP*, the investigation of the community health impacts of air pollutants, especially smoke, is highlighted. Actions proposed for modelling, monitoring and research, in common with Melbourne's *Draft AQIP*, emphasize the importance of a well-maintained air emissions inventory, a representative monitoring network, and a modelling capability to research the nature and sources of emissions.

A notable action that has achieved success is Perth's TravelSmart program, which encourages the use of public transport, cycling, and walking. A pilot project in South Perth achieved a significant reduction in car use on an average day. The overall result was a 14% decline in VKT and included a 25% increase in public transport journeys (March–June 1999 compared to March–June 2000). These programs support regional centres and major workplaces across the Perth metropolitan region, and are a means of minimizing travel requirements and long-term parking. Other TravelSmart programs include workplace and school programs. TravelSmart Workplace aims at people employed in the Perth Central Business District (CBD), raising awareness and understanding of air pollution issues and transport choices. TravelSmart school programs raise awareness among children in primary schools of transport alternatives to the car. Research indicates that social conditioning of children in the "car culture" is already evident in children as young as seven, and is well established by the time they are 13. A trial resulted in a 22% reduction in car trips to the school. The trial produced a resource kit for school Years Five to Seven and extended the initiative to many other schools. State government subsidies provide TravelSmart officers to eight local councils (PAQMP [17]).

The *PAQMP* describes the strategy development processes for prioritizing, scheduling and implementing the actions, reflecting similarities with Brisbane's *SEQRAQS*. The *PAQMP* acknowledges that each of the 126 actions will require an individual implementation plan. These plans will consider relevant issues, resources, timing and costs for each step of the action. A ranking process assessed the relative effectiveness of actions, identifying cost-effective priorities that present short-term and long-term benefits to air quality (Table 7). This process identified 6 priority actions (Table 8).

Table 7: Ranking criteria to assess air quality actions in Perth (PAQMP [17]).

Ranking Criteria	Issues
Direct emission reduction potential	Estimated unit reduction in pollution per year, per capita, and total emission reduction
Indirect emission reduction potential	Increased information allows future pollutant reduction. An action that precedes another action
Timing of effectiveness	Time frames for an action to have an effect and to be fully implemented
Cost effectiveness	Costs of an action to the individual, to the community, to industry; to Government
Equity	Confirmation that an action targets the original source of pollution. Identification of stakeholders who would bear the costs of the action; and individuals or groups within the community who may be disadvantaged
Technical feasibility	Availability and suitability of current technology. Likelihood of new technology being developed to solve the problem
Enforceability	Availability and suitability of enforceable regulatory mechanisms. Administrative barriers to the introduction of the action. Need for enhanced or new enforceable mechanisms
Public acceptance	Community opinion of an action. Public perception barriers to the introduction of the action
Public health benefits	Reduced hospital admissions, morbidity and mortality

The Perth *AQMP* describes for each of the 126 actions: an identified output; expected outcomes; agencies responsible for coordination and support; a time frame for commencement and completion; and expected progress after 5 years (Table 8). Implementation of the *PAQMP* is managed by the Air Quality Coordinating Committee (AQCC), comprised of one state parliament representative, seven state government senior executives, and seven non-government members representing local government (2), business and industry (2), and community groups (3). The AQCC will consider new information and results as they become available, and annually review the *PAQMP* to ensure actions remain appropriate and effective.

4.5 Discussion

As outlined in the preceding section, the air quality management plans for the four major urban areas address common concerns (Table 9). The most important issues are reducing emissions from transport, industrial and commercial premises, and domestic activity, as well as controlled burning in Perth and Brisbane. This section discusses similarities, differences and unique issues of local concern.

Table 8: Perth AQMP key actions (DEP [17]).

Action	Output	Expected outcome	Air quality impact	Time	Expected status
Continue with travel behaviour programs such as TravelSmart	Campaigns, reports, annual review	Improvement to system, behaviour change	Direct	1999 to 2010	Reports, campaigns, reviews in progress
Implement program to reduce in-service emissions from motor fleet	Program	Improvement to system	Direct	2002 to 2005	Program in prgress
Regulate fuel quality standards	Regulation, review 2005	Basis for future decisions	Direct	2000	Regulation completed, review in progress
Encourage national action on cleaner fuels	Regulation	Basis for future decisions	Indirect	2000 to 2029	Regulation completed, review in progress
Promote CNG and LPG for passenger and freight transport	Report	Basis for future decisions	Indirect	2001 to 2002	Report completed
Reduce industrial No_x and VOC emissions	Report	Basis for future decisions	Indirect	2003	Report in preparation

Table 9: Number of actions and percentages (approximate) by strategic areas in air quality management plans.

Key strategic area / emissions	*Sydney*		*Melbourne*		*Brisbane*		*Perth*	
	No.	*%*	*No.*	*%*	*No.*	*%*	*No.*	*%*
Transport	33	59%	29	34%	53	30%	36	30%
Industrial and commercial	10	18%	20	23%	27	15%	14	10%
Domestic	4	7%	14	16%	15	8%	13	10%
Open burning	3	6%	3	4%	18	10%	19	15%
Monitoring, modelling, research	6	10%	15	17%	30	17%	21	17%
Health research					3	2%	13	10%
Communication, awareness, education			5	6%	8	4%	10	8%
Land-use planning					18	10%		
Implementation, review					7	4%		
Total	56	100%	86	100%	179	100%	126	100%

4.5.1 Transport emissions

Reducing transport emissions is the most important issue identified by each of the relevant air quality management plans, constituting over half of all actions for Sydney, and approximately one third of all actions for Melbourne, Brisbane and Perth. Common themes for reducing transport emissions target two broad objectives: integrating air quality goals with urban and transport planning; and improving the emissions performance of new and in-service vehicles. The relative weighting of actions in the AQMPs addressing these objectives is generally in the ratio of 2:1 for Sydney, Melbourne and Brisbane, although more evenly weighted for Perth. The weighting emphasises the links between urban form, travel demand and air quality and is consistent with the AATSE ([4]), which recognised the need for greater consideration to the shape of urban development, encouraging lower levels of car-dependency.

Integrating air quality goals with urban and transport planning The Melbourne *Draft AQIP* suggests that increasing the population of a city, irrespective of urban form, would be expected to lead to an increase in traffic densities, at least in activity/transport centres and along arterial routes. This possibility validates the integrated approach of the AQMPs, requiring a package of actions, including strong disincentives to using the car, the provision of attractive alternatives and supportive planning. The experience of overseas cities (Kenworthy *et al.* [59]), questions the ambitious goals set by Sydney's *Action for Air* [14] for zero growth in per capita VKT by 2011, and zero growth in total VKT by 2021. The later AQMPs for Melbourne, Brisbane and Perth are less prescriptive in setting goals in relation to VKT. While rapid urban growth is expected to continue, infrastructure investments are planned to reshape metropolitan Perth into a more compact form, with growth occurring in regional centers (PAQMP [17]).

Improving the emissions performance of new and in-service vehicles The need for long-term reductions in motor vehicle emissions from new and in-service vehicles is recognised as crucial in all four AQMPs. Statutory vehicle emission standards for new vehicles in Australia are set at the national level as Australian Design Rules (ADRs), as noted by each of the AQMPs. All plans advocate tighter national emission standards for new cars. Sydney's *Action for Air* (EPAN [14]) notes the NSW coordination of a national working group supporting new standards equivalent to those set by the US in 1994 and Europe in 1996. The new standards are estimated to achieve major reductions in emissions of CO (20%), VOCs (30%) and NOx (70%), (EPAN [14]). Melbourne's *Draft AQIP* (EPAV [15]) notes the Commonwealth Government's 1999 commitment to harmonise ADRs with the European standards by 2007. The *Draft AQIP* advocates progress in ADRs to maintain consistency with international standards.

Melbourne's Draft AQIP (EPAV [15]) is the only AQMP to advocate the introduction of zero-emission vehicles, clearly communicating the positive impact on air quality by improving the emissions performance of in-service vehicles. The most poorly maintained vehicles are estimated to emit five to 10 times the exhaust emissions of well-maintained vehicles of the same age. The results of a Commonwealth study (FORS [60]) predicts emission improvements of 16% for VOCs, 25% for CO and 9% for NOx, from regular servicing of vehicles on a fleet-wide basis. Regular servicing also has fuel savings benefits, at least around 5% for the most poorly maintained cars (EPAV [15]).

The Melbourne *Draft AQIP* emphasises that a significant improvement in vehicle maintenance is more depended on low-cost measures, strong regulations and incentives for regular servicing, rather than the random testing proposed in most major urban areas. The plan proposes that proof of vehicle servicing within the past 12 months be provided before a vehicle may be re-registered annually. Vehicle owners who provide evidence that their vehicle has been serviced would be charged less for registration than those who do not (EPAV [15]).

4.5.2 Industrial emissions

Strategies to reduce industrial emissions in the four major urban areas generally propose a mix of regulations to achieve higher standards, and cooperative programs for introducing cleaner technologies and environmental management systems. The relative proportion of actions targeting industrial emissions varies from generally one fifth to one tenth of all actions (Table 9).

Economic incentives to reduce emissions were introduced in 1990 as a movement away from the command and control approach to managing industrial emissions by government imposed regulation. Although economic instruments take effect through government legislation or regulation, they usually allow for choice and decision-making involving those whose industrial operation is to be modified. The *Intergovernmental Agreement on the Environment* (CoA [2]), required the polluter pays principle to be incorporated in development and implementation of environmental policies and programs in Australia. A national audit concluded that there was scope for economic instruments to be used more extensively (Productivity Commission [62], in James [61]).

All four AQMPs propose a greater role for economic incentives to contain emissions through market processes or other financial incentives. The NSW EPA's load-based license fee system was introduced in 1991 and strengthened in 1997 by Sydney's *Action for Air* (EPAN [14]), representing one of the most positive innovations in pollution management in Australia. The scheme incorporates emissions targets with fees dependent on the discharge load and the potential impact on the receiving environment. Improving the load-based license fee system and the links with NPI reporting is recognised in the Melbourne *Draft AQIP* (EPAV [15]).

Sydney's *Action for Air* (EPAN [14]), is representative of the major urban AQMPs for action proposed to reduce industrial NO_x emissions from major industrial sources, such as power stations. NSW hosts major power generators, with five coal-fired power stations located north of the central metropolitan area. The proposed NO_x policy sets a limit or cap on total emissions for the greater metropolitan region and issues tradable discharge permits. The proposal was based the results of earlier modelling (MAQSa [7]), suggesting that emissions from industrial facilities at the edge or beyond the greater metropolitan area may exacerbate photochemical smog in the major urban area of Sydney. However, recent modelling focusing on inter-regional transport of air pollutants from the power stations, suggests that the proposed NO_x cap and trading scheme may not be effective in controlling O_3 levels in Sydney (Nelson [63]). The example demonstrates not only the role of airshed modelling to inform debate and evaluation of management options, but also the complexity of formulating appropriate policy.

4.5.3 Area-based sources (domestic activity, controlled burning)

All AQMPs recognise the role of high national standards for design and operation of heating systems to reduce particulate emissions from domestic wood burning, a significant source of winter pollution events, particularly in Melbourne and Perth. More energy-efficient homes and buildings reduce common urban air pollutants as well as greenhouse gases.

Brisbane and Perth, and more recently Sydney, propose a national rating scheme to indicate the relative air quality effects for various consumer items including solvent-based products, lawn mowers, outboard motors and jet skis.

Melbourne and Brisbane plan to follow Sydney's lead in banning domestic waste incineration. Perth, emphasising planned burning as a management tool in agriculture, forestry and native ecology, proposes weather and smoke modelling to improve burn decision processes.

5 Conclusion

Table 1 illustrates the components of air pollution management: merging the best available air pollution science with policy and regular review, at a range of levels. The process is illustrated in Australia since 1990.

Air quality management in Australia over the last 20–30 years has moved progressively away from 1960s notions of pollution as the price of progress, through 1980s command and control regulations, towards actions based on the principles of ecologically sustainable development.

Since 1990, national leadership in decision-making that links social, economic and environmental responsibility has created policy formulation networks between researchers, industry, the community and governments at national, state and local government levels. The result is a nationally consistent approach that links science, policy and air quality management. Mechanisms for monitoring, review and community reporting operate via the national pollution inventory and the ambient air quality monitoring network. National goals for common urban air pollutants set explicit standards which guide urban air quality management planning.

AQMPs for the major urban areas identify methods for reducing emissions and for future research for the next 20–30 years. Health effects research, emissions inventories and atmospheric dispersion modelling have improved understanding of the nature, sources and impacts of air pollutants. Knowledge of air pollution science and trends at the local and regional level, have been applied in the development of management options. These plans incorporate community involvement in decision making and a partnership approach to implementation. A range of cost-effective actions address common priorities. Integration of urban and transport planning, new vehicle emission standards, new fuel quality standards, and alternative fuels, target the reduction of transport-related emissions. New regulations on solid-fuel burning for domestic heating, agriculture, and forest and bushfire management aim to reduce particle pollution. Economic incentives aim to reduce major industrial emissions. Regular review and reporting of air quality management is included in local, state and national reporting on the state of the environment.

As the actions in the current air quality management plans reduce common urban air pollutants to meet national standards, processes to extend national goals continue. Draft discussion and issues papers for national protocols on $PM_{2.5}$, air toxics and indoor air quality have been released for public comment (CASOE [11], CAZANZ [44]), indicating the focus for action in the future.

Acknowledgement

The authors acknowledge: Olivier Rey-Lescure, Cartographer, University of Newcastle, NSW, Australia, for providing Figures 1 and 4; and Dr Peter Manins, Chief Research Scientist, Australian Commonwealth Scientific and Industrial Research Organisation, for providing data for Figures 2 and 3.

References

[1] Bridgman, H., Air pollution management in Australia: The example of Newcastle, NSW., *Air Quality Management*, Eds. J.S.W. Longhurst, D.M. Elsom & H. Power, WIT Press: Southampton, UK, pp. 159–198, 2000.

[2] CoA, *Intergovernmental Agreement on the Environment,* Commonwealth of Australia, Canberra, 1992, http://www.ea.gov.au/esd/national/igae/index.html

[3] NEPC Act, *National Environment Protection Council Act 1994,* Commonwealth of Australia, Canberra, 1994, http://www.austlii.edu.au/au/legis/cth/consol_act/nepca1994432/index.html

[4] AATSE, *Urban Air Pollution in Australia,* Australian Academy of Technical Sciences and Engineering, Parkville, Australia, 1997 http://www.ea.gov.au/atmosphere/airquality/urban-air/index.html

[5] Bridgman, H., Manins, P. & Whitelock, B., *Assessment of the Cumulative Emissions of Air Pollution from Kooragang Island and the Inner Suburbs of Newcastle,* a report submitted to the New South Wales (NSW) Department of State Development, TUNRA, University of Newcastle, NSW, Australia, 1992

[6] Boyle, R., Dewundege, P., Hazi, J., McIntosh, C., Morrell, A., Ng, Y. & Serebryanikova, R., *Report on the Air Emissions Trials for the National Pollutant Inventory,* Volumes 1 and 2, Environment Protection Authority of Victoria, Melbourne, Australia, 1996

[7] MAQS, *Metropolitan Air Quality Study*, Volumes 1–4, NSW Environment Protection Authority, Sydney, Australia, 1997

[8] NEPMa, *National environment protection measure for ambient air quality*, National Environment Protection Council, Commonwealth of Australia, Canberra, 1998, http://www.ephc.gov.au/php/list_documents.php?pdfType=Air_Quality_NEPM

[9] NSWSOEa, *New South Wales State of Environment Report 2000*, NSW Environment Protection Authority, Sydney, Australia, 2000, http://www.epa.nsw.gov.au/soe/soe2000.htm

[10] NEPMb, *National environment protection measure, and impact statement, for ambient air quality*, National Environment Protection Council, NEPC Service Corporation, Adelaide, 1998

[11] CASOE, *Australia State of the Environment Report, Atmosphere*, Commonwealth of Australia, 2001, http://www.ea.gov.au/soe/

[12] WPC and DEP, *The Perth Photochemical Smog Study*, Western Power Corporation and the Department of Environmental Protection, Perth, Australia, 1996, http://aqmpweb.environ.wa.gov.au:8000/air_quality/Publications/

[13] DEP, *Air Quality in Perth: 1992–1999*, Technical Report 109, Department of Environmental Protection, Perth, Australia, 2001, http://aqmpweb.environ.wa.gov.au:8000/air_quality/Publications/?AltContent=content_html

[14] EPANAA, *The NSW Government's 25-year Air Quality Management Plan, Action for Air,* NSW Environment Protection Authority, Sydney, Australia, 1998, http://www.epa.nsw.gov.au/air/actionforair/index.htm

[15] EPAV, *Draft Air Quality Improvement Plan, Port Phillip Region,* Environment Protection Authority of Victoria, Melbourne, Australia, 2000, http://www.epa.vic.gov.au/Air/aqip.asp

[16] EPAQ, *The South-East Queensland Regional Air Quality Strategy,* Environment Protection Agency of Queensland, Brisbane, Australia, 1999, http://www.epa.qld.gov.au/cgi-bin/w3-msql/environment/environment/airfeature/msqlwelcome.html?page=introduction.html

[17] PAQMP, *Perth Air Quality Management Plan,* Department of Environmental Protection, Government of Western Australia: Perth, Australia, 2000, http://aqmpweb.environ.wa.gov.au:8000/air_quality

[18] NEPMc, *Ambient Air Quality – Revised Impact Statement for the Ambient Air Quality National Environment Protection Measure,* National Environment Protection Council, Commonwealth of Australia: Canberra, Australia, 1998, http://www.ephc.gov.au/php/list_documents.php?pdfType=Air_Quality_NEPM

[19] Beer, T., Exposure assessment and health risk assessment—the Australian context, in *Air pollution and health risk,* Ed. T. Beer, CSIRO and Clean Air Society of Australian and New Zealand, Melbourne, Australia, pp 1–26, 2000.

[20] Morgan, G., Air quality and health risk in Sydney, In: *Air pollution and health risk,* Ed. T. Beer, CSIRO and Clean Air Society of Australian and New Zealand, Melbourne, Australia, pp. 101–116, 2000

[21] Simpson, R.W., Williams, G., Petroeschevsky, A., Rutherford, S., Mitchell, C. & Thalib, L., Air quality and health risk studies in Brisbane, In: *Air pollution and health risk,* Ed. T. Beer, CSIRO and Clean Air Society of Australian & New Zealand, Melbourne, Australia, pp. 79–90, 2000

[22] EPAVM, *Melbourne mortality study: Effects of ambient air pollution on daily mortality in Melbourne 1991–1996*, Publication 709, Environment Protcction Authority of Victoria, Australia, 2000, http://www.epa.vic.gov.au/Air/health.asp

[23] Hearn, D., Health risks due to motor vehicle emissions in Melbourne, *Clean Air*, **29**, 37–441, 1995

[24] USEPA, *The Benefits and Costs of the Clean Air Act, 1970–1990*, US Environmental Protection Agency, Draft, Washington, DC, USA, 1997
[25] Simpson, R. & London, J., *Report to the Brisbane City Council, An Economic Evaluation of the Health Impacts of Air Pollution in the BCC Area,* Brisbane, Australia, 1995.
[26] Keywood, M., Ayers, G., Gras, J. & Cohen, D., Size-resolved chemistry of Australian urban aerosols, *Proceedings of the 15th International Clean Air and Environment Conference*, Clean Air Society of Australia and New Zealand, Mitcham, pp. 550–555, Australia, 2000.
[27] NPI, *National Environment Protection (National Pollution Inventory) Measure, as varied 2000,* NEPC Service Corporation, Adelaide, Australia 2000, http://www.ephc.gov.au/nepms/npi/npi_downloads.html
[28] Rae, I.D., *Report of the Review of the National Environment Protection Measure for the National Pollutant Inventory,* NEPC Service Corporation, Adelaide, Australia 2000, http://www.ephc.gov.au/nepms/npi/npi_downloads.html
[29] Carnovale, F., Tilley, K., Stuart, A., Carvalho, C., Summers, M. & Eriksen, P., *Metropolitan Air Quality Study – Air Emissions Inventory*, NSW Environment Protection Authority, Sydney, Australia, 1996.
[30] DEPEI, *Perth Airshed Emissions Inventory Update 1998 - 1999,* Technical Series 110, Department of Environmental Protection, Perth, Australia, 2002, http://aqmpweb.environ.wa.gov.au:8000/air_quality/Publications/
[31] EPAVTS, *Air Quality Management Plan: Taking Stock*, Publication 535, Environment Protection Authority of Victoria, Melbourne, Australia, 1997
[32] Hyde, R., Young, M., Hurley, P. & Manins, P., *Metropolitan Air Quality Study – Meteorology – Air movements*, Consultants Report to the NSW EPA, NSW Environment Protection Authority, Sydney, Australia, 1997.
[33] EPANMP, *National Environment Protection Measure for Ambient Air Quality, Monitoring Plan for NSW*, NSW Environment Protection Authority, Sydney, Australia, 2001.
[34] NSWSOE, *NSW State of Environment 1997,* NSW Environment Protection Authority, Sydney, Australia, 1997, http://www.epa.nsw.gov.au/soe/97/index.htm
[35] EPAVEI, *Air emissions inventory: Port Phillip region*, Publication 632, Environment Protection Authority of Victoria, Melbourne, Australia, 1998, http://www.epa.vic.gov.au/Air/Emissions_Inventory
[36] QSOE, *State of the Environment Queensland 1999,* Environmental Protection Agency of Queensland, Brisbane, Australia, 1999, http://www.epa.qld.gov.au/cgi-bin/w3-msql/environment/science/environment/msqlwelcome.html
[37] DEPSK, *Perth air quality management plan: State of knowledge*, 2000, Department of Environment, Perth, Australia, http://www.environ.wa.gov.au/DEP/aqmp
[38] EPAVMP, *Victorian Air Quality Monitoring Plan –*, Environmental Protection Authority of Victoria, Melbourne, Australia, 2001, http://www.epa.vic.gov.au/Air/Monitoring/vicaqmp.asp

[39] QDE, *Air emissions inventory – south-east Queensland airshed*, Department of Environment, Brisbane, Australia, 1997.

[40] EPAQMP, *Ambient air monitoring plan for Queensland,* Environment Protection Agency of Queensland, Brisbane, Australia, 2001, http://www.epa.qld.gov.au/cgi-bin/w3-msql/environment/science/air/msqlwelcome.html?page=introduction.html

[41] Newton, P., *Re-shaping cities for a more sustainable future*, Research Monograph No. 6, Australian Housing and Urban Research Institute, Melbourne, Australia, 1997.

[42] Hurley, P. *The Air Pollution Model (TAPM) Version 1:Technical Description and Examples,* Atmospheric Research Technical Paper No. 4, Commonwealth Scientific and Industrial Research Organisation, Melbourne, Australia, 1999, www.dar.csiro.au/res/aq/TAPM

[43] DEPH, *The Perth Haze Study 1994–1996: Summary and Major Findings,* Department of Environmental Protection, Perth, Australia, 1996, http://aqmpweb.environ.wa.gov.au:8000/air_quality/Publications/

[44] CASANZ, *A summary of air quality management in Australia and New Zealand*, Clean Air Society of Australia and New Zealand, Mitcham, Australia, 2002.

[45] NEPC, *Annual Report 2000–2001,* National Environment Protection Council, Commonwealth of Australia, Canberra, 2001.

[46] SSROC, *Innovative Ways of Working Together*, South Sydney Regional Organization of Councils, Sydney, Australia, 1994.

[47] EPAVAQM, *State Environmental Planning Policy for Air Quality Management,* Environment Protection Authority of Victoria, Melbourne, Australia, 2000, http://www.epa.vic.gov.au/Air

[48] EPAVAAQ, *State Environmental Planning Policy for Ambient Air Quality,* Environment Protection Authority of Victoria, Melbourne, Australia, 2000

[49] RPI, Regional pollution index, NSW Environment Protection Authority, available at http://www.epa.nsw.gov.au/envirom/regpollution.htm

[50] PVAE, *Protecting Victoria's Air Environment, Draft Variation to SEPP (AQM) and SEPP (AAQ) and Draft Policy Impact Statement*, Environment Protection Authority of Victoria, Melbourne, Australia, 2000.

[51] DEEI, *Air emissions inventory – south-east Queensland airshed,* Department of Environment, Brisbane, Australia, 1997.

[52] DEMA, *The movement of air pollutants in the south-east Queensland airshed,* Department of Environment, Brisbane, Australia, 1997.

[53] DLPG, *The South-east Queensland Regional Framework for Growth Management,* Department of Local Government and Planning, Brisbane, Australia, 1998.

[54] Macintosh, K., Wainwright, D., Patterson, C. & Riese, R., Air Quality Management in South-east Queensland/ implementing SEQRAQS – the EPA Experience, *Proceedings 16th International Clean Air and Environment Conference*, 19–22 August 2002, Christchurch, New Zealand, CASANZ, Mitcham, Victoria, Australia 2002.

[55] DCE, *The Kwinana Air Modelling Study*, Report Number 10, Department of Conservation and Environment, Perth, Australia, 1982.

[56] Dockery, D.W., Pope, C.A., Xu, X., Spengler, J.D., Ware, J.H., Fay, M.E., Ferris, B.G. & Speizer, F.E., An Association between Air Pollution and Mortality in Six U.S. Cities, *New England Journal of Medicine,* **329**, pp. 1753–59, 1993.

[57] Dockery, D.W. & Pope, C.A., Acute Respiratory Effects of Particulate Air Pollution', *Annual Review of Public Health*, **15**, pp. 107–132, 1994.

[58] LA, *Report of the Select Committee on Perth's Air Quality* Legislative Assembly State Law Publisher, Parliament of Western Australia, Perth, Australia, 1998.

[59] Kenworthy, J., Laube, F., Newman, P. & Barta, P., *Indicators of Transport Efficiency in 37 Global Cities,* A report for the World Bank, Institute for Science and Technology Policy, Murdoch University, Perth, Australia, 1997.

[60] FORS, *Motor Vehicle Pollution in Australia - Report on the National In-Service Vehicle Emissions Study*, Federal Office of Road Safety, Commonwealth of Australia, Canberra, 1996.

[61] James, D., *Environmental Incentives: Australian Experience with Economic Instruments for Environmental Management*, Department of Environment, Sport and Territories, Commonwealth of Australia, Canberra, 1997.

[62] Productivity Commission, *Stocktake of Progress in Microeconomic Reform,* Australian Government Publishing Service, Canberra, Australia, 1996.

[63] Nelson P., Azzi, M., Carras, J., Lilley W., Cope M., Hurley, P., and Hyde R., *Inter-Regional Transport of Air Pollutants Study (IRTAPS),* CSIRO Energy Technology Investigation Report Et/IR 570R *Prepared for Pacific* PHB Hagler Bailly 2000, 'Assessment of the Proposed NOx Cap and Trade Scheme for NSW'. Vol. 1 Review and Assessment of Policy Options' *Final report prepared for Pacific Power International Power International on behalf of: Eraring Energy, Delta Electricity & Macquarie Generation* http://www.dar.csiro.au/publications/cope_2003a.pdf

CHAPTER 3

A critical evaluation of the local air quality management framework in Great Britain. Is it a transferable process?

J.W.S. Longhurst[1], C.I. Beattie[1], T.J. Chatterton[1], D.M. Elsom[2], N.S. Leksmono[1] & N.K. Woodfield[1]
[1]*Air Quality Research Group, Faculty of Applied Sciences, University of the West of England, Bristol, United Kingdom*
[1]*Faculty of Social Sciences & Law, Oxford Brookes University, Oxford, United Kingdom*

Abstract

The Air Quality Management (AQM) framework in Great Britain is an effects-based, risk management process designed to provide a dynamic solution to public health issues associated with elevated concentrations of seven air pollutants. The statutory basis of the AQM process is provided by The Environment Act 1995, which required an Air Quality Strategy to be published by Government and enabled the Government to introduce a series of air quality objectives for lead, carbon monoxide, 1,3-butadiene, sulphur dioxide, nitrogen dioxide, benzene and particulate matter (PM_{10}) in Regulations. The Act provides a framework in which national and local actions are required to identify and remediate areas of poor air quality. The Act places a series of duties and responsibilities upon Local Authorities to review and assess local air quality against air quality objectives set on the basis of best available epidemiological and public health information. The objectives, specified in Regulations, represent the Government's judgement of achievable air quality between the years 2003–2010 on the evidence of costs and benefits and technical feasibility. The objectives apply in areas where the public may be exposed for the averaging time of the relevant objective such as building facades, public open spaces, pavements and gardens. The objective will not be exceeded if there is no relevant public exposure. Local Authorities are required to identify those areas in which air quality objectives will be exceeded by a target date and to declare these areas as Air Quality Management Areas (AQMA). In such locations a Local Authority must develop an Action Plan detailing both the measures to be taken and the time-scale for implementation in order that the air quality objectives are achieved within the AQMA. The Act places a duty on Government to support Local Authorities through the provision of guidance and other initiatives. This has included the development of high quality national monitoring networks, the development of emissions inventories, the provision of training for local authority personnel and the provision of additional financial resources to assist local authorities in purchasing equipment and technical resources. A key support element has been the development of web, telephone and email help desks and support materials.

The Air Quality Strategy is dynamic and subject to rolling review to reflect developments in European legislation, technological and scientific advances, improved air pollution modelling techniques and an increasingly better understanding of the socio-economic issues involved. Similarly the Air Quality Regulations, which give statutory weight to the air quality objectives, are reviewed periodically to reflect new developments in epidemiological and scientific understanding of the relationship between human health and specific air pollutants. Local Authorities began the process of review and assessment in 1999 and the first round of the process concluded in 2001 following which some 129 Local Authorities have or expect to declare AQMAs. Review and Assessment is considered by Government to be a continuous process and Local Authorities are encouraged to consider it as such although there are distinct phases of work during which the Government exercises its power as a Statutory Consultee to appraise the work of Local Authorities and to assure itself that the statutory duties are being undertaken in an appropriate manner. The review and assessment elements of the framework have recently been subjected to an evaluation and the essential elements of this process have been confirmed as fit for purpose. Thus the procedures will continue in use for the next round of reviews and assessments that are due to be concluded by the end of 2003 except in London where the Mayor has agreed a deadline of 2004.

1 Introduction to historical context and conceptual framework

Great Britain (all areas of the UK excluding Northern Ireland) has a long history of air pollution and regulatory responses to its control [1]. Air quality problems have been an inescapable partner of the economic development of Great Britain and such problems predate the industrial age. Air pollution can be conceptualised as an outcome of political, economic and societal decision making. The impact of air pollution is mediated by the availability of technology and the interaction between these issues is a dynamic one. Social, political and economic responses to air pollution have often lagged behind awareness and understanding of the impact of air pollution and the public's willingness to accept a particular amount of pollution changes over time. The ability and willingness of political institutions to tackle air pollution varies with the level of public concern, the economic costs of change, the impacts of air pollution and the availability of technological solutions. The approach taken in Great Britain to manage air quality should be considered in this context. Considered in a 200-year historical context, the dynamic interaction of air pollution with British social, economic and political considerations can be viewed as a process as follows.

- Adoption of new technology leads to observation of unintended consequences;
- Raised public concern leads to scientific inquiry and identification of the major components of the problem;
- Public concern and disquiet;
- Institutional inability or unwillingness to act sufficiently to manage the problem;
- Risk of catastrophe, public outcry, institutional reaction and legislation;
- Implementation of regulation and adaptation of society.

Economic, political, social and technological actions and reactions have sought to manage pollution whilst minimising the adverse consequences of control on business and society at large [2]. However, the reaction has not always been appropriate or timely particularly in the case of air pollution in urban areas where a city's form and function and often its topography predisposes it to air pollution. In Britain's air pollution history much of the concern has been with the use of coal as a fuel where the incomplete combustion of coal in industrial furnaces and kilns or domestic grates and fireplaces created air pollutants including smoke and sulphur dioxide. There are good descriptions of air pollution and its impacts in Great Britain from the 17th Century onwards and these observations included an analysis of potential solutions [1]. It required a catastrophe - the London Smog of 1952 - before the UK addressed urban air pollution effectively. Even then it took two decades before the last severe coal based smog was eliminated from London. The 1952 Great Smog of London was a seminal event in the reaction it provoked [3]. Although smogs had been experienced for nearly 200 years, the scale and impact of this smog exceeded any that had preceded it [2]. On Friday December 5th 1952 an unusually severe fog descended upon London as a slow moving anticyclone came to a halt over London [1,3]. On Friday morning the fog was the thickest in living memory. It grew thicker throughout the day and people began to experience discomfort in breathing. Those outside rapidly became smoke blackened and aware of the choking smell. That evening respiratory cases, especially bronchitis, were twice the normal rate. Some million chimneys added more smoke to the fog. On Saturday visibility was near zero, prize animals at Smithfield show took ill and some had to be destroyed. At Sadlers Wells an opera performance had to be stopped due to poor visibility in the theatre. People continued to suffer health effects. On December 8th the fog was as thick as ever, respiratory problems continued to be reported. Transport was at a complete standstill. The fog cleared by the 10th of December as the anticyclone moved away [1]. It is estimated that the smog caused 4000 excess deaths. The smog of 1952 was a catastrophe with a very high increase in respiratory mortality and morbidity. As is often the case in environmental matters it was the acute impact that was the catalyst for legislation. It is appropriate to ask why a catastrophe was necessary for action to be taken? The sources, impacts and controls were well known before the 1952 smog, indeed had in part been known since the 17th Century [1,2]. Many reasons could be offered of which indifference and concerns about the cost would figure highly. What is clear is that mitigating the public health impact was not afforded a sufficiently high priority. What is often overlooked is that prolonged exposure to high wintertime concentrations of smoke reduced the quality of life, causing illness for many and brought an untimely death to many more. The total of these deaths foreshortened will far outweigh the total deaths caused by the 1952 smog but they were not enough to bring effective change.

In 1953 the Beaver committee was established to examine the nature, causes and effects of air pollution and the efficacy of present preventative measures; to consider what further measures are practicable; and to make recommendations for action by government. The report recommended action on domestic smoke and the effective use of fuel in industry. The Beaver Committee's recommendations were neither novel nor new but they catalysed the government into action. The recommendations suggested the clearance of the log jam of obsolete laws and administrative practices and created a policy framework for clean air, an act to cover domestic and industrial emissions of smoke from new and existing premises. The 1956 Clean Air Act banned dark smoke emissions from chimneys, railway engines and vessels, required new furnaces to be smokeless and required the emissions of grit and dust from furnaces to be minimised [4]. Importantly, the Act gave local councils' powers to introduce smokeless zones. The 1956 Act was later amended and extended by the *Clean Air Act, 1968* [3, 4]. The 1956 and 1968 Acts were consolidated and their provisions re-enacted in the *Clean Air Act, 1993.* The 1956 and 1968 Acts constituted the operative legislation against pollution by smoke, grit and dust from domestic fires and other commercial and industrial processes not covered by the Alkali Acts and other subsequent legislation [4]. They regulated the combustion of solid, liquid and gaseous fuels and controlled the heights of new industrial chimneys that are not scheduled elsewhere. The Acts prohibited the emission of dark smoke from any chimney, provided for government funding for the conversion of domestic grates to smokeless operation, and regulated the fuels that could be burnt on them [3, 4].

The Clean Air Acts dealt with visible pollution especially smoke. They had a dramatic effect in reducing the concentration of visible black smoke particles in the air. In Manchester the annual mean concentrations of smoke fell by some 90% between 1959 and 1984. The bronchitic death rate – an indicator of the health effects of air pollution – reduced in line with the concentration of smoke [2, 5]. During the 1980s, attention again turned to the regulation of industrial sources of emission and led directly to the *Environmental Protection Act 1990* which sought to build upon the long standing British philosophy of using the best practicable means to render harmless such emissions as are released from industrial activities [4]. The Act introduced integrated pollution control (IPC) to cover all major releases to air, land and water from the most polluting and complex industrial processes, known as Part A processes, covering some 2000 processes in chemical works, power stations, oil refineries and other industries. The Act introduced the concepts of Best Available Technology Not Entailing Excessive Costs (BATNEEC) and Best Practicable Environmental Option (BPEO). The enforcing authority sets conditions so as to ensure that the BPEO is used. Part 1 of the Act also introduced a new range of responsibilities for local authorities in respect of air pollution control from less complicated processes, described in the Act as Part B processes [4]. The regulation of such processes by the local authorities is known as Local Air Pollution Control (LAPC). Whilst both IPC and LAPC are concerned with managing and minimising the release of pollutants to the environment, the integration between them did not extend to an encompassing effects-based programme for urban air pollution minimisation. Nor even when considered in combination with the Clean Air Acts can a holistic approach be said to have occurred.

The Clean Air Acts cleared the air of visible pollutants but did not clean the air of all pollutants. The Acts did not specifically target sulphur dioxide or very fine particles and although substantial reductions in annual mean concentrations did occur, this was a by-product of other processes set in train by the acts. The emphasis on visible pollutants was appropriate at the time the Acts came into statute and their success was due to the focus on one fuel - coal - and on the emission of one pollutant, black smoke. Their success bred complacency and fostered a view that urban air pollution had been conquered. Black smoke had been dealt with but the problem of gaseous pollutants and fine particles would return later as the location of sources, the intensity of the sources of emission and the composition of air pollution changed. Concentrations of pollutants regulated by the Acts fell, whilst those from motor vehicles increased substantially in the period from 1985 onwards [6]. Thus the elimination of "pea soupers" created a prevailing ideology that urban air pollution had been conquered. Resources were directed towards other problems and the onset of health-based concerns about the impacts of the growth of road traffic in the 1980s and 1990s left the UK woefully ill prepared to tackle an emergent problem of traffic-sourced air pollution [6]. The press directed a powerful current of public concern over child health, asthma and pollution from motor vehicles that repositioned urban air pollution and public health at the forefront of the political agenda and led to *the Environment Act*, 1995 [6]. This time there was more than one fuel and one pollutant to confront.

2 Developing the new legislative framework for air quality management

In 1990, the Government White Paper *'Our Common Inheritance. Britain's Environmental Strategy'* [7] introduced a new direction for air quality control.

This was to build upon the existing technology-based controls by adding an *effects-based*, risk management approach through the formulation of a series of air quality standards [6]. The new framework was termed *Local Air Quality Management* and led to the formulation of the Environment Act 1995 (Part IV, Air Quality) [8], the primary legislation for the process of air quality management. Before the passing of the Act, the then Department of the Environment (DoE) published two discussion papers for consultation, entitled '*Improving Air Quality*' (1994) and '*Air Quality: Meeting the Challenge*' (1995). The first paper discussed the policies and procedures required for the setting, achievement and maintenance of air quality standards. It recognised the need for an effects-based approach to air quality management expressed in terms of air quality standards and management regimes, tailored to local circumstances [9]. The second discussion paper, 'Air Quality: Meeting the Challenge' [10] proposed the placing of a statutory duty on all local authorities to review air quality regularly, not only those with designated areas of poor air quality. The paper also set out the Government's aim to achieve air quality standards by 2005. The various consultation papers recognised that an approach solely based on point-source control was no longer appropriate for the multi-sourced, multi-pollutant, air pollution situation of the 1990s and thus the Act introduced a radical shift from sources to effects-based control and introduced human-health-based objectives and standards [6]. The Act applies only to Great Britain. Northern Ireland will introduce a bill in 2002 with the intent of bringing into force an Act in 2003. The bill in Northern Ireland will be based on the experience in Great Britain. The key sections of the Environment Act, 1995 that relate to air quality management are as follows:

- **Section 80** Obliges the Secretary of State (SoS) to publish a National Air Quality Strategy.
- **Section 81** Obliges the Environment Agency to take account of the Strategy.
- **Section 82** Requires local authorities, unitary or district, to review air quality and to assess whether the air quality standards and objectives are being achieved. Areas where standards fall short must be identified.
- **Section 83** Requires a local authority, for any area where air quality standards are not being met, to issue an order designating it an air quality management area (AQMA).
- **Section 84** Imposes duties on a local authority with respect to AQMAs. The local authority must carry out further assessments and draw up an action plan specifying the measures to be carried out and the timescale to bring air quality in the area back within limits.
- **Section 85** Gives reserve powers to cause assessments to be made in any area and to give instructions to a local authority to take specified actions. Authorities have a duty to comply with these instructions.

- **Section 86** Provides for the role of County Councils (a higher tier of government operating at sub-regional level) to make recommendations to a district on the carrying out of an air quality assessment and the preparation of an action plan.
- **Section 87** Provides the SoS with wide-ranging powers to make regulations concerning air quality. These include standards and objectives, the conferring of powers and duties, the prohibition and restriction of certain activities or vehicles, the obtaining of information, the levying of fines and penalties, the hearing of appeals and other criteria. The regulations must be approved by affirmative resolution of both Houses of Parliament.
- **Section 88** Provides the SoS with powers to make guidance that local authorities must pay due regard.

The National Air Quality Strategy (NAQS) was published in March 1997 [11]. The Strategy outlined the methods and targets to be pursued by Government based on health effects standards for eight pollutants. The standards were established through the work of a government-commissioned Expert Panel on Air Quality Standards (EPAQS, see http://www.defra.gov.uk/environment/airquality/aqs/index.htm.). The Air Quality Regulations 1997 [12] subsequently provided the legal framework for standards and objectives for seven pollutants, excluding ozone from local AQM, to be achieved by 2005. The NAQS was reviewed in 1998/99 to reflect developments in European legislation, technological and scientific advances, improved air pollution modelling techniques and an increasingly improving understanding of the economic and social issues involved. The revised strategy also slightly changed the UK air quality approach. Prior to the revision, the NAQS was exclusively based on protecting human health. However the new UK Air Quality Strategy for England, Scotland, Wales and Northern Ireland (UK AQS) [13] (http://www.defra.gov.uk/environment/airquality/airqual/index.htm) included the protection of the environment (i.e. vegetation and ecosystems), although these objectives are not under local regulation. Of the eight pollutants included in the NAQS, five objectives were tightened, two stayed static and one (PM_{10}) was relaxed. Whilst the relaxation of the particulate objective was seen as a pragmatic response to uncertainty, the Government has acted in the case of benzene, 1,3-butadiene and carbon monoxide by changing the target date to 2003 (http://www.defra.gov.uk/environment/airquality/airqual/index.htm). In 2002 further amendments to the air quality objectives have been consulted on in England and the devolved administrations. This consultation has resulted in (for the first time) different objectives being set in England, London, Wales and Scotland. In Scotland, the Air Quality (Scotland) Amendment Regulations 2002 were formalised in June 2002 and the Air Quality (England) Amendment Regulations in late 2002. In Wales the proposals were expected to be in place by the end of 2002. The current position is presented in Table 1.

Table 1: Air quality objectives for the purposes of air quality management in the United Kingdom

Pollutant	**Air Quality Objective**		**Date to be achieved by**
	Concentration[1]	**Measured as**	
Benzene All authorities	16.25 μg/m^3	running annual mean	31.12.2003
Authorities in England only[1]	5 μg/m^3	annual mean	31.12.2010
Authorities in Scotland [Wales] and [Northern Ireland] only[2]	3.25 μg/m^3	running annual mean	31.12.2010
1,3 Butadiene	2.25 μg/m^3	running annual mean	31.12.2003
Carbon monoxide	10.0 mg/m^3	running 8-hour mean	31.12.2003
Lead	0.5 μg/m^3 0.25 μg/m^3	annual mean annual mean	31.12.2004 31.12.2008
Nitrogen dioxide[3]	200 μg/m^3 not to be exceeded more than 18 times a year 40 μg/m^3	1-hour mean annual mean	31.12.2005 31.12.2005
Particles (PM_{10}) (gravimetric)[4] All authorities	50 μg/m^3 not to be exceeded more than 35 times a year 40 μg/m^3	24-hour mean annual mean	31.12.2004 31.12.2004
Particles (PM_{10}) (gravimetric)[4] Authorities in [Wales] and [Northern Ireland] only[2]	50 μg/m^3 not to be exceeded more than 7 times a year 20 μg/m^3	24-hour mean annual mean	31.12.2010 31.12.2010
Particles (PM_{10}) (gravimetric)[4] Authorities in Scotland only[5]	50 μg/m^3 not to be exceeded more than 7 times a year 18 μg/m^3	24-hour mean annual mean	31.12.2010 31.12.2010

Sulphur dioxide	350 µg/m^3 not to be exceeded more than 24 times a year	1-hour mean	31.12.2004
	125 µg/m^3 not to be exceeded more than 3 times a year	24-hour mean	31.12.2004
	266 µg/m^3 not to be exceeded more than 35 times a year	15 minute mean	31.12.2005

1. Regulations 2002 have not yet been laid for England and Wales
2. The Welsh Assembly Government has yet to make a decision on the objectives to be included in the Regulations 2002 and the values stated in this table are subject to confirmation. In Northern Ireland the objectives are to be given legal effect in Regulations following the introduction of the Local Air Quality Management Bill, in June 2003
3. The objectives for nitrogen dioxide are provisional
4. Measured using the European gravimetric transfer sampler or equivalent.
5. These 2010 air quality objectives for PM_{10} apply in Scotland only, as set out in the Air Quality (Scotland) Amendment Regulations 2002.

The Air Quality Strategy sets out how the Government and the devolved administrations (these are the administrations in Scotland, Wales and Northern Ireland that are responsible for the setting and implementation of air quality policy) will secure the Air Quality Objectives. Local Air Quality Management is designed to identify those local areas (hotspots) where national actions on their own will not secure the air quality objectives. The Strategy sets out a clear and simple policy framework, with realistic objectives that are underpinned by regulation and financial incentives to help achieve objectives [13]. The strategy provides an analysis of costs and benefits in order to prioritise actions and the achievements of strategy targets are to be tracked by high quality air pollution monitoring, the results of monitoring are made publicly available. Government have described the purpose of the Air Quality Strategy as "to make sure everyone can enjoy a level of ambient air quality in public places which poses no significant risk to health or quality of life" [13].

The Air Quality Strategy has set out a range of measures and policies to reduce emissions of regulated pollutants. Concentrations of lead, carbon monoxide, benzene, sulphur dioxide (in most cases) and 1,3 butadiene are unlikely, in the Government's view [13], to exceed objectives in the relevant year because of the measures and policies introduced nationally and enforced by government and its national agencies. In some industrial locations, benzene and lead exceedances are predicted but there is no public exposure. However, in the case of nitrogen dioxide, PM10 and in some cases, sulphur dioxide, local action is needed to complement and enhance the effect of national polices and actions. Such action is defined by the Act and Strategy as Local Air Quality Management

[13]. Government, in the publication of the revised Air Quality Strategy and Air Quality Regulations in 2000 has made it clear that the responsibility for securing the Air Quality Objectives rests with Government and the devolved administrations, noting that Local Government has a duty to work towards securing the Air Quality Objectives. This is a significant change from 1997 when the duty to secure the air quality objective lay with the local authority.

Section 82 of the Environment Act, 1995 required Local Authorities in Great Britain to undertake a review of local air quality and to assess, using a risk management based approach, the likelihood of achieving a number of health based air quality objectives set out in Table 1. Where air quality objectives are not likely to be achieved, local authorities are required to designate Air Quality Management Areas (AQMAs) and put in place action plans to improve the local situation. The objectives defined in the Strategy apply at locations which are situated outside buildings or other natural or man made structures above or below ground and where members of the public are regularly present and might reasonably be expected to be exposed over the relevant averaging period [13]. In the Government view "to pursue polices to achieve objectives at places where the highest measurable concentrations occur without regard to whether or not the public might be exposed would be inappropriate and highly inefficient." [13]. An Air Quality Objective specified in the Air Quality Regulations, 2000 may be exceeded only when public exposure occurs. This is the heart of the risk management approach to Local Air Quality Management, for the process of Review and Assessment is designed to identify those areas where poor air quality coincides with public exposure. In reaching a decision to declare an AQMA the Local Authority must be convinced that relevant public exposure exists in relation to the averaging time of the Objective.

Objectives with an annual mean, such as benzene or lead, apply at all background locations where members of the public might regularly be exposed and at building facades of residential properties, schools and hospitals. They do not apply at the facades of buildings or places of work where the public does not have regular access. Kerbsides, gardens and other locations where exposure is expected to be short term also are not applicable sites. In the case of pollutants such as sulphur dioxide or PM_{10} which have a 24 hour objective or carbon monoxide with an 8 hour objective, then the objectives apply at all locations where the annual objectives apply and in addition will apply in the gardens of residential properties. They will not apply at kerbsides or other locations where the public exposure is judged to be short term. The sulphur dioxide and nitrogen dioxide 1-hour objectives apply to locations where the 24-hour objectives apply and in addition at kerb sides of busy shopping streets, those parts of car parks and railway stations that are not fully enclosed and to any location where the public might reasonably be expected to be exposed. The objectives will not apply to kerbside locations where the public would not be expected to have regular access. The 15 minute sulphur dioxide location will apply in those locations where the public might reasonably be expected to be exposed for a period of 15 minutes or longer.

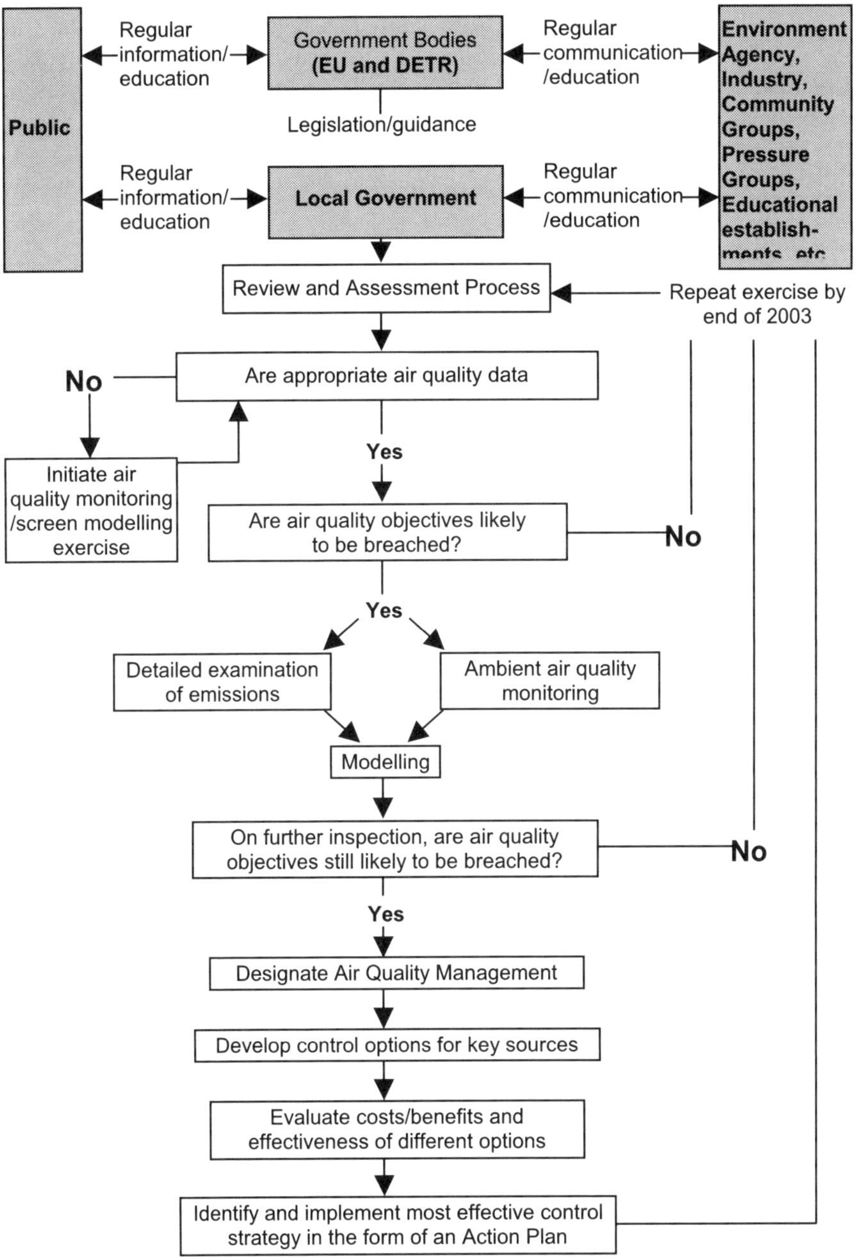

Figure 1: Process of Review and Assessment as undertaken by Local Authorities in Great Britain.

As well as the system of Local Air Quality Management the Government and the devolved administrations are bound by the requirements of the European Union Directives. Whilst the EU Limit Values do not have the same emphasis on exposure as do the Air Quality Objectives, there are common principles between the EU and British approach including the assessment against target dates and the requirement to produce action plans in areas where Limit Values will not be met [14].

Local authorities have now finished the first round of reviews and assessments and those local authorities who identified a problem and designated an air quality management area(s) are currently undertaking the action planning stage of the process. This remainder of this paper is a review of the first round of air quality reviews and assessments and reflects on some of the issues and lessons learnt prior to the next round of reviews and assessments, which is due for completion before the end of 2003 in Great Britain, outside of London. Figure 1 shows the process of Review and Assessment as undertaken by local authorities in Great Britain.

3 Local air quality management

The emphasis of delivering the requirements of the AQS was action at a local level. Local government has a statutory duty to work towards the air quality objectives, and has subsequently been provided with new powers to assist in delivering the strategy (for example vehicle-emissions testing). Controlling air pollution and thereby quality of air is not an entirely new responsibility for local authorities. Local government has been responsible for air emissions from those less complex industrial processes (or Part 'B' processes) identified by the Environmental Protection Act 1990 [15]. The situation is slightly different in Scotland where the Scottish Environmental Protection Agency is responsible for all prescribed processes. However, for the first time, local government in Great Britain is required to consider air pollution and subsequently air quality in an effects-based approach. Emissions to air clearly arise from a variety of sources over varying spatial and temporal scales, and it is becoming increasingly challenging to integrate solutions into existing policy packages, particularly through the use of transport plans, land use plans, economic development plans and sustainable development plans.

3.1 Review and assessment

The first aspect of the LAQM regime is a review and assessment of air quality in the local authority's area. The government recommended a three-stage approach, whereby each stage of the review and assessment process increases in depth and complexity. This should be consistent with the risk of failing to achieve the air quality objectives [11]. The first stage of the process is essentially a desktop study of emissions sources within, and impacting on, the local authority area. Stage 2 introduces screening models and interpretation of available monitoring data. Stage 3 is a complex modelling and monitoring exercise assessing air quality concentrations against the specified objectives. Examples of reports produced at each stage of the process can be found at www.uwe.ac.uk/aqm/review/examples/index.html. At each stage pollutants can be omitted from the process where it can be shown that the air quality objectives are likely to be achieved. On completion of a third stage review and assessment, and in areas where air quality objectives are predicted to exceed by their target date, an Air Quality Management Area (AQMA) must be designated. Where AQMAs have been designated, local authorities are required to prepare a written action plan illustrating how the air quality standards and objectives are to be achieved and improved upon in the designated area [16].

Prior to the publication of the National Air Quality Strategy, the probable timetable for the review and assessment process was that assessments would be complete by 1998 with action plans implemented thereafter. After 1997 and the late publication of the NAQS, Central Government proposed that local authorities should complete their Stage 1 review and assessments by the end of 1998, with the initial review and assessment phase (to Stage 3 where necessary) to be completed by the 31 December 1999. This timetable was initially extended to June 2000 and subsequently to the end of 2000. Where an AQMA is designated, a 12-month period for a further review and assessment of air quality in the AQMA is to be undertaken. All local authorities (except in London) must undertake a further round of review and assessment of air quality before 31 December 2003.

It was initially anticipated that many (particularly rural) authorities would not proceed further than a Stage 1 assessment and that large metropolitan areas and cities would be most likely to declare AQMAs [17]. It was not anticipated therefore that the majority of local authorities would require a Stage 3 assessment. The pollutants causing greatest challenges were initially anticipated to be nitrogen dioxide and PM_{10}, although the level to which the annual mean nitrogen dioxide objective is predicted to be exceeded was never anticipated. Transport has long been recognised as the major contributor of emissions in locations requiring air quality management areas. However, the wide scale coincidence of concentrations above objective values and exposed populations was not anticipated.

3.2 Support mechanisms for local authorities

Following publication of the NAQS, local authorities expressed concern about the lack of necessary tools or indeed expertise to undertake their AQM responsibilities effectively. Central government responded by making available national resources that local authorities could draw upon [14]. Resources included specific guidance documents, Internet-based information including monitored data, emissions inventories, and telephone and e-mail help desks [18].

Paper-based guidance documents were the initial method of helping local authorities in the UK with their AQM tasks. They were published in 8 volumes, with G1–G4 covering the framework of the management process [16], developing local air quality action plans and strategies [19], air quality and traffic management [20] and air quality and land use planning [21]. These were subsequently followed by a series of 4 technical guidance notes (TG1-4) offering advice on monitoring [22], emissions inventories [23], dispersion modelling [24] and a pollutant-specific guidance document [25], which outlined specific methodologies for reviewing and assessing each pollutant.

Following the updating of NAQS to the UK Air Quality Strategy, the guidance documents were also updated in the same format as G1-4 [26–29] and TG1-4 [30–33]. The Scottish Executive at the same time produced their own G series of guidance notes, SG1-4. These documents are available on www.uwe.ac.uk/aqm/review/links.html. The technical guidance has been prepared by a consortium of leading organisations in the field of air quality and thus the guidance represents a technical consensus on the approach to be taken by a local authority. This consensus underpins the review and assessment process.

The technical guidance has been summarised in the checklists local authorities are advised to use in their review and assessment work. The checklists may be viewed at www.uwe.ac.uk/aqm/review/checklists/ and a summary is provided in Appendix 1. These checklists are used in the independent appraisal, commissioned by Government, of local authority review and assessment work.

The next round of review and assessment will be different from the previous round. It has been amended on the basis of the evaluation of the process utilised in the first round. A two-step process is to be utilised and a revised set of guidance documentation is currently being prepared. These will incorporate the latest scientific and technical considerations and the format, following feedback in the evaluation process, will be more prescriptive than that used in the first round. The four technical documents have been amalgamated into one document [34] and likewise the four policy guidance documents have also been combined into one document [35].

A variety of other documents have also been provided for local authorities for example guidance on air quality management consultation [36] and advice on the designation of Air Quality Management Areas [37] and land use planning and air quality considerations [38]. More recently guidance on air quality action plans and strategies has been provided by the National Society for Clean Air [39, 40]. These can be found at www.uwe.ac.uk/aqm/centre/aqaps/index.html.

Other sources of technical information for review and assessment include a government co-ordinated Automatic Urban and Rural Network (AURN) of automated analysers (http://www.airquality.co.uk). Whilst a substantial number of AURN sites are part funded by local authorities in combination with central government, these stations become part of the AURN once instrumentation and site operation procedures are approved. Central government provides telemetry to link the AURN sites with a central data collection point. The government provides a quality control service, but the cost of establishing and operating these sites remains largely with local government.

Pollution data from all AURN sites and other pollution monitoring networks, such as the national NO_2 diffusion tube network comprising over 1100 tubes, are available via the National Air Quality Information Archive internet site established in April 1997 (see http://www.airquality.co.uk). The AURN and other pollution monitoring networks have enabled background pollution concentrations (in 1×1 km grid squares maps) to be compiled for both current and future target years, which are also available at the above internet site [14]. In response to limited emissions inventory information available for local authorities, the National Atmospheric Emissions Inventory (NAEI) was provided on the internet (see http://www.naei.org.uk). This inventory provides estimates of emissions of the key pollutants on a 1×1 km level basis. The internet site offers other key information such as emission factors for vehicles (disaggregated by type and speed) and industrial processes emission factors. The Government also commissioned ten detailed emissions inventories of major urban areas [see, for example, 41] including inventories for London, Manchester, Birmingham, Merseyside and Bristol amongst others. Academic work also contributed to the development of tools and techniques for estimating local emissions [see for example, 42]. Also in response to local authority needs, an existing screening modelling tool, the Design Manual for Roads and Bridges (DMRB), which had been developed by the Department of Transport primarily for use in assessing the environmental impact of road schemes, was updated for use by the air quality community to provide direct comparisons with air quality objectives. The updated version was then produced in spreadsheet format for use by local authorities and made available free on the internet (www.stanger.co.uk). In addition to this tool, the model AEOLIUS used to assess air pollution in low-dispersing canyon streets, pioneered by the Meteorological Office was also updated to reflect its new role in the review and assessment process (http://www.meto.gov.uk/environment/aeolius1.html).

Helpdesks and associated web sites were also set up to support local authorities in their review and assessment duties. These include:

- Review and Assessment helpdesk (operated by Air Quality Consultants Limited and University of the West of England) (www.uwe.ac.uk/aqm/review/index.html)
- Monitoring and emission inventory helpdesk (operated by NETCEN) (www.airquality.co.uk)
- Modelling helpdesk (operated by Casella Stanger) (www.stanger.co.uk/modelhlp)
- Action Planning helpdesk (operated by Casella Stanger in conjunction with Transport and Travel Research Ltd) (www.stanger.co.uk/actionplan)

These same organisations collaborate in the preparation of technical guidance for the local air quality management process.

In order to ensure that local authorities discharged their duties in an appropriate manner and also to ensure a consistency of approach, Government introduced an appraisal system linked to Government's statutory consultee role. The appraisal system provides an objective, independent report to Government on the appropriateness of the approach and conclusions reached by a local authority (for details see www.uwe.ac.uk/aqm/review) The authors of this paper are part of a team contracted by the Department for Environment, Food and Rural Affairs, and devolved administrations to appraise all UK air quality review and assessment reports and maintain the national archive of assessment reports.

4 How have local authorities fared in the review and assessment process?

This section will briefly outline some of the main outcomes of the review and assessment process. The data includes that generated by the University of the West of England authors on behalf of DEFRA and the devolved administrations (www.uwe.ac.uk/aqm/review) and also results from a 5-year research programme looking at the implementation of the Environment Act Part IV by urban local authorities in England (for more information on research currently being undertaken at the University of the West of England, Bristol, see www.uwe.ac.uk/aqm/research). The survey work is designed to identify, track and evaluate capacity in local authorities to manage local air quality. The methodology allows both technical (e.g. modelling, monitoring) and managerial (e.g. communication strategies, consultation, policy integration and joint working) capabilities to be considered. The surveys have included environmental professionals, who have been responsible for pollution legislation within local authorities and are, in most cases, co-ordinating their air quality management efforts, as well as other local authority professionals. Contemporary air pollution sources are diverse and hence solutions to tackle problem locations will need to come from a variety of sources including transport planning, land use planning and economic development sectors as well as agencies and bodies external to the local authority such as health authorities, the Environment Agency and Highways Agency. Hence each of these sectors have been surveyed annually since 1998 with regard to their involvement in the management of local air quality (see http://www.uwe.ac.uk/aqm/research/urban). Case studies of authorities identified as advanced in their approach to air quality management responsibilities have been undertaken. A*dvanced* as defined by the authors is where local authorities have shown an integration of air quality issues into other planning processes as well as competence in the technical aspects of monitoring and modelling.

Table 2 illustrates the pollutants taken further than stage 1 of the process by all local authorities in the UK. This indicates the pollutants causing greatest problems in relation to achieving the health based objectives and highlights some of the regional differences. In all cases, nitrogen dioxide, PM_{10} and sulphur dioxide have involved the most work by local authorities within the review and assessment process. Sulphur dioxide has been a particular problem in Northern Ireland where there are still large areas of domestic coal burning. Carbon monoxide has been more of an issue in London, which has generally higher traffic flows and low traffic speeds in the congested urban centres.

Table 2: Local Authorities in Great Britain and Northern Ireland Intending to proceed to Stage 2/3 Following Completion of Stage 1, specified by pollutant (as of January 2002)

Country	**Total for 1 or more**	**Pollutants to be taken to Stage 2/3**							
		CO	Benz	1,3-but	Pb	NO_2	PM_{10}	SO_2	All 7*
England (exc. London)	291	174 (60%)	82 (28%)	44 (15%)	103 (35%)	285 (98%)	285 (98%)	212 (73%)	27 (9%)
London	33	27 (82%)	7 (21%)	1 (3%)	11 (33%)	33 (100%)	33 (100%)	26 (79%)	-
Scotland	21	5 (24%)	2 (10%)	1 (5%)	3 (14%)	17 (81%)	15 (71%)	11 (52%)	1 (5%)
Wales	20	6 (30%)	3 (15%)	1 (5%)	5 (25%)	20 (100%)	18 (90%)	12 (60%)	1 (5%)
N. Ireland	22	6 (27%)	3 (14%)	-	2 (9%)	21 (95%)	19 (86%)	21 (95%)	-
Total	387	218	97	47	124	376	370	282	29

* The numbers within this column are also included in the individual pollutant columns.

4.1 Air Quality Management Areas (AQMAs) designated

By July 2003, 122 local authorities have declared AQMAs with a further 9 anticipating declaring in due course (an up-to-date record of declared AQMAs can be found at www.uwe.ac.uk/aqm/centre/aqmas/index.html). Nitrogen dioxide (NO_2) is the pollutant for which predicted exceedences are greatest, particularly with respect to the annual mean objective. Coupled with this is a widespread anticipated exceedance of the short-term PM_{10} objective, both of which are largely as a result of traffic emissions. Over 95% of AQMAs are to be designated as a result of traffic emissions. Traffic congestion exerts the greatest influence on the AQMA decision-making process within local authorities, and is therefore likely to be a significant target for action in air quality policy across England, Scotland and Wales.

Four of the seven regulated pollutants are predicted to remain below their objective concentrations or to have no exceedances in areas of public exposure, and only predicted exceedences of the sulphur dioxide (SO_2) objectives will require a small proportion (10%), of authorities to declare AQMAs other than for nitrogen dioxide and particulates. Those AQMAs to be declared as a result of SO_2 are due to industrial point source emissions, or combinations of emissions from a number of such point sources, or other sources such as shipping. Of all anticipated AQMAs in the UK, a quarter of declaring local authorities will require a single area of designation for one pollutant, with 16% of authorities likely to require more than one AQMA for a single pollutant, and 11% anticipating a single AQMA for more than one pollutant. Local authorities have not received prescriptive guidance on the designation of AQMAs in the UK, although informal guidance has been developed to assist authorities with the process [37] (see http://www.uwe.ac.uk/aqm/review/aqap2.pdf).

Of the AQMAs declared and anticipated, a suite of common AQMA typologies has emerged. Most AQMAs will include urban road networks, or a number of individual streets collectively, and authorities are choosing to declare individual roads and specific receptors exposed to exceedences along them, as well as pockets and areas of land in between road networks. A number of authorities, in smaller conurbations or semi-urban or rural locations have identified, through monitoring and in some cases less advanced modelling techniques, small areas of predicted exceedences along individual high streets [43]. Some rural authorities, such as Salisbury DC in Wiltshire, South Lakeland DC in Cumbria and Harborough DC in Leicestershire have designated local canyon-like streets with residential property close to the carriageway as AQMAs, where insufficient dispersion as a result of the street topography and slow moving traffic is predicted to cause future exceedences. The size of AQMAs varies from a single property to several square kilometres in area.

The National Air Quality Strategy (1997) anticipated that major air quality problems would exist adjacent to motorways as traffic flows and HGVs proportions are amongst the highest on the UK road network [11]. However this fear has not been realised due to a combination of good dispersion conditions in the vicinity of motorways and the absence of public exposure. AQMAs as a result of traffic emissions from motorways are mainly within the South and South East regions of England. Concentrations of NO_2 and PM10 quickly decrease from the kerbside, and public exposure is not so prevalent in close proximately to motorways in comparison to urban networks of roads and minor highways (see http://www.uwe.ac.uk/aqm/review/monitor.pdf for details). However, a number of authorities anticipate AQMA designations along specific corridors or certain distances either side of motorways.

4.2 Local authority practice

4.2.1 Technical capabilities

The results of the questionnaire surveys show a gradual increase in air quality monitoring practice within local authorities over the four years of the study. By

2001, all urban local authorities were undertaking some form of monitoring with the pollutants monitored reflecting those causing greatest problems in achieving the air quality objectives. In the survey undertaken by Elsom and Crabbe in 1994 [44], 55% of the urban sample of local authorities did not have permanent real-time monitoring equipment and hence did not have the capacity to assess with confidence air pollution concentrations within their area. By 2001 91% of urban local authorities had real time analysers.

Modelling practices by local authorities show even more categorically the increases in use of tools for air quality management. In the 1994 survey of local authorities [44], 5% had utilised some sort of modelling. In 2001, this had increased to 99% with a gradual increase in the intervening years. The greatest increases in types of models used by local authorities were in ADMS urban, a complex dispersion modelling system and in DMRB, a screening model used most often in Stage 2 of the review and assessment.

Apart from some issues of obtaining funding and support for air quality management tools, the technical capabilities of urban authorities for the implementation of review and assessment has increased considerably. Local authorities technical capabilities are (with exceptions) viewed by the local authorities involved in case study analysis as sufficient for the successful implementation of the technical aspects of air quality management.

4.2.2 Within authority collaboration

Questionnaire surveys have shown that under a half of urban local authorities have an internal group or project team that spends at least some of its time dealing with air quality issues. However, the remit of these groups varies significantly, which cannot be identified in detail through questionnaire surveys. The remit of the majority of the groups that environmental professionals are involved in, included review and assessment, or action planning or strategies within their remit. From the case study authorities it can be shown that the smaller authorities had less need to meet formally. Departments were generally much smaller, and often just one person within transport or land use planning was responsible for all air quality implications of their profession. However, in larger authorities it is difficult to foresee how different policy areas could integrate on a day to day basis without formally meeting up and discussing each other's agendas and requirements.

4.2.3 Between local authority collaboration

Questionnaire surveys have shown that particularly for environmental health professionals, regional working of some kind is widespread. Other professions are not as involved in regional groups undertaking some air quality work. There is no clear trend as to whether regional initiatives are increasing or decreasing. The case studies attempted to decipher which initiatives may be more successful in integrating different local authority working practices. Regional working was shown to have two distinct facets, at a policy level, and at an officer working level.

Information systems or the dissemination of information to the public, between departments within local government and between local authorities and outside agencies have not really evolved at all over the four years of the study. There have been slight increases in most of the methods of disseminating information to the public over the four years, but no major change in approach. The only method that has shown substantial increases in use is local authority internet sites. The majority of local authorities now have their own internet site with some form of air quality information upon it. There has been even less progress in the methods of disseminating information to other departments within local authorities, and to neighbouring authorities and outside agencies.

4.2.4 Collaboration between local authorities and stakeholders

Outside agencies and bodies have been involved in the review and assessment process to varying degrees. Health authorities' involvement seems to have tailed off in 2000 and 2001, although there has been an increase in the amount of health advice given out relating to air pollution. There has also been an increase in the use of information from air quality studies being made available at regional groups and also used in the review and assessment process.

The Environment Agency, who regulate large industrial sources in England and Wales, has in some areas taken a proactive approach to managing air pollution sources in their jurisdiction. Throughout the four years of the survey their input into local air quality management has remained static, with a base of monitoring and modelling to build on. Over 90% of Environment Agency respondents had supplied information to local authorities even in 1998. The Highways Agency, who oversee the management of motorways and trunk roads in England and Wales, has, however, been less involved. In contrast to the Environment Agency, the Highways Agency as a whole had little previous knowledge or expertise in the field of air quality monitoring or management. In many cases, exceedances of the air quality objectives are caused by motorways or trunk roads and, with little involvement by the Highways Agency, it is difficult to foresee how solutions can be found and implemented in time for the objective deadlines.

5 Discussion

Local authorities have gained capacity and confidence in the review and assessment process having started from a heterogeneous position. In most cases local authorities have discharged their review and assessment functions in an effective and efficient manner. Surprisingly however, the survey evidence points to a series of managerial as opposed to technical difficulties being faced by local authorities. Difficulties have been reported in policy integration and communication within local authorities. Questionnaire surveys have shown that local authorities do not appear to have the capability to manage air quality issues between local government departments. Of the non-environmental health local authority functions, transport planners have generally been more involved in the process than other professionals. This may be because of funding incentives through the process of Local Transport Plans (LTPs) and related processes in the devolved administrations). There were also problems in communicating the outcomes of the review and assessment process to the public.

Local authorities have made substantial progress in managing a complex and dynamic atmospheric environment against a legislative framework that has seen substantial changes since the publication of the first NAQS in 1997. Government Guidance issued for the process is intended to provide consistency by way of specifying the minimum expectations for the AQM process. However, survey data has illustrated a divergence in capability and achievement in LAQM practice within urban local authorities in England. A related study investigating rural local authorities [45] has shown urban areas to be in advance of rural authorities generally, suggesting an even greater divergence in practice between local authorities in the UK. A number of reasons may have caused this divergence, notably the greater experience of urban authorities in addressing more acute air pollution problems. Urban authorities are more likely to operate as a single tier authority (i.e. as a unitary authority), with more efficient communication mechanisms. Rural authorities, in contrast, are more financially constrained and have a potential absence of historical air pollution monitoring and modelling data and officer capability. With AQMAs being declared in smaller, rural locations these constraints are likely to hinder the subsequent action planning process.

The amended time-scale for reviewing and assessing air quality has caused problems for integrating air quality objectives into transport planning processes seen by many as the delivery mechanism for solutions to identified problems. The UK has, over the time-scale of the first round of reviews and assessments, experienced a rapid change in relation to transport policy. The Government White Paper 'A New Deal for Transport: Better for everyone', published in 1998, promised innovative approaches to transport planning in England. It emphasised the responsibility of local authorities to produce Local Transport Plans (LTPs), designed to integrate different transport modes and respond to local circumstances. LTPs replaced the former Transport Policy and Programmes (TPP) system for allocating resources for local transport capital expenditure. LTPs require environmental and sustainability consideration, of which air quality is a major component [46]. The first complete round of Local Transport Plans (LTPs) was completed in July 2000. The LTP process bids for funding for the next 5 years and in many cases the reviews and assessments were not completed in time. It should be noted that there are parallel processes being implemented within the Devolved Administrations and London.

There has been limited involvement of the planning community in the air quality review and assessment process as a whole. The potential success of future air quality action plans and their effective implementation to improve local air quality within designated AQMAs will require active participation and action on the part of the land-use planning community, working collaboratively with transport planning and other disciplines within the local authority. This will be assisted by the development of local air quality strategies for individual local authorities and the development of protocols for ensuring that air quality considerations become an integral aspect of local authority planning and development processes and functions.

At the conclusion of the first round of review and assessment in Great Britain over 30% of the 407 local authorities will require one or more AQMAs within their area of jurisdiction.

6 Conclusions on the British approach

Despite some initial difficulties in the first round of air quality review and assessments in Great Britain, it is evident from the data collected that the LAQM process is likely to deliver improved air quality. The great quantity of air quality information assembled and disseminated, together with the monitoring and modelling undertaken in direct response to the AQM review and assessment process has led to a position where the air quality of Great Britain is better understood now than at any time in history. The analysis of source contribution and the spatial location of exceedances has provided some surprising outcomes.

Traffic emissions are the most significant source of predicted exceedences, and AQMAs are being declared alongside motorways and trunk roads, across road networks, as well as along individual road sections where low traffic speeds, high percentages of HGV traffic and canyon-like street topographies lead to poor air dispersion. An unexpected outcome has been the identification of areas of exceedance in smaller towns where traffic flows have been modest, but an exposed population has been identified.

Local authorities are at the stage of writing and implementing action plans following designation of AQMAs. The review and assessment process up to designation has been mainly a scientific process, although collaboration has been required with other departments and outside agencies, particularly for information such as traffic and health data. However, as local authorities move into implementing solutions through air quality action planning, the collaborative aspects of the process will be central to the success of air quality management.

In the authors' opinions the LAQM process in Great Britain is an effective one. The British experience and process is, in part, a transferable model and this is a useful and effective process for others to consider in meeting their air quality obligations and requirements. An evaluation of the review and assessment process in Great Britain has recently been undertaken by the University of the West of England authors (in collaboration with others) on behalf of the UK Government and the devolved administrations. The outcomes of this work can be found at http://www.uwe.ac.uk/aqm/review/index.html. The authors intend to continue to monitor the progress of AQM implementation as the review and assessment process moves into its second round and the action planning process unfolds. This will provide further insight into the often complex solutions required in improving local air quality.

7 A transferable system?

The AQM framework and process in Great Britain is characterised by health-effects-based objectives and a risk management approach. It provides a model against which other industrial states could evaluate their existing procedures. Clearly no system of environmental management is directly transferable between different countries without adjustment for local circumstances, history and philosophy. The air quality management process is no different. However, in the recent independent evaluation of the British system of review and assessment (http://www.uwe.ac.uk/aqm/review/ereport.pdf) there were certain elements that were judged to work well. These include national support actions for local authority review and assessment such as training, internet support, help desks, prescriptive guidance and a high-quality national monitoring network. The existence of a technical consensus about the way in which guidance for review and assessment should be implemented is a significant contributory factor to the success of the British system. In addition to the formal technical guidance commissioned and issued by Government, organisations such as the National Society for Clean Air and Environment Protection provided informal guidance on aspects of the process and this was supported by training activities. Important to the success of these activities was the support of Government for the approach being taken. The risk-management-based approach to the setting of objectives based on sound science and an assessment of the costs and benefits is a pragmatic one, and of equal importance is the absolute requirement that relevant public exposure should exist in areas where air quality objectives are exceeded. In order that local authorities are both consistent and correct in their risk management approach to decision making the Government introduced an appraisal system to confirm the decisions reached by local authorities.

The support elements and the check on the approach and conclusions reached by a local authority have helped ensure that a consistent approach to local air quality management has been developed in Great Britain. This is despite the very large number of local authorities undertaking review and assessment work. These features of the British system are recommended as valuable contributions to an effects based, risk management approach to air quality.

8 Recommendations

The following recommendations and conclusions are offered:

1) Good air quality is a critical parameter for public and environmental health protection and a strategy for air quality management should be an integral component of a sustainable development pathway in a city.

2) The location and size of areas of poor air quality in a city, the polluted status of air within the area and the relative contribution of sources causing poor air quality are not always well understood especially at the

local level. Proportionate and cost effective remedial action requires that such information is available.

3) The combination of local and national actions is recommended to provide a comprehensive policy and implementation framework in which air quality objectives can be identified and remedial action targeted in a proportionate and cost-effective manner.

4) A local review and assessment process is recommended in order that peak or local "hot spot" concentrations are identified and improved. In the medium term the focus can shift from reducing "hot spot" concentrations to managing a reduction in the mean concentration of objective pollutants in areas of public exposure.

5) The reality or perception of poor air quality in a city can impact negatively on inward investment decisions and high peak and /or mean concentrations of pollutants will increase the burden placed on public health services. Conversely, investment in air quality management initiatives will reduce the cost impact on health services and, dependent upon the relative source contributions, can bring cost savings through capital investment in new plant, pollution control technology or in reducing congestion experienced by road users.

References

[1] Brimblecombe, P., *The Big Smoke*, Routledge, London, 1987.

[2] Longhurst, J.W.S., Lindley, S.J., Conlan, D.E., Rayfield, D.J. & Hewison, T., Air quality in historical perspective, In: *Urban Air Pollution III*. eds. H. Power & N. Moussiopoulos, CMP: Southampton, pp. 69–109, 1996.

[3] National Society for Clean Air and Environmental Protection (NSCA), The Clean Air Revolution: 1952–2002, *Clean Air and Environmental Protection,* **32(4)**, pp. 1–75, 2002.

[4] National Society for Clean Air and Environmental Protection (NSCA), *Pollution Handbook 2002*, NSCA, Brighton, 2002.

[5] Elsom, D.M., Atmospheric pollution trends in the United Kingdom, In: *The State of Humanity*. ed. J.L. Simon, Blackwell: Oxford, UK & Cambridge, USA, pp. 476–490, 1995.

[6] Longhurst, J.W.S., Lindley, S.J., Watson, A.F.R. & Conlan, D.E., The introduction of local air quality management in the United Kingdom. A review and theoretical framework, *Atmospheric Environment,* **30(23)**, pp. 3975–3985, 1996.

[7] Department of the Environment, *This Common Inheritance. Britain's Environmental Strategy*. CM1200, The Stationery Office, London, 1990.

[8] HM Government, *Environment Act 1995*. The Stationery Office, London, 1995.

[9] Department of the Environment, Scottish Office, Welsh Office and Department of the Environment for Northern Ireland, *Improving Air Quality. A Discussion paper on Air Quality Standards and Management*, Department of the Environment, London, 1994.
[10] Department of the Environment, Scottish Office, Welsh Office and Department of the Environment for Northern Ireland, *Improving Air Quality. Air Quality: Meeting the Challenge. The Government's Strategic Policies for Air Quality Management.* Department of the Environment, London, 1995.
[11] Department of the Environment, Scottish Office and Welsh Office, *Improving Air Quality. National Air Quality Strategy.* CM 3587, The Stationery Office, London, 1997.
[12] HM Government, *The Air Quality Regulations 1997*, The Stationery Office, London, 1998.
[13] Department of the Environment, Transport and the Regions, Welsh Office, Scottish Office and Department of the Environment for Northern Ireland, *The Air Quality Strategy for England, Scotland, Wales and Northern Ireland*, CM 4548, The Stationery Office, London, 2000.
[14] Elsom, D.M., Development and implementation of strategic frameworks for Air Quality Management in the UK and European Community. *Journal of Environmental Planning and Management,* **42**, pp. 103–121, 1999.
[15] HM Government, *Environmental Protection Act 1990,* The Stationery Office, London, 1990.
[16] Department of the Environment, Transport and the Regions, *LAQM.G1(97) Framework for Review and Assessment of Air Quality.* The Stationery Office, London, 1997.
[17] Bartlett, T., Brown, T., McDonald, T. & Price, F., Local Authority progress in Air Quality Management. *Clean Air,* **27**, pp. 97–98, 1997.
[18] Beattie, C. I., Longhurst, J. W. S. and Woodfield, N. K., Air Quality Management: evolution of policy and practice in the UK as exemplified by the experience of English local government. *Atmospheric Environment,* **35**, pp. 1479–1490, 2001.
[19] Department of the Environment, Transport and the Regions, *LAQM.G2(97) Developing Local Air Quality Action Plans and Strategies: the principal considerations*, The Stationery Office, London, 1997.
[20] Department of the Environment, Transport and the Regions, *LAQM.G3(97) Air Quality and Traffic Management*, The Stationery Office, London, 1997.
[21] Department of the Environment, Transport and the Regions, *LAQM.G4(97) Air Quality and Land Use Planning*, The Stationery Office, London, 1997.
[22] Department of the Environment, Transport and the Regions, *LAQM.TG1(98) Monitoring for Air Quality Review and Assessments*, The Stationery Office, London, 1998.

[23] Department of the Environment, Transport and the Regions, *LAQM.TG2(98) Preparation and Use of Atmospheric Emissions Inventories*, The Stationery Office, London, 1998.
[24] Department of the Environment, Transport and the Regions, *LAQM.TG3(98) Selection and Use of Dispersion Models*, The Stationery Office, London, 1998.
[25] Department of the Environment, Transport and the Regions, *LAQM.TG4(98) Pollutant Specific Guidance*, The Stationery Office, London, 1998.
[26] Department of the Environment, Transport and the Regions and National Assembly for Wales, *LAQM.G1(00) Framework for Review and Assessment of Air Quality*. DETR, London, 2000.
[27] Department of the Environment, Transport and the Regions and National Assembly for Wales, *LAQM.G2(00) Developing Local Air Quality Action Plans and Strategies: The Main Considerations*, DETR, London, 2000.
[28] Department of the Environment, Transport and the Regions and National Assembly for Wales, *LAQM.G3(00) Air Quality and Transport*. DETR, London, 2000.
[29] Department of the Environment, Transport and the Regions and National Assembly for Wales, *LAQM.G4(00) Air Quality and Land Use Planning*. DETR, London, 2000.
[30] Department of the Environment, Transport and the Regions, National Assembly for Wales and Scottish Executive, *LAQM.TG1(00) Monitoring Air Quality*. DETR, London, 2000.
[31] Department of the Environment, Transport and the Regions, National Assembly for Wales and Scottish Executive, *LAQM.TG2(00) Estimating Emissions*. DETR, London, 2000.
[32] Department of the Environment, Transport and the Regions, National Assembly for Wales and Scottish Executive, *LAQM.TG3(00) Selection and Use of Dispersion Models*. DETR, London, 2000.
[33] Department of the Environment, Transport and the Regions, National Assembly for Wales and Scottish Executive, *LAQM.TG4(00) Pollutant Specific Guidance*. DETR, London, 2000.
[34] Department for Environment, Food and Rural Affairs, Scottish Executive, Welsh Assembly Government and Department of Environment in Northern Ireland, *LAQM.TG(02) Review and Assessment: Technical Guidance*, DEFRA, London, 2002.
[35] Department for Environment, Food and Rural Affairs and Welsh Assembly Government, *LAQM.PG(02) Policy Guidance*, DEFRA, London, 2002.
[36] National Society for Clean Air and Environmental Protection, *The How to Guide: Consultation for Local Air Quality Management*, NSCA, Brighton, 1999.
[37] National Society for Clean Air and Environmental Protection, *Air Quality Management Areas: Turning Reviews into Action*. NSCA, Brighton, 2000.
[38] Royal Town Planning Institute, Air *Quality and Land Use Planning*, NSCA, Brighton, 1999.

[39] National Society for Clean Air and Environmental Protection, *Air Quality Action Plans: Interim Guidance for Local Authorities*, NSCA, Brighton, 2000.

[40] National Society for Clean Air and Environmental Protection, *Air Quality: Planning for Action*, NSCA, Brighton, 2001.

[41] London Research Centre, *Atmospheric Emissions Inventories for Four Urban Areas. Merseyside, Bristol, Southampton/ Portsmouth, Swansea /Port Talbot.* London Research Centre, London, 1997.

[42] Lindley, S.J., Longhurst, J.W.S., Watson, A.F.R. & Conlan, D.E., Procedures for the estimation of regional scale atmospheric emissions - an example from the northwest region of England. *Atmospheric Environment*, **30(17)**, pp. 3079–3091, 1996.

[43] Woodfield, N.K., Longhurst, J.W.S., Beattie, C.I. & Laxen, D.P.H., Designating Air Quality Management Areas (AQMAs) in the UK: Implications for Securing UK Air Quality Objectives. *Water Air and Soil Pollution*, in press.

[44] Elsom, D.M. & Crabbe, H., Local Air Quality Management in the UK Survey. *Clean Air,* **25**, pp. 95–107, 1995.

[45] Ing, C., Beattie, C.I. & Longhurst, J.W.S., Progress with implementing local air quality management in rural areas of England. *Journal of Environmental Management,* **6**, pp. 137–147, 2001.

[46] Beattie, C.I., Longhurst, J.W.S. & Woodfield, N.K., A Preliminary Analysis of the Linkages between Air Quality Management and Transport Planning Policies in a Sample of English Highways Authorities. *Journal of Environmental Planning and Management,* **44**, pp. 391–408, 2001.

Appendix 1
Review and Assessment Checklists

The technical guidance has been summarised in the checklists local authorities are advised to use in their review and assessment work. The checklists may be viewed at www.uwe.ac.uk/aqm/review/checklists/. These checklists are used in the independent appraisal, commissioned by Government, of local authority review and assessment work.

Stage 1

General

- Is an assessment of all seven pollutants presented?
- Have localities that will be the focus of stage 2 assessment been identified?
- Have the sources of information used to compile the report been identified?
- Has due consideration been given to the potential for exposure of individuals in relevant locations?
- Have details of part A processes been provided?
- Have details of part B processes been provided?
- Has the first stage report concluded by listing those pollutants requiring further examination?
- Are current concentrations described?

Carbon Monoxide

- Have roads with traffic flows above the guidance levels been identified?
- Have junctions been considered?
- Have relevant Part A and Part B processes been identified?
- Has the nomogram been used correctly?
- Have sources in neighbouring authorities been considered?
- Have planned developments been described?

Benzene

- Have relevant Part A and Part B processes been identified?
- Has the nomogram been used correctly?
- Have sources in neighbouring authorities been considered?
- Have planned developments been described

1,3-Butadiene

- Have relevant Part A and Part B processes been identified?
- Has the nomogram been used correctly?
- Have sources in neighbouring authorities been considered?
- Have planned developments been described?

Lead

- Have relevant Part A and Part B processes been identified?
- Has the nomogram been used correctly?
- Have sources in neighbouring authorities been considered?
- Have planned developments been described?

Nitrogen Dioxide

- Have NOx background concentrations been correctly estimated?
- Have nomograms been used correctly to identify potentially significant roads?
- Have other transport sources been considered?
- Have relevant Part A and Part B processes been identified?
- Has the nomogram been used correctly?
- Has consideration been given to sources acting in combination?
- Have sources in neighbouring authorities been considered?
- Have planned developments been described?

PM_{10}

- Have PM_{10} background concentrations been correctly estimated?
- Have nomograms been used correctly to identify potentially significant roads?
- Have other transport sources been considered?
- Has consideration been given to domestic solid fuel?
- Have relevant Part A and Part B processes been identified?
- Has the nomogram been used correctly?
- Has adequate consideration been given to uncontrolled or fugitive sources?
- Have sources in neighbouring authorities been considered?
- Have planned developments been described?

Sulphur Dioxide

- Have relevant Part A and Part B processes been identified?
- Have solid fuel or fuel-oil combustion systems >5 MWt been considered?
- Has the nomogram been used correctly?
- Is the density of coal burning household >300 per km^2
- Have sources in neighbouring authorities been considered?
- Have planned developments been described?

Stage 2

The detail required in Stage 2 is more complex and the example of nitrogen dioxide is used to illustrate the range of issues to be covered.

Nitrogen Dioxide

MONITORING

- Have worst-case locations been used?
- Have 'relevant-exposure' locations been used?
- Are the monitoring period and data capture appropriate?
- Has monitoring identified existing exceedances?
- Are monitoring data used properly to predict both objectives?
- Have future concentrations been estimated correctly?
- Is sufficient detail of QA/QC procedures provided?
- Have diffusion tube data been validated against a continuous monitor?
- Have diffusion tube data been scaled?

MODELLING

- Has an appropriate model been used?
- Has background been properly added?
- Is modelling used properly to predict both objectives?
- Has modelling identified 2005 exceedances?

Road traffic

- Has DMRB been used correctly?
- Have appropriate traffic data and speeds been used?
- Have junctions been considered?

Industrial

- Has EA GSS modelling been applied correctly?
- Has 'other' modelling been used properly
- Has correct approach been used for background?
- Have multiple sources been considered?

GENERAL

- Have combined impacts been considered?
- Has due consideration been given to the potential for exposure of individuals in relevant locations?
- Does the decision to proceed to Stage 3, or not, appear to be soundly based?

Stage 3

The final stage, Stage 3, is more complex and again the example of nitrogen dioxide is used to illustrate the detail and sophistication of analysis that is required.

Nitrogen Dioxide

MONITORING

- Has continuous monitoring been undertaken?
- Is the monitoring period and data capture appropriate?
- Have diffusion tube data been validated against a continuous monitor?
- Have diffusion tube data been scaled?
- Are monitoring data used properly to predict both objectives?
- Has monitoring identified 2005 exceedances?
- Is sufficient detail of QA/QC procedures provided?

MODELLING

- Has background been properly added?
- Has the model been appropriately validated?
- Are adjustments to predicted concentrations clearly documented?
- Is appropriate modelling used?
- Is modelling used properly to predict both objectives?
- Has modelling identified 2005 exceedances?

Road traffic

- Are road traffic data detailed and accurate enough?
- Has the NO_2:NO_x ratio been properly addressed?
- Has any traffic modelling been validated against traffic counts?

Industrial

- Have actual emissions data been used?
- Have uncertainties regarding emissions been documented (i.e. batch processes, fugitive emissions)?

GENERAL

- Have both the magnitude and geographical extent of any exceedance been defined?
- Have combined impacts been considered?
- Has due consideration been given to the potential for exposure of individuals in relevant locations?
- Does the decision to declare or not declare an AQMA appear to be soundly based?

CHAPTER 4

Monitoring and modelling air quality in Mendoza, Argentina

E. Puliafito & C. Puliafito
Institute for Environmental Studies (IEMA)
University of Mendoza, Argentina
Also CONICET members

Abstract

Like many big urban centres, Mendoza, a city of about 900,000 inhabitants, has increasing air pollution mostly from mobile sources. To assess its air quality, two main strategies have been developed: setting up a network of monitoring stations and the development of a comprehensive model. Continuous monitoring and daily measurements of the main pollutants had been performed for more than 10 years, at different sites, combined with several measuring campaigns. On the other side, a model, based on a geographical information system has been developed to evaluate the air quality derived from stationary and mobile sources. Once the model has been calibrated with the given measurements, the system is then used to test the compliance to local standard air quality, or to study the environmental impact of new industries or to determine the changing conditions when vehicle circulation is increased. This allows us to characterise areas with potential increase in air pollution or improvement in air quality. It is also possible to compute past or present conditions by changing such basic information as traffic flow, or stack emission rates. In this article a review of the monitoring and f efforts, together with a prognosis of future possible scenarios will be presented. Also a case study during a high pollution event occurred in May 1995 is described here.

1 Introduction

The city of Mendoza is located in western Argentina in an arid to semi-arid zone of low rainfall, as will be described below. Its people have a clear oasis culture, where great care is given to managing its limited resources, especially water and space, and developing an appropriate habitat based on a steady effort.

Therefore environmental preservation and sustainable development is a consequence of its long cultural survival process. The General Water Law of 1884, still in force, and the Irrigation Department created in 1916, which is responsible for managing the irrigation water resources, is a clear example of this environmental culture. We should also mention, as important milestones of the end of the nineteenth and beginning of the twentieth century, the development of sanitary works, the creation of a 450 ha park and public tree fostering in the urban area. Its many trees and green spaces could be identified as a natural way to moderate desert climate and to improve urban life quality. A century later, this visionary urban planning proved insufficient to stop the progressive degradation of natural environment, and to solve the uprising of the new conflicts. The increasing urbanisation problems, with traffic congestion, air pollution, and loss of green or agricultural space, placed additional pressure on the administration of natural resources and existing infrastructures.

This situation finally pushed the society to a broad discussion. As a consequence, new environmental laws were passed by the Provincial Congress and in 1989 an Environmental Ministry was incorporated into the Provincial Executive Power. Lots of effort was made to rearrange the institutional relations within the same Provincial State, the Municipalities and the Federal State, regulating the application of Provincial and Federal Laws or national decrees.

The main objective of this Ministry was the action and coordination with Municipalities, Enterprises and Scientific institutions, in the following fields: a) environmental Information; b) development and up-dating of legal, institutional and techno-economics instruments for environmental protection; c) development of public environmental awareness; d) natural flora and fauna preservation; e) environment control and sanitation; f.) urban regulation and environmental development; Urban services with direct environmental incidence (drinkable water, sewers, parks, and public transportation); and g) housing. The environmental policy was based on prevention (instead of acting on the consequences), participation (protection and improvement do require permanent consultation with the population, in order to know their hopes and expectation), and political and technical cooperation (the solution of environmental problems requires inter disciplinarity and new technology). Only through a vast cooperation among public offices, research institutes and non-governmental organisations is it possible to solve many of these problems and to prevent for continuous deterioration of urban life quality. Environmental legislation enforcement, has been very active in the area of pollution reduction and control, especially in the following areas: a) atmospheric pollution by mobile sources; b) pollution by urban solid domestic residues; c) contamination by dangerous and special waste; d) atmospheric pollution due to stationary sources; e) industrial water pollution; f) contamination by oil exploitation; and g) contamination by mining exploitation. However, new regulations mandate the monitoring and tracking of numerous issues, introducing extra demands on agencies at the local government level.

Since the early 1990s, the University of Mendoza through its Institute for Environmental Research (IEMA), in cooperation with the Centre for Environmental Research UFZ-Leipzig-Halle, and the Max-Planck Institut for Aeronomy MPAE-Lindau (Germany), among other local universities and research centres, have joined the effort of the public office, implementing a broad environmental research program. Relevant aspects of this program included research in the area of urban pollution monitoring [1–5] ecosystem modelling, environmental standards and legislation [6], health effects due to ambient air, [7, 8] and sustainable architecture.

In this article we will concentrate only on aspects related to the air pollution situation in the city of Mendoza, showing the modelling efforts performed by our Institute, aimed at contributing to local authorities to implement an effective air pollution control strategy.

The air in an urban area is influenced by many factors. They range from local meteorology and orography, natural and anthropogenic emissions, which sends to the atmosphere, among others, pollutants like carbon monoxide, Oxides of Sulfur and nitrogen, photochemical oxidants and hydrocarbons, and dust. The major source of man-made pollution comes typically from industry, commerce, domestic cooking and heating, and transportation activities. Urban environment is stressed by a number of common problems such us population, industrial and commercial growth, increasing energy and transportation demands. Worldwide, before 1990 most of the concern and control efforts was given to the industrial sources, but lately it was very clear that the major source for the degradation of the air quality was no longer factories, but the road vehicles – buses, passengers cars, light and heavy duty trucks or motorcycles. Additionally concern has been given in the last decade to understand and evaluate the impact of such emissions on the population health.

A program to control the air pollution is generally based on several goals: a) setting air quality targets, based normally on health effects; b) data monitoring of principal pollutants and meteorology; c) completion of a sources inventories, through measuring and surveys; d) setting an appropriate model, to control the air quality targets and to correlate air quality standards with emission-reduction targets; and e) implementation of legal and socio-economic measures to achieve the desired air quality targets. The selected model should help to evaluate and compare transport policies and plans, to determine air quality standards, to assess the environmental impact or health risk. The air quality modelling segments should include at least: a) emission forecasting models to predict the monitoring measurements; b) study and forecast of wind patterns and transport models; c) dispersion models to correlate emission with ground ambient concentrations in order to understand the contribution of each of the sources. With the exception of ozone, which is a secondary pollutant, the present models represent approximately the pollutant situation, therefore it is possible to model NO_x, CO and hydrocarbons based on the emission rate at the source.

Different approaches may be adopted to handle these large amounts of information, depending on the available technology, skills and environmental programs. Moreover, this information needs to be handled not only in temporal data formats for some monitoring stations, but over a wide spatial distribution.

The actual efforts in improving the air quality in Mendoza, and why a model based on a Geographical Information System (GIS) can meet the required needs, will be presented in this article. GIS is a computer-based systems that helps to translate environmental science into policy by using relevant geographically located information, combined with algorithms, procedures and models that represent general scientific understanding. Data sets created and organised under GIS can be shared between various government entities and the public. GIS applications allow any user, department or agency to access common geographical data and make it easy for the public to browse the data, providing an excellent software for development and analysis. Today, GIS is surely one of the most used software platform for planning all over the world. These features justify choosing GIS as an appropriate management tool for air pollution control.

2 Materials and methods

2.1 Meteorology and orography

The city of Mendoza, (33°S, 68°W, 750 m. a. N. s. l.), is located at the east side of the Andes Mountains in western Argentina. The Cordillera close to Mendoza, reaches an average height of 5,000 m with peaks up to 7,000 m. This natural barrier influences the meteorological conditions determining a strong dependence in the air pollution situation. The region is characterised by its low precipitations: 120 to 400 mm/a, (mean 230 mm/a), which occurs especially during the summer months (November to March). Only 3% of the total province territory (250,000 km^2) is irrigated by five main rivers with annual modules less than 50 m^3, conforming three main oases. Almost 70% of the population is concentrated in the northern oasis, conforming the Mendoza metropolitan area with approximate 900,000 inhabitants.

This region is characterised by passing cyclones developing from cold polar air masses at the polar front of the southern hemisphere. Consequently the weather is disturbed by passing fronts, especially during the winter. However, during the summer season the polar front is pushed to the south, developing a stable area of high pressures in the north-east of Argentina. Unlike the northern hemisphere, the southern hemisphere is covered by more sea than land. The circulation is dominated by the planetary system. Therefore, the NE winds arise from the semi-permanent anticyclone on the atlantic east of the South American continent at approximately 30° S and thus belong to an “Atlantic wind system”. This subtropical circulation brings warm and moist air into the region. Since the polar front is shifted to the north in the winter, the metereological conditions of Mendoza, during this time, are determined by Pacific air masses (cyclones). Consequently in this season there is little rainfall.

Due to its closeness to the mountains, Zonda winds similar to Föhn or Chinook winds, prevail in the higher layers mostly during the winter months

(May to October) with high probability of frost, which contributes to a higher degree of pollution due to the occurrence of strong thermal inversion layers. Damp, warm tropical air masses pass over The Andes cordillera from west to east, and therefore, the air masses ascends saturated adiabatically, losing their moisture in the western hillside of The Andes. This wind descends dry adiabatically on the eastern hillside provoking strong increases of the temperature of about 4°C per each kilometer of mountain height, and with speeds ranging 5-6 m/s, and gusts up to 12 m/s.

The mean wind intensity is around 2 m/s, with mostly calm days (< 0.5 m/s) 35%, and predominant wind from S 13%, NE and W 12%, SW 10%, SE 8 %. Low relative humidity (50%), low incidences of fog and few days with covered skies (65-75 days/annum). The solar global radiation varies from 270 MJ/m^2 in winter to 780 MJ/m^2 in summer, with annual mean daily heliophany of 7.9 hours. The climate can be classified according to the Köeppen scale as BWakw: dry dessert, with mean temperature of the hottest month > 22°C, cold and dry winter with a mean temperature below 18°C. The area presents low relative humidity (50%), low incidences of fog and few days with covered skies (65–75 days/annum). Another important feature for the description of the air pollution is the day-night variation, characterised by a typical valley-mountain circulation. From the first hours after sunset to the early hours after sunrise, there is a clear wind flow from WSW, while in daylight hours the circulation is ENE. Strong solar heating on the valley side causes an up-slope wind flow at daytime (ENE). At night due to rapid radiational cooling on the valley slope, the circulation switches over causing the air masses to move down the mountain from WSW. This strong night cooling and day heating produces an important variation in mixing heights and inversion layers. More on meteorological and climatic aspects relevant to the region can be seen in [1, 2].

2.2. Monitoring and sampling

Ambient concentration measurements of main pollutants have been monitored since 1970 in the metropolitan area of Mendoza by the Ministry of Environment and Public Works through its Direction for Environmental Control. In the urban area of Mendoza, around 15 stations monitors daily mean values of Total Suspended Particles (TSP) by filter capture and reflectometry, daily mean values of nitrogen oxides NO_x by colorimetric Griess–Saltzmann, once a week 24 h mean values of lead Pb by colorimetric ditozone, and sulfur dioxide SO_2 by colorimetric West and Gacke modified by Pate. Also, since 1995, our Institute IEMA has measured black carbon and PAH poliaromatic compounds (Aethalometer GIV reflectometry), surface ozone (O_3), nitrogen oxides (NO_x), carbon monoxide (CO), using a set of Horiba instruments (series APO350E for O_3 and APN360 for NOx and CO) and meteorological parameters, including solar global radiation. Besides the mentioned monitoring data, paper filter samples of particulate matter have been collected using a high volume pump at several measuring points. At the beginning of the project, little organised information was available over the emission of the fixed sources, thus it was necessary to set up an emission inventory of the main sources. These forms contain relevant information on type of fuel, yearly consumption, physical data of boilers, chimneys and emission measurements if available. Additionally various stack measurements were performed by independent consultants using an isokinetic stack samplers, following U.S. EPA Standard Methods. The main industrial areas are located on the periphery of the city: oil refineries, petrochemical industries and energy production are situated SW of the city, two cement industries are on the north edge and finally the agro-industrial production is mainly on the east side of Mendoza. Figure 1 shows a satellite picture of the city and surroundings, showing also the main industrial sources and the monitoring stations.

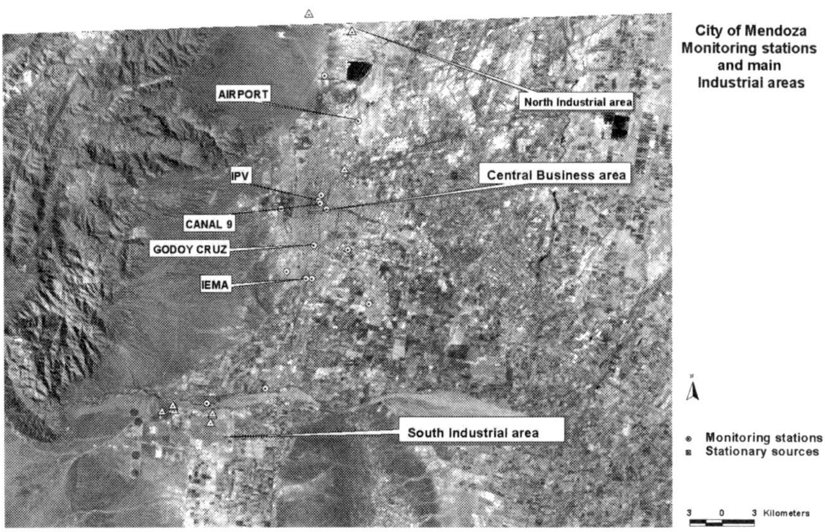

Figure 1: Satellite image of the area under study. Monitoring and meteorological stations are shown with triangles, main industrial sources are represented with circles.

2.3 Cycles of the airborne pollutants

The described meteorological and orographic conditions of the city, together with man's activity produce a characteristic air-pollution pattern. A detailed analysis of the gathered ambient data shows several well-defined cycles.

A main cycle is the seasonal variation (winter–summer), with higher values in winter and lower values in summer. Figure 2 shows monthly mean values of daily measurements of total suspended particles, monitored at the downtown area. This cycle is directly coupled to the meteorological conditions described above, namely calm winds, little precipitation and high occurrences of low inversion layers in winter. Infrequent polar fronts from SE bring snowfalls to mountain and valley, breaking the accumulation cycle of air pollution. Another clear cycle is the day–night variation, characterised by a typical valley–mountain circulation as mentioned above.

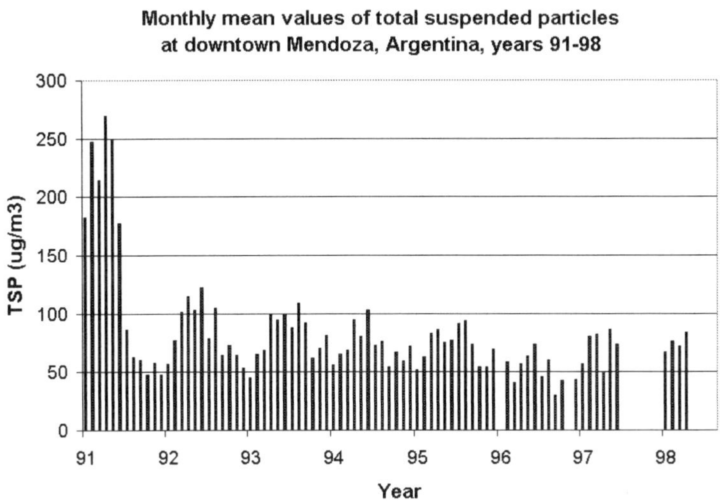

Figure 2: Monthly average of 24h values of total suspended particles at a downtown monitoring station.

Figure 3 shows a typical day, with solar radiation, wind direction, surface ozone (O3), carbon monoxide (CO), and nitrogen oxides (NO_x) measurements, monitored at the IEMA Institute. This graphic shows a strong traffic-related emission, where it can be clearly seen the beginning of normal day activities from 7 to 9 am and back-to-home rush hour after 8 pm. It may also be noted in Mendoza that despite an important flux of back-to-home traffic at noontime, mainly for lunch and because of the end of school activities, the vehicle pollution is not heavily shown in the ambient concentration values. This is an indication of the high convective dispersion present around midday. Note that Mendoza has a very dry (low humidity) climate, and then there is less heat exchange due to water vapor. Analyzing the monitoring data, it is possible to distinguish a typical weekly cycle, which confirms the relevance of mobile sources emissions in the air quality of Mendoza.

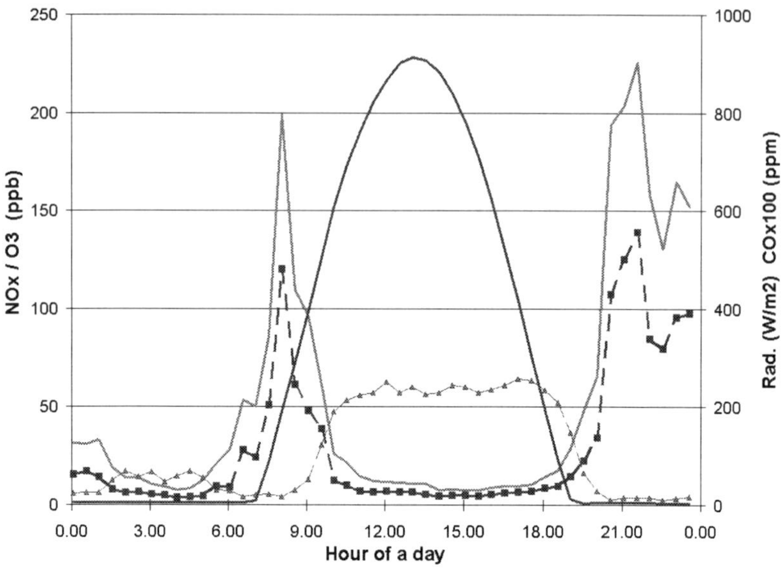

Figure 3: Typical daily variation of global solar radiation, and main pollutants monitored at the IEMA Institute, sited in Benegas, Mendoza. NOx: gray line (ppb); solar radiation: dark line (W/m^2); ozone: thin line with triangles (ppb); CO: dashed line with squares (×100 ppm).

Figure 4 shows the annual daily mean values for TSP and NOx according to the day of week, from a data series of several years taken at one of the fixed monitoring station. So, it can be seen that from Monday through Thursday the values are very similar, Friday is lower, and a strong decrease is present on weekend days.

The ozone data also shows a very interesting pattern, according to our monitoring data. Ozone is a major secondary air pollutant in many urban and rural areas of the world. High concentrations of surface ozone have become a significant problem affecting human health, vegetation, and numerous man-made and natural materials. Ozone in the lower atmosphere is produced during hydrocarbon oxidation via the catalytic action of the nitrogen oxides NO and NO_2. Although high ozone pollution of the lower troposphere has been widely investigated in the northern hemisphere, ozone has only been recorded to a limited extent in South America. Since 1994, the Institute IEMA has been measuring surface ozone from different locations in the Mendoza urban and suburban areas. The photochemical origination of ozone from pollutants (primarily from vehicles engines) and its strong dependence on the sun's ultraviolet (UV) irradiation would suggest a high level of ozone production and accumulation in Mendoza. But this is not, in fact, the case. In winter the night-time temperature may drop below 0°C until the air is heated again after sunrise at about 8:00 a.m. In the semi-arid region of Mendoza, the air is very dry (especially in winter), and with temperatures of 15°C at most, there are very few clouds. Hence, ozone formation takes place parallel to the temperature (radiation profile).

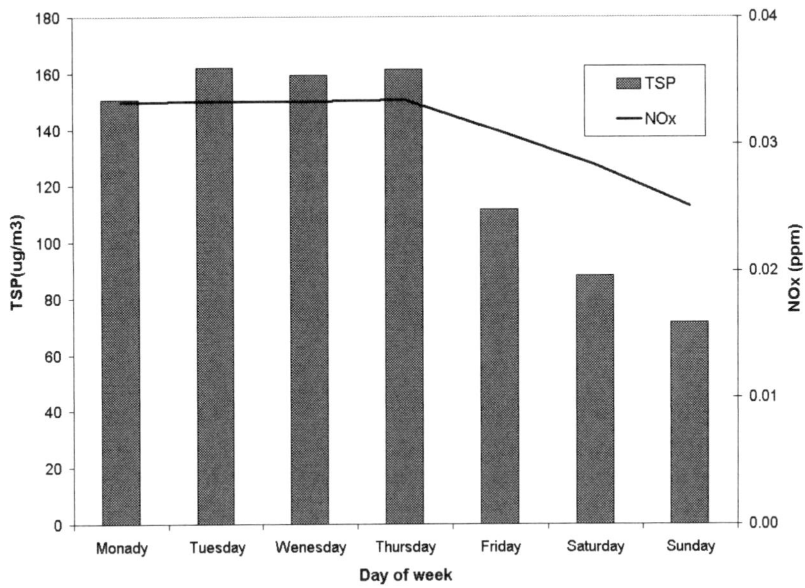

Figure 4: Annual mean values of TSP and NOx in μg/m^3 as a function of the day of the week measured at a downtown area in Mendoza.

The time period between sunrise and sunset is shorter in winter than in summer. This causes very sharp diurnal ozone peaks. Minor transport by wind causes the midday ozone concentration to reach the highest values of the year. The lower the daily maximum temperature, the lower is the transport and the higher is the midday ozone peak. In the morning and evening rush hours, vehicle emissions are able to enrich and reduce ozone. The ozone concentration immediately drops to zero. If any transport occurs, it mostly takes the form of the mountain-valley circulation systems. It lowers the midday ozone peak and produces ozone peaks during the dark night-time when warm air rich in ozone arrives from hills. During fall, with temperatures above 10 °C ozone and temperature peaks follow the diurnal radiation cycle, as they do in winter. During normal conditions, the temperature continuously increases after sunrise, reaching a maximum at about 2:00 p.m. and then generally declining until the next morning. This pattern is superimposed by the concentrations changes due to the mountain-valley transport system, which are much stronger than in the winter (up to 0.11 mg/m^3). The weak but continuous circulation starts at approximately 10:00 a.m. and reduces the ozone concentration. This explains the lower ozone maxima in the fall (peaks up to 0.07 mg/m^3) and even in the summer (peaks up to 0.055 mg/m^3) than in the winter. During the summer, as well as during the fall, the circulation system leads to additional minima in the diurnal ozone profile. The onset of circulation is connected with the transport of clean air and produces a morning minimum of concentration. At sunset, ozone formation is terminated and an evening minimum appears. However, the circulation system reverses and increases the ozone concentration anew by the downhill transport of ozone-rich air from the west. The strong circulation system causes the midday peak to be lower and the night-time peak to be higher in summer than in winter. This explains the findings of the highest ozone variations during winter and the generally low–medium ozone pollution [2]. One other possible cause that should be considered, in addition to meteorological influences, is the chemical reduction of ozone by particles and other pollutants from vehicle exhausts. This reaction, the probability of which depends on the temperature, leads to a drop in the ozone concentration, especially during summer.

2.4 Geographical Information System

GIS is a computer-based system for capturing, storing, checking and manipulating data that are spatially referenced. For this study we used the ArcView® application due to its relative user friendliness and its wide use by local authorities and many research institutes. New applications of GIS are been developed to explore dynamic environmental models. Like in every computer language, GIS uses data types, which are stored and manipulated in the form of a vector (point, line or polygon) and or raster (cells or grid). So it is possible to associate each of this shapes with consistent information in their databases.

In our study, industries and monitoring sites are examples of point shapes, which correspond to records in the respective database. In each of these records, information regarding time series of monthly, annual mean values of ambient concentration are associated to the monitoring site database. Stack dimensions, emission rates and coordinates are associated to the industry database. In the GIS system, each street is represented by a segment corresponding to a record in the database. We have characterised streets according to three hierarchies: a) primary or main city access and inter-county freeway; b) secondary or intra-county roads; and c) tertiary or residential streets. The hierarchies have been selected according to traffic intensity, hourly variation, street width, dominant use, such as: private or public transport; light or heavy duty; commercial or residential; and other urban criteria. Streets axis, number and emission of vehicles are included in records of line-shape-type database. Each of these segments also includes length, street type, county and so on. Several other indicators such as inhabitants per car use, inhabitants growth rate, main city attractors, and a mobility source-destination public survey, have been used to complement the knowledge of the city mobility.

A particular display of the different shapes and options are called themes and can be viewed or selected in any order, e.g. localisation of monitoring sites, emission patterns, etc. These sites are sorted according to the modeller criteria, and the most relevant features may be highlighted on individual maps (or views) and displayed on different scales. It is important to note that GIS is not only used as a map viewer, but more as an integrated tool to handle data from many sources.

2.5 Data collection

Any air quality model requires good information on emission sources, energy consumption and meteorological data. There are several categories of sources: a) large stationary sources, like steelworks, petrochemical plants, power generation; b) medium-small stationary sources, including industrial boilers and process plants, commercial and residential heating, laundries, dry cleaners; c) domestic activities, such as heating and cooking; and d) road vehicles like gasoline or diesel passengers cars, public buses, light or heavy duty vehicles, and two-wheel vehicles. The emission inventories may be prepared according to the type of sources. For example, for large sources, by sending questionnaires to large energy consumers and checking it with in-situ stack measurements; and for small energy consumers, it can be treated as area sources derived from fuel consumption and calculating it through average emission factors. To estimate the mobile sources it can be calculated thorough traffic emissions, using average emission factors for different fuel and power, number of vehicles, average speed and travelled distances, complemented with traffic counting and public registers and surveys.

2.6 Calculation of industrial emissions

To treat the stationary sources, the model calls the program ISC3 (EPA: U.S. Environmental Protection Agency) that calculates the ambient concentration for multiple industrial sources at a receptor grid. As input for the ISC3 program we used the data of the prepared emission inventory: stack dimensions, output stack temperature, emission flow and velocity, source relative position and so on. The dispersion program also needs as input the local meteorological data from the monitoring stations. The output of this program is then calculated for each cell of the area under study. These concentrations are organised as a raster shape database inside the GIS system.

It is important to note, that the Gaussian dispersion models and the available wind data have certain limitations, which can be summarised as: a) the dispersion models suppose linearity between source emission and ambient concentration values at the receptors when enough averaging data are present; the minimum significant correlation is a one-hour average; b) the model does not consider chemical combination, that is, the gases and particulate matter are considered inert, and the range leeward must not exceed 25 km; c) most of the anemometers record wind in 16 sectors of 22.5° degrees each, where an hourly mean value indicates the most frequent direction (mode); d) local micro dynamic at the receptor, due to the influence of trees, buildings, etc., are normally not considered. Despite these limitations, the calculations represent adequately daily, monthly and annually mean values, and are suitable for compliance with the air quality standards. The ambient concentration at a particular cell located at coordinates (x,y,z) from a fixed point source is generally calculated using a bidimensional Gaussian plume moving in the wind direction x. At ground level ($z=0$) the concentration C(g/m^3) is:

$$C(x,y,0) = \frac{Q}{u}\frac{1}{\pi\sigma_y\sigma_z}\exp(-\frac{H^2}{2\sigma_z^2})\exp(-\frac{y^2}{2\sigma_y^2}) \qquad (1)$$

where Q (g/m.s) is the emission rate of the source, H(m) is the effective stack height, y is the distance transversal to the wind direction in the horizontal plane, z is the altitude above ground, and u (m/s) is the wind speed. The lateral and vertical dispersion coefficients σ_x σ_y (m) depend on the stability class (and indirectly on the wind speed) and increase with increasing distance to the source x. The equation is well described in many books [9, 10].

2.7 Calculation of the vehicle emissions

Two complementary approaches are generally proposed to estimate the emissions from road vehicles: the top-down approach or the bottom-up approach. The selection of one of these methods will depend on the availability of the main input data and the desired spatial and temporal resolution. In the top-down approach, the total annual emission from mobile sources is calculated using the number of road vehicles actually running for a given region. At any time and street, the vehicle flux is estimated through indirect information such as population density, structure of the automotive park, fuel consumption, and number of cars per inhabitants. Three other estimation should be also considered, i.e., the average speed; the annual mean traveled distance; and the annual emission factors, normally based on fuel consumption. This information has acceptable spatial distribution, but a poor temporal resolution, usually on an annual base. To estimate the emissions from road vehicles in the bottom-up approach, traffic counting and speed recording in many streets are required. Normally this has a high temporal resolution (an hourly base) but poor spatial resolution. Also it is convenient to determine the vehicle fleet distribution according to power, size, fuel, and typical vehicle use. It is clear, then, that this approach requires high data density. To calculate the total emissions, it is necessary to estimate, average emission factors and fuel consumption for each vehicle category together with the mean traveled distances. Usually this latter approach is selected in urban areas when the information is available. Although the bottom-up design seems to be more accurate, the randomness of the involved variables degrades the accuracy of the detailed information. To compare both approaches it is necessary to calculate the total annual emissions on the same temporal scale and for a given region. After doing this, one must check for fuel consumption, number of total traveled distances per vehicle, etc. The advantage in using GIS is that most of the gathered or estimated information can be geographically distributed, simplifying the association with temporal data.

In this study, the bottom-up scheme was selected and downloaded to the GIS system. Traffic counting in main streets, average driving speed, and used emission factors, were recorded and grouped for different fuel and vehicle types. Also, this information was cross checked using a top-down methodology. In this case, relevant information from public registers and surveys was obtained: the structure of the automotive park, the average traveled distances the fuel consumption, the inhabitants density, car per inhabitants, etc. The emission information from the vehicle factories and relevant international regulations and legislations was also gathered [11].

The emission factor e (g/km) due to street vehicles belonging to a group i is expressed as the mass of pollutant per unit length as a function of the traveled speed v. The total pollutant q for each street (or segment) with traffic flow n (veh/h) belonging to I vehicular groups is calculated as:

$$q = \sum_{I} \frac{p_i}{100} e_i(v).n_i \tag{2}$$

where p is the percentage of vehicular group i with respect to the total flow. As emission factors we used Tables 8.1 to 8.3 of the Emission Inventory Guidebook [12]. Table 1 shows typical values of the emission factors used for Mendoza. Primary street ambient concentrations, where the roads present no obstacles to dispersion, are treated as a line source using a Gaussian dispersion model. Small flux residential streets (tertiary) are considered as area sources, and for downtown streets a canyon plume box model is used. These ambient concentrations C (μg/m^3) may be calculated as [13]:

$$C = a\left[\frac{q/3.6}{W(u+0.5)}\right] * F + bT + dH \tag{3}$$

where a is a constant ≈7, q is given in g/(km.h), u is the wind speed (m/s) at roof level, F is a shape factor depending on the wind direction, W is the street width (m), T is the ambient temperature (in Celsius) and H is the mixing height (m); b and d are the regression coefficients.

The total emissions of air pollutants from vehicles, using the top-down approach, for a given area, and for an (yearly) average period, it is characterised by three main factors:

$$E = N \times e \times l \tag{4}$$

where E (g/unit time) is the total emission in the time considered, N is the number of average circulating vehicles in the period, e is the specific average emission factor measured in g/km per vehicles, and l is the mean traveled distance in km. The emission factor is an empirical function of the fuel consumption, the speed and age of the vehicle. This information is useful to understand the problems associated with road transport: i) The number of vehicles N: the monitoring data clearly show a rise of nitrogen associated to an increase in the automotive park. A reduction in N may be interpreted not only in a reduction of number of vehicles, but better as a decrease in the number of daily trips. Some logical implications to reduce the effective number of road vehicles are: a) to promote and improve public transportation; b) to increase the occupational rate of vehicles; c) to improve the coordination of time schedules, in order to balance the transport demands. ii) The emission factor e: this factor is directly related to the available technology standard, and the driving style. A reduction in e is slowly incorporated as new vehicles with better technology replace old vehicles, but since Mendoza is situated in an arid area, the average vehicle age is >10 years, and mostly not fitted with catalytic converters. Regular technical vehicle inspection and carefully planning of new avenues and traffic light synchronisation may contribute to reduce the effective emission factor. iii) The driving distance l. To reduce l it is necessary to coordinate with city urban planners and decision makers some of the following aspects: a) decentralisation of the public offices, which actually are very much concentrated in the downtown area; b) improvement of public educational activities at the peripheral areas; c) decentralisation of shopping centres and bank offices. This latter has only been done by market needs, but unfortunately with not enough urban planning. Decentralisation will promote and strengthen the local communities, will increase their cultural, educational and commercial offers, and certainly will contribute to reduce the needs of transportation. At present, the increase in use of the Internet may contribute positively to increase the efficiency in the transportation of goods.

Table 1: Example of emission factors in g/km for1999 for gasoline and diesel fuels for given vehicle speed.

Speed	TSP		NO_x		HC		CO	
	G	D	G	D	G	D	G	D
12 km/h	1.10	2.50	1.50	1.40	5.40	4.50	58.70	6.60
23 km/h	0.90	2.30	1.30	1.20	3.45	2.55	38.90	4.20
40 km/h	0.80	2.20	1.20	1.10	2.35	1.55	27.50	2.80
60 km/h	0.80	2.20	1.20	1.10	1.75	1.10	21.30	2.10

TSP: Total suspended particles, NO_x Nitrogen oxides, HC: hydrocarbons, CO carbon monoxide, G: Gasoline, D: Diesel.

2.8 Description of the model

The diagram in Fig. 5 represents the principal steps of the used model, which will be briefly described. First, it is necessary to gather the necessary source information such as: industry inventory, number of vehicle, street types; secondly, emission factors for both stationary and mobile sources should be measured or estimated. Secondly, an emission pattern can be prepared, which will be computed and stored as either line shape (vehicle) or grid shape (industry) databases.

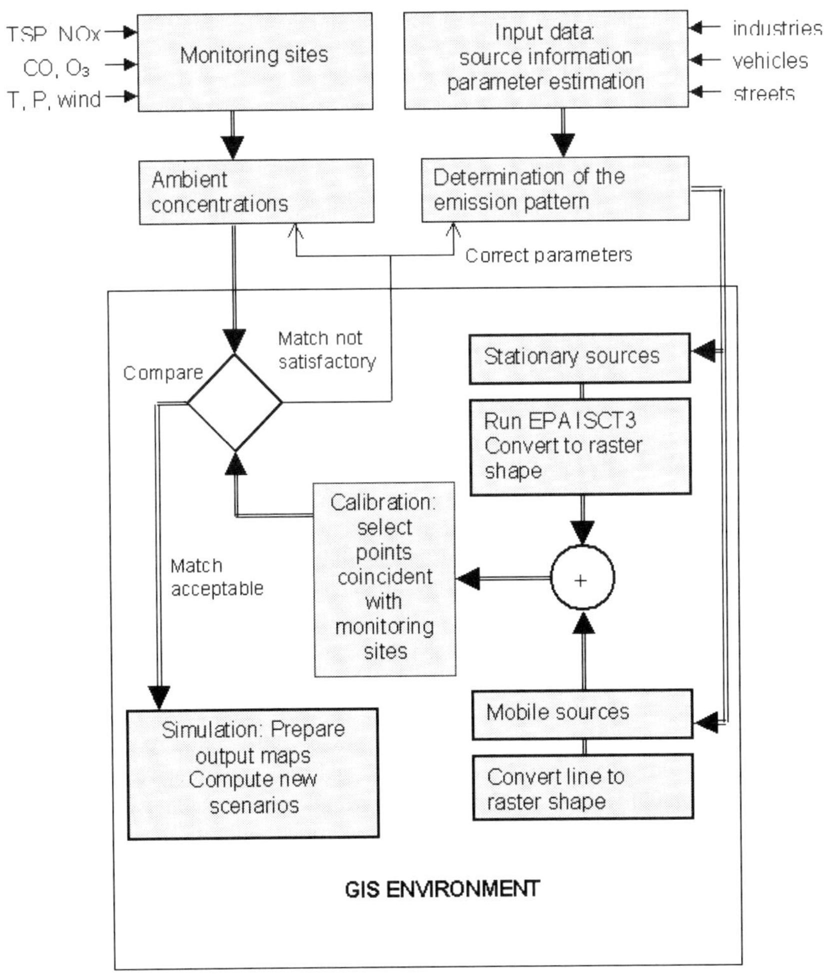

Figure 5: Diagram of the described model.

To handle this information, the city has been divided into cells of 350 m × 350 m, where each cell represents a record in the database. This database contains information such as, coordinates of the receptor grid and ambient concentration for the main pollutants. For areas close to the industrial sources smaller grid of 100 m × 100 m are also available.

In order to compare, at the street level, both types of database, the pollutant coming from the industrial sources is added as a background concentration to the ambient concentrations of pollutants provided by the mobile source calculations.

Once the model is calibrated, then it is possible to use the model to simulate different scenarios, and prepare the output maps. If no acceptable match is obtained between calibration and measurements, then it is necessary to return to the first step and check for errors in the parameter estimation or to perform model corrections.

2.9 Calibration of the model

Data from monitoring stations and from various measuring campaigns were collected to calibrate the model. For areas characterised mainly by mobile sources, three data sets corresponding to three urban monitoring sites were selected. One set of two weeks of 3-min continuous measurements was used to set the coefficients in (3), and two series with several years of daily mean values were used as control. Additionally, several ad-hoc measuring campaigns were collected for those areas where the industrial sources are predominating. We will here briefly present the calibration results.

2.9.1 Calibrating the model for mobile sources

One urban downtown monitoring station was selected to measure one month continuously of NO_x, CO, O_3, black carbon, temperature, wind speed, and pressure close to a sidewalk. Also total suspended particle (TSP) measurements were captured with daily high volume filter samples. Additionally, vehicle counting in the surrounding streets of the monitoring station was performed regularly. Then, TSP, CO and NO_x concentrations were calculated (using eqns. 2 and 3) and compared with the measured data. This comparison allowed the estimation of the unknown parameters.

The selected parameters were tested in a two-control data set at different monitoring sites corresponding to several years of daily average values of NO_x and TSP. Consequently, we simulated the distribution of ambient concentration of NO_x and TSP. Since there were no traffic counting available for every day, it was necessary to suppose a Gaussian distribution around the daily mean of the number of vehicles. Following the same concept, the emission factors are only known as an average process, thus again a Gaussian distribution should be adopted. For the emission factors, we use as a mean, the values adopted in the first calibration test and we included a small variance (25% for n and 20% for e). The wind intensities were taken from the available measurements.

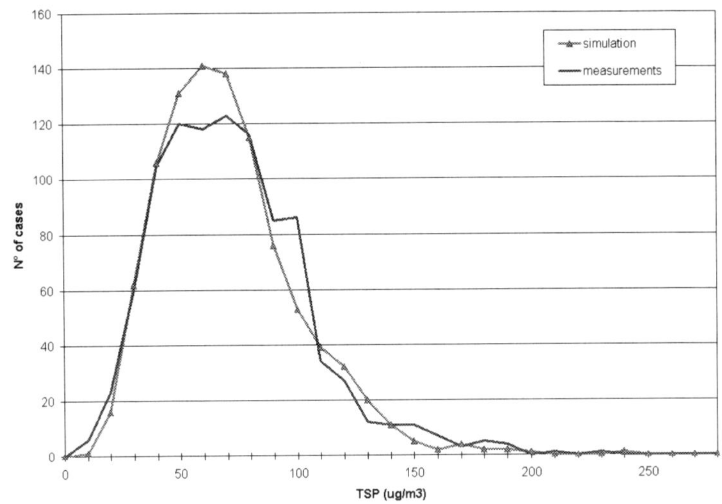

Figure 6: Frequency density distribution of 5 years (1993–1998) of daily mean measurements and calculated total suspended particles (μg/m^3) in a downtown area (station "IPV"): measurements (dark line), calculated data (gray line with triangles).

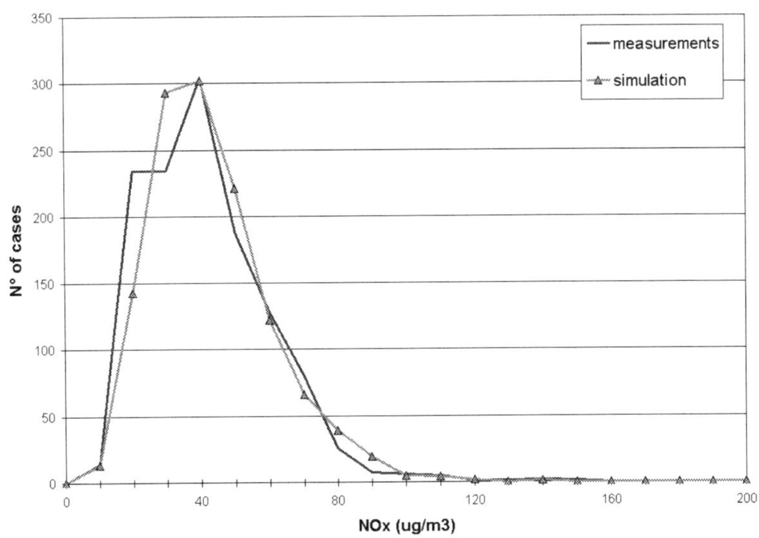

Figure 7: Frequency density distribution of 5 years (1993–1998) of daily mean values measured and calculated nitrogen, (NOx) in μg/m^3, at station "Godoy Cruz". Measurements (dark line), simulated data (gray line with triangles).

Figures 6 and 7 show the resulting frequency distribution for 5 years of daily measurements (1993–1998) at station "IPV" and "Godoy Cruz", respectively. These control tests show that not only the mean value and variances are matched, but also their distributions are well fitted. These results are not possible if the Gaussian variation were not included in the traffic flow and emission factors.

2.9.2 Calibrating the model for industrial sources

In the Mendoza area, it is possible to identify three main groups of stationary sources: a) an oil refinery and a petrochemical industry as main emitters of particles and SO_2; b) an energy power plant, mainly based on natural gas, with significant NO_x emissions; and c) two medium companies responsible for black smoke fumes. These latter sources produce metal and ferrous alloy and coal briquettes for heating. Each of these two industries has also open piles for the storage of iron and coal respectively, which were included in the dispersion model as area sources of 400 m × 400 m each. The geographical location of industrial sources was positioned using a GPS receiver.

To calibrate the industrial source emissions, several monitoring campaigns were performed close to an industrial complex situated at the SW of the city. In these campaigns, continuously (3-min sampling) NO_x, CO, O_3, black carbon, temperature, wind speed, and pressure at different sites were taken. The black carbon data were compared with a gravimetric method, by taking several samples on a paper filter (48 hours each) using a high vacuum pump. The data were computed using eqn. (1) and then compared with the measured values at the test monitoring sites (Fig. 8), showing a correlation factor of 0.85.

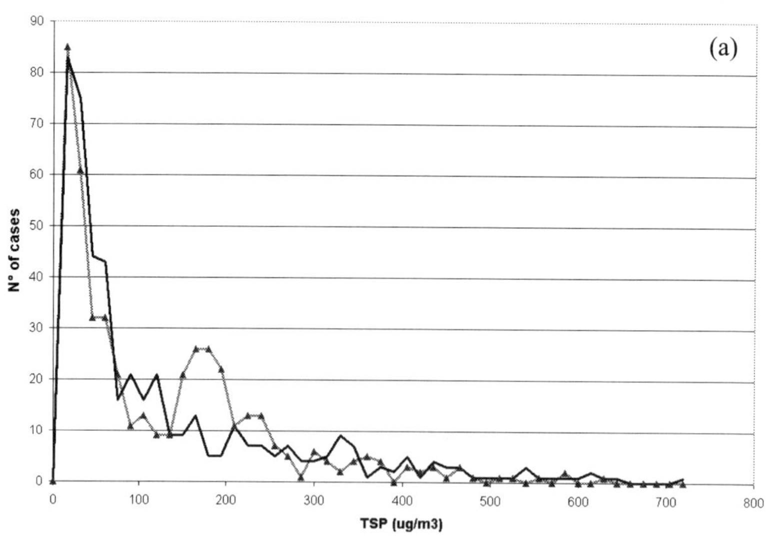

(b)

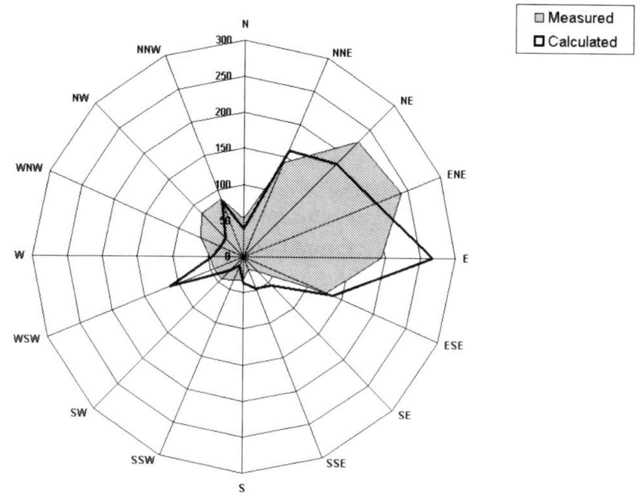

(c)

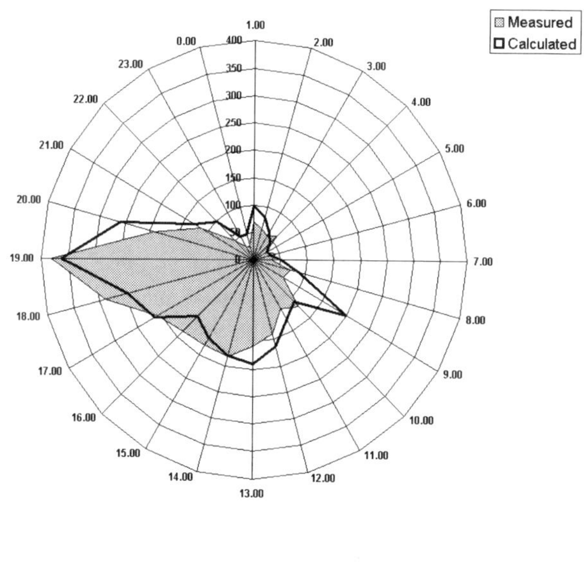

Figure 8: (a) Opposite: Frequency distribution for measured and calculated data of nitrogen oxides (μg/m^3) taken at the south industrial area (dark solid line: measured; gray line with triangles: calculated). (b) Measured and simulated mean hourly values of NO_x with respect to wind direction; and (c) same data as a function of hour of a day.

3 Results

3.1 Model outputs

After the calibration process was concluded, it is possible to study the spatial distribution of the ambient concentration for the entire city. As an example of these model outputs, two years (1997–98) of meteorological data, has been calculated.

Different temporal integrations are possible, but normally, the following averaging conditions are usually computed: maximum hourly values, 24 h, 30 days, and annual mean values. The next figures show several types of outputs, which illustrate the calculated conditions. Figure 9 shows a map with the spatial distribution street hierarchy and figure 10 displays the respective average traffic flow for each street. Based in this information, it is possible to establish a pollutant emission inventory for the mobile sources. Figure 11 plots, for example, the emission pattern for NO_x. Figure 12 shows the calculated ambient concentration for NO_x produced by the mobile sources, where it is possible to recognise areas with a mayor impact on the air quality, especially close to the central business area.

As examples of the industrial areas, Figure 13 shows the influence of the refinery located south of the city as the main source of SO_2, reaching maximum hourly values at the urban areas of the city of 100 $\mu g/m^3$, and more than 400 $\mu g/m^3$ in areas close to the source. Most of it is vented through the torch, although lately the refinery has incorporated a desulphuration Claus plant to reduce their SO_2 emissions.

At the south industrial area, the ferrous alloy and the coal-burning processes are the main emitter, especially because they function without any filtering device to their dust emissions. Maximum hourly ambient values may reach 1500 $\mu g/m^3$ and more than 200 $\mu g/m^3$ daily mean values close to the sources. Ten kilometres away, at a residential part of the city, annual mean values of 30 to 40 $\mu g/m^3$ can be obtained, especially during the winter months. According to the calculations and the monitoring measurements, the influence of the industrial sources over the city do not exceed the air quality standard, with the exception of areas close to the industrial area.

The power plant at Lujan de Cuyo is the main NO_x emitter, but the concentration values do not exceed the local standard (100 $\mu g/m^3$ for 1 year and 200 $\mu g/m^3$ for 24 h). Mean daily values reaches de 20 to 40 $\mu g/m^3$ and annual mean do not exceed 5 to 10 $\mu g/m^3$, due to industrial sources. NO_x values at the urban areas are mainly produced by vehicle exhaust. Figure 14 show a daily mean concentration map of nitrogen oxides emitted from the southern industrial area.

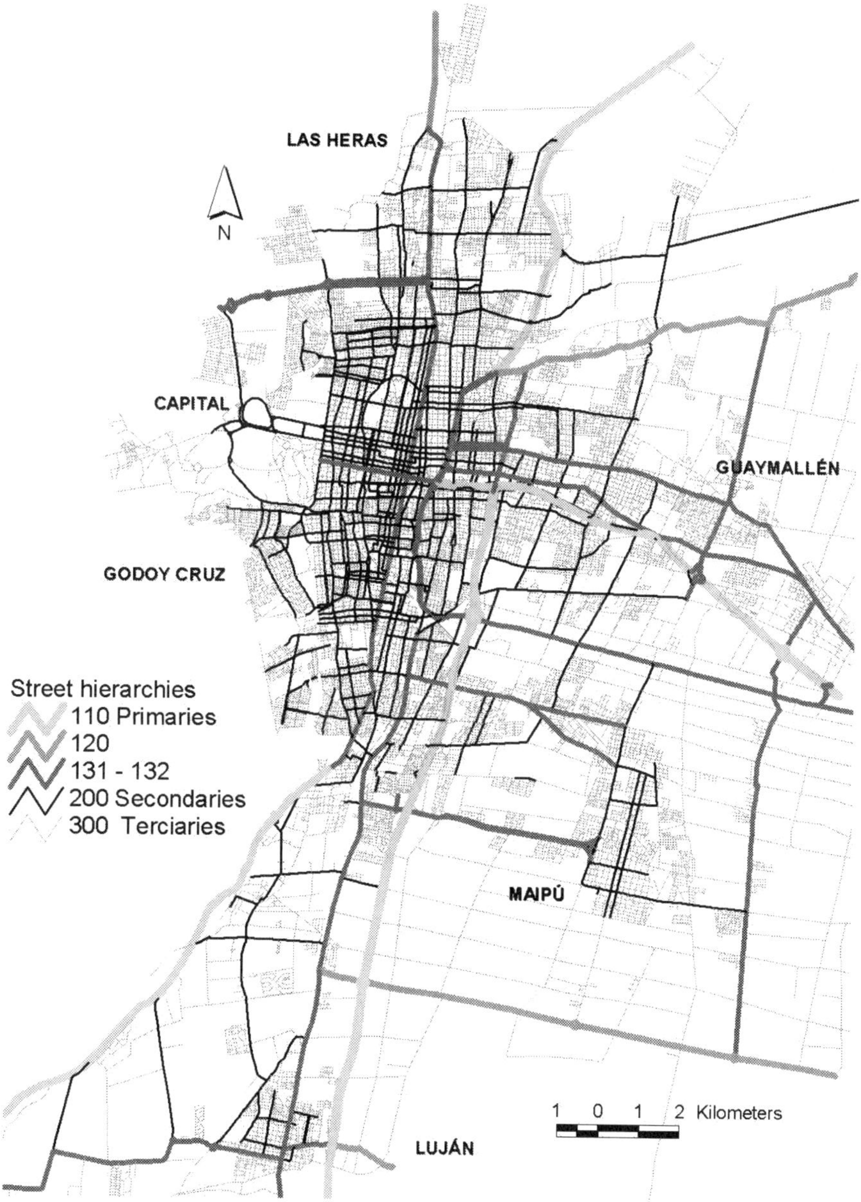

Figure 9: Street hierarchy for the Mendoza metropolitan area.

Figure 10: Average traffic flow in the city metropolitan area.

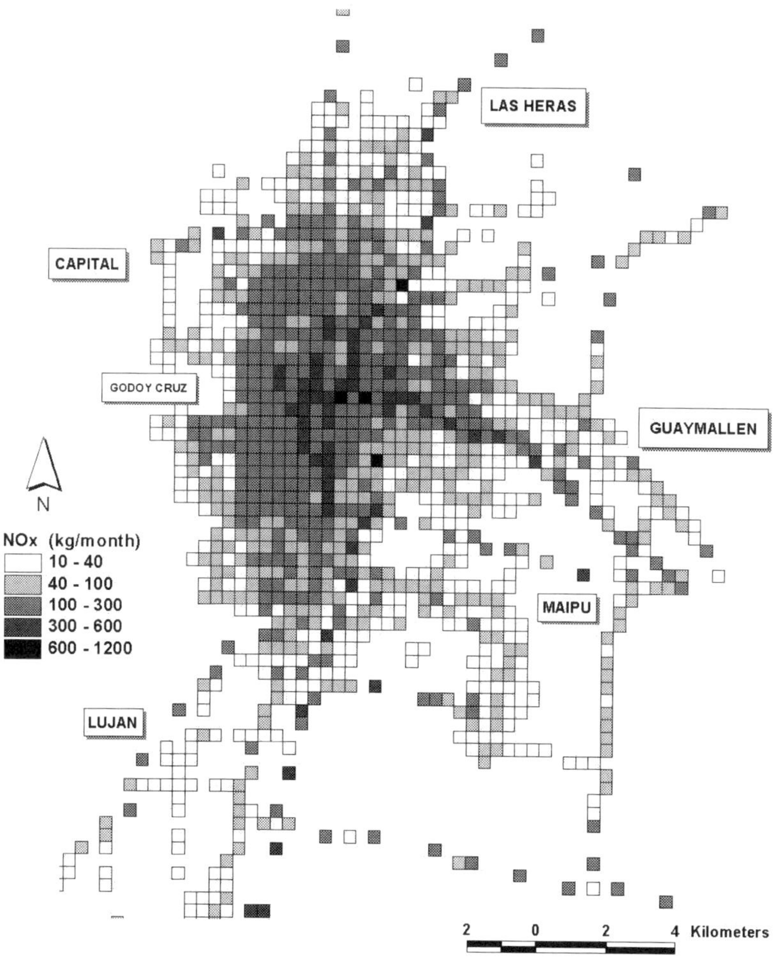

Figure 11: Emission pattern for mobile sources in Mendoza.

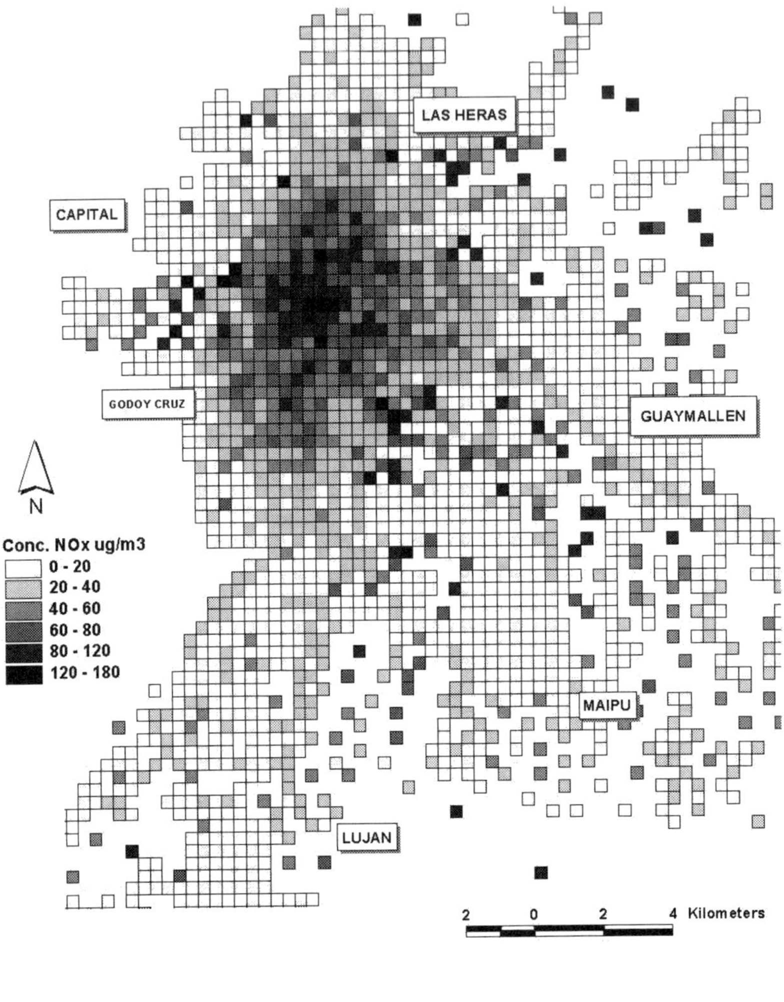

Figure 12: Ambient concentration values of NO_x ($\mu g/m^3$) produced by the road vehicles in Mendoza.

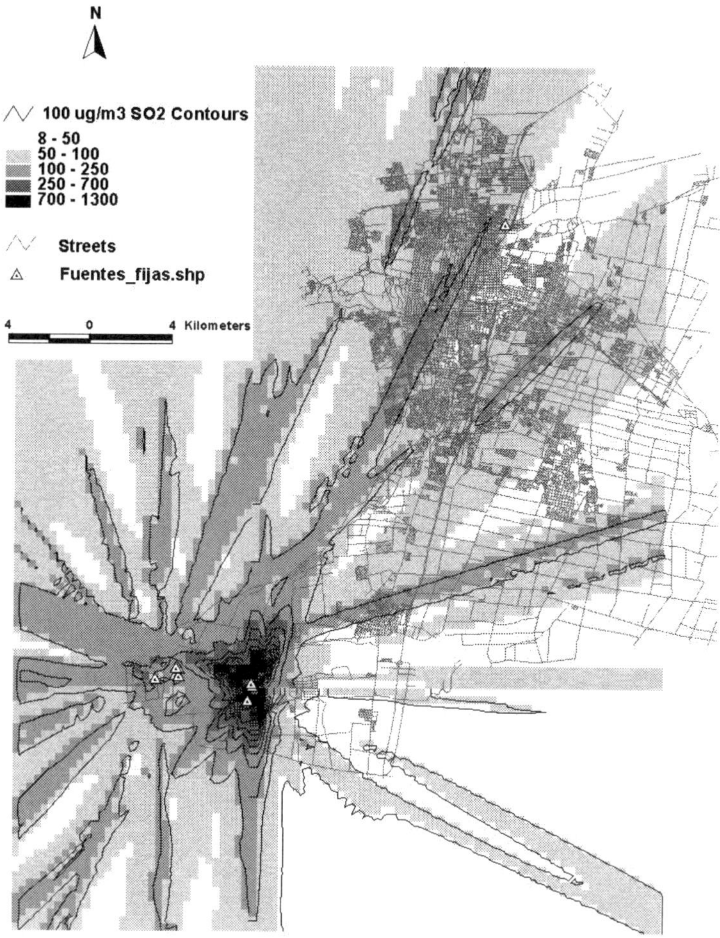

Figure 13: Annual average of 8 hr. Sulphur dioxide concentrations.

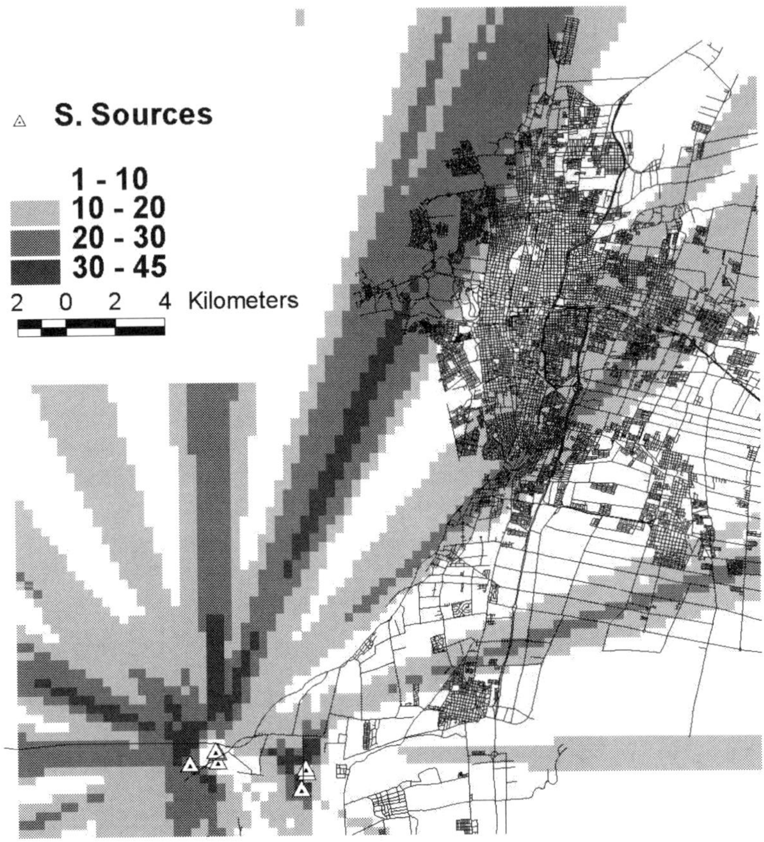

Figure 14: Nitrogen daily mean concentration (μg/m^3) from the south industrial area.

3.2 Future scenarios

The model may be also used to analyze the air quality situation under a changing scenario. To test this situation, we will suppose that the public transport system may be changed. At present, important discussions are carried out among authorities, enterprise managers and users of the public transportation system to propose and introduce a new electric train and tram system to replace the current diesel buses. A basic topic for the discussion is the environmental conditions the new system should have, that is, how much reduction of air pollution would this system allow. One of the proposals was based on a multi-modal system. This system will reuse the railway to commute at mid-distances, coupled with electric tram and trolley buses, which will run downtown and for short distances. However, some diesel buses will co-exist for those areas where no electric transportation is available. To assess this future scenario, the actual situation needs to be first evaluated. The second step requires the study of the city development for the next decade maintaining the present bus system, that is, it is necessary to determine the status for the principal indicators: population growth, number of vehicles, mobility, etc. The third step requires the recalculation of the latter situation, but when the transportation system is changed by a multi-modal electric tram system. Finally, it is possible to compare both situations.

An important task of the simulation is to select an appropriate number of indicators that will well represent the expected changes. For this task, four parameters where selected: changes in mobility and means of transportation, automotive park, and population growth. There are also some secondary indicators like: preferences in the public transportation means, and changes in vehicle technology, which are shown to be very useful.

From the source and destiny survey of 1998, it is known that 1,400,000 trips are performed daily within the metropolitan area. Public diesel buses carry around 510,000 passengers daily, implying 35% of the daily trips, while another 610,000 trips (45% of the daily trips) are done in private cars. However, the ratio public-to-private transportation is decreasing, showing a preference for using one's own car. Table 2 records the main parameters and indicators used to estimate years 1999 and 2010. Analyzing the present trend, a 5% yearly increase in new vehicles is expected, with a 1.5% yearly withdrawal. With these values, it is possible also to estimate the actual vehicle technology and also to approximate the variation of the emission factors. In Table 3 we summarise the expected emission factors for year 2010, considering the estimated composition of the on-road vehicles. It is foreseen, a reduction in the emission factors for 2010, due to technological improvement, however the impact of these improvements is weighted by the age of the older fleet.

As an important parameter scenario for year 2010, we considered 25% of private car users using public transportation and a second scenario with only 10% of car users switching to the electric system. Under this basis, daily traffic, emissions and ambient concentration values for air pollutants were estimated.

Table 2: Number of daily trips vs. means of transportation.

Year	1999		2010
Means of transportation	No. Of Trips	Relative %	No. Of Trips
Public diesel bus	493600	34.9%	623100
Trolley bus	14900	1.1%	18800
Contracted (diesel) bus	5900	0.4%	7500
School bus	13400	0.9%	16900
Taxi or remise	23000	1.6%	29000
Own car, as driver	354700	25.1%	447800
Car, as passenger	201000	14.2%	253700
Motorcycle	35200	2.5%	44400
Bicycle	102700	7.2%	129600
Walking more than 1000 m	111200	7.9%	140300
Other means (not specified)	59700	4.2%	75400
Total	1415300	100.0%	1786500

Table 3: Principal indicators for 1999 and 2010.

Year	1999	2010
Population (inhabitants)	895000	1050000
Mobility (number of daily trips per inhabitants)	1.58	1.7
Growth in the number of trips (1999-2010)		1.26
Population growth rate (1999-2010)		1.21
Private vehicle growth rate 1999-2010		1.36
Motorisation rate (inhabitants per car)	3.1	2.7
Passengers per private car (per trip)	1.45	1.50
Number of private vehicles	291000	381000

3.3 Results of the simulation

Once the scenario is chosen, the principal calculations are the geographical distribution of daily traffic flow, emissions and ambient values for air pollutants. Normal outputs for this model are: a) daily vehicle fluxes, b) daily mean emission for TSP: Total suspended particles, NO_x: Nitrogen oxides, HC: hydrocarbons and CO carbon monoxide; c) total emissions for given areas of the city, discriminated according to the vehicle type and fuel; d) expected immissions for the present situation (1999), for year 2010 under several scenarios and for year 2010 incorporating the electric tram system.

The total daily emissions for the city, for an approximate area of 25 × 20 km, can be calculated as a function of the type of fuel consumption, which can also be interpreted in terms of private (mainly gasoline) and public (mainly diesel) transport. This information is displayed in Table 4, while Table 5 shows this data as a ratio of emission to transported passenger. These two tables clearly remark that the public transport has a better overall performance than private cars, considering total daily emissions. Only for particles (total mass) diesel buses are worst than the gasoline vehicles. Table 6 shows a comparison of the estimated total mean daily emission of the city, calculated for: a) year 1999; b) year 2010, no change (NP); and c) year 2010 with the proposed tram system on going (WP).

Table 4: Daily emission of pollutants by fuel type in kg for year 1999.

Emission (in kg) / Transport mode	Public (diesel)	Private (gasoline)
Total suspended particles (TSP)	600	400
Nitrogen (NO_x)	850	2500
Hydrocarbons (HC)	950	2900
Carbon monoxide (CO)	5250	18400

Table 5: Ratio of the daily emissions to carried passengers for year 1999.

Emission per passenger (in g/pass.)	Public (diesel)	Private (gasoline)
Total suspended particles (TSP)	1.15	0.60
Nitrogen (NO_x)	1.65	3.65
Hydrocarbons (HC)	1.80	4.25
Carbon monoxide (CO)	10.25	26.65

Table 6: Total daily emission of pollutants by mobile sources in tons.

Contaminant (in tons)	1999	2010 (NP)	2010 (WP)
Total suspended particles (TSP)	1	1.4	1.0
Nitrogen (NO_x)	3.4	4.3	3.2
Hydrocarbons (HC)	3.9	4.7	2.7
Carbon monoxide (CO)	23.6	30.3	18.0

An increase in air pollutant emissions from 20 to 40% is foreseen for the year 2010, under the scenario presented in Tables 2 and 3. Consequently a worsening of the air quality is expected, both in absolute concentrations values as well as an extension of the affected area. As an example, Fig. 15(a) shows the calculated suspended particle concentrations for the year 2010 and Fig. 15(b) plots how the area with critical ambient concentrations values could be reduced in extension and in absolute values, when the expected tram project is carried out. A higher positive impact is expected on carbon monoxide and hydrocarbons emissions, because of their nonlinear dependence on the vehicle velocity. This double effect will improve or at least stabilise the present air quality in the city, despite the expected population and traffic growth. As traffic density increases, lower speeds are developed, hence producing higher emissions. On the contrary, as more users prefer a clean and safe electric public transport, traffic intensity will diminish considerably, and higher regular speeds can be achieved for those who use private transportation.

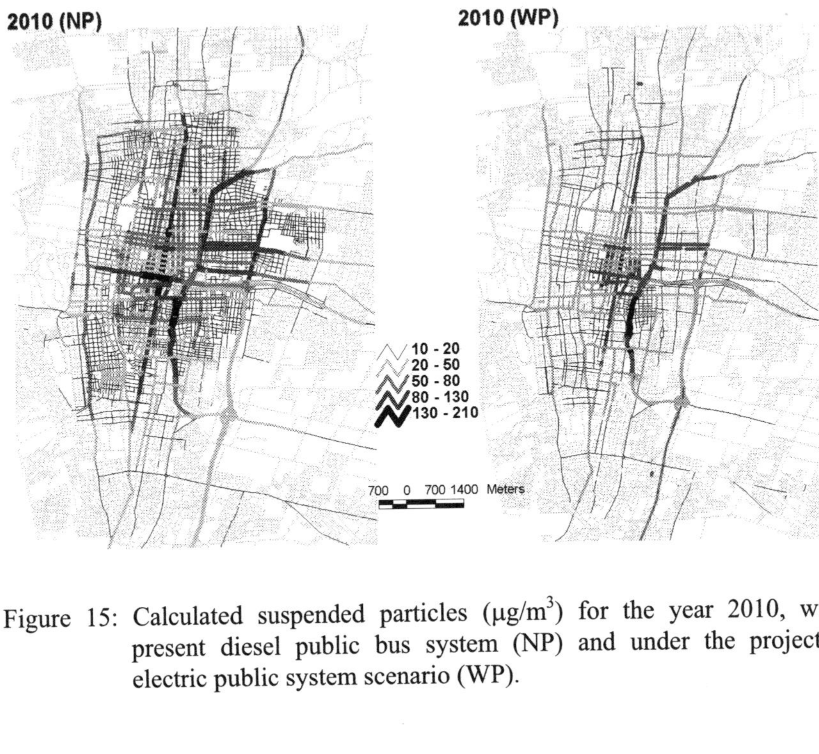

Figure 15: Calculated suspended particles (μg/m^3) for the year 2010, with present diesel public bus system (NP) and under the projected electric public system scenario (WP).

4 Case study

In the early morning of May 15, 1995, the population of the city of Mendoza was affected by a high pollution event, which produced serious pulmonary affections. During several hours hundreds of persons, especially children, needed to be hospitalised. Since no Enterprise or Company claimed to be responsible for the situation, our Institute for Environmental research was called by the Federal Justice to study the case and to produce an environmental assessment. To study this case, we followed three lines of action: 1) we performed an epidemiological survey on the symptoms of the affected persons to see what kind of pollutant could be the responsible; 2) We collected the meteorological data and, according to the suspected pollutant, we ran a dispersion model for possible emissions; 3) We tested the dispersion model with continuous monitoring measurements. Finally the federal judge used our written assessment as an important proof for the respective investigation and trial.

4.1 Epidemiological survey

The main symptoms were associated in two groups: a) respiratory affections and dyspnea; b) headache, migraine, nausea and vomiting. According to the patients, the event had a duration of between two and four hours. Given also the strong smell witnessed by the affected persons and the intensity of the respiratory affections, we preliminary concluded that it could be a combination of sulfur oxides and hydrogen sulfide (SO_2 and H_2S) in the presence of particulate matter [14–16]. The reached ambient concentration of the gases could approximately be between 0.05 and 1 ppm. Note that the lower olfactory threshold of H_2S is approx. 0.025–0.05 ppm. Unfortunately there were no SO_2 or H_2S measurements available that day.

4.2 Meteorological conditions

During the first hours of May 18, the meteorological conditions were: cold, calm winds (less than 1.5 m/s) from SW, overcast with high pressure, characterised as stable with low inversion layer less than 200–300 m. By the midday the conditions improved to convective with soft winds.

4.3 Dispersion models

Given the meteorological conditions and the intensity of the epidemiological aspects, only a few possible stationary sources could be responsible of the emission. We simulated a short high emission from the Refinery located 16 km SW from the affected area of the city. Supposing a worst-case scenario, with 500 kg/h of sour gases emitted at an 80 m high stack, and considering that 10 km downwind we may have had at least 10 % of H_2S and 90 % of SO_2, we obtained:

Conc. $SO_2 = 2830\ (\mu g / m^3).\ 0{,}9 = 2547\ (\mu g / m^3) = 1{,}14$ ppm

Conc. $H_2S = 2830\ (\mu g / m^3).\ 0{,}1 = 283\ (\mu g / m^3) = 0{,}24$ ppm

which confirms that our first guess on possible concentrations could have been possible.

4.4 Monitoring measurements

As stated before, there were no SO_2 or H_2S measurements available during this episode. But at our Institute IEMA we monitored surface ozone, detecting an anomalous situation. The monitoring station is located around 16 km SW downwind from the refinery, and is close to the affected area. We studied how

O_3 could be affected by H_2S, concluding that, after neglecting other smog components, the following reaction could take place [10]:

$$H_2S + O_3 \rightarrow H_2O + SO_2 \tag{5}$$

Considering that the reaction time of 1 ppb of H_2S in the presence of 50 ppb of O_3 with 15 000 particles / cm^3 is about 2 hours.

Then the dynamic variation of the ozone concentration may be expressed by:

$$[O_3]^M = [O_3]^M(t_0)e^{-K_1(t)} + e^{-K_1(t)}\int_0^t \frac{d}{d}\frac{[O_3]^E}{T}e^{K_1(T)}dT \tag{6}$$

where $K_1(t) = \int_{t_0}^{t} k_r[H_2S]dT$ and $K_1(T) = \int_{t_0}^{T} k_r[H_2S]dT'$, $[O_3]^M$ is the measured concentration for any t, $[O_3]^M$ (t_0) is the initial value, and $[O_3]^E$ is the due value if $[H_2S]$ were not present; t_0: initial time, t_1 is end time of the reaction, k_r reaction rate for H_2S 1.11×10^{-5} [ppb^{-1} seg^{-1}], for T = 25°C (289,15 K); $[H_2S]$: hydrogen sulfide concentration.

4.5 Results of the simulation model

The resolution of the continuity equation yields in a hydrogen sulfide $[H_2S]$ pulse as it can be seen in Fig. 16. The $[H_2S]$ concentration, which is represented as a continuous line, is the input in our program that reproduces the measurement of May 18. The mean square root of the olfactory threshold (50 ppb) is drawn as a dotted line. This pulse begins approximately at 8:00 a.m. with a wave front of 100 ppb. This curve surpasses the olfactory threshold some time before 9:00 a.m., and around 12:30 p.m. the smell is no longer perceived.

Figure 17 shows, with the pointed line, the average value of ozone for May $[O_3]^E$ (t), this value represents the expected value of ozone if the H_2S pulse were not present. The continuous line is the true measured ozone value for May 18 $[O_3]^M$ (t). The dashed line represents the simulation value $[O_3]^C$ (t). Finally, the point-dashed line represents the measured value; compare this curve with the measured value.

The value measured on May 18 had a different feature compared to ozone values of many days with similar meteorological characteristics, which gave a strong evidence of the presence of an anomalous situation. This could not be explained only by the presence of typical smog situation. The strong decay measured between 11:30 am and 12:00 pm gave us an indication of the reaction time of eqn (6).

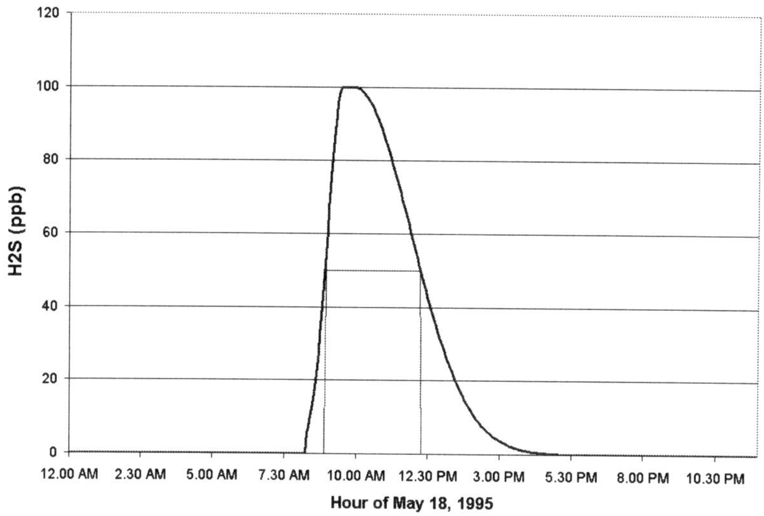

Figure 16: Emission of hydrogen sulfide H_2S during an pollution episode in May 18, 1995.

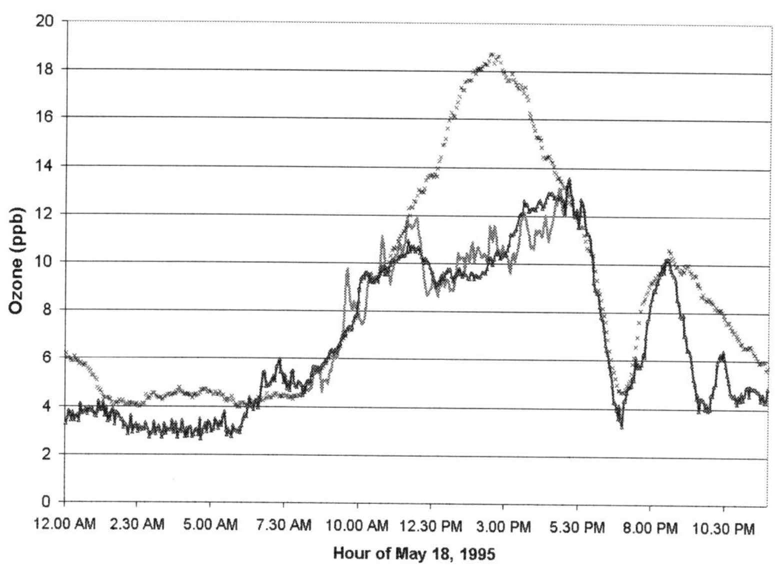

Figure 17: Reconstructed ozone concentrations at Benegas during May 18, 1995. O_3 Measurements (continuous dark line), expected ozone values if the H_2S pulse of Fig. 16, were not present (dotted line with X), simulated ozone concentrations (gray line).

4.6 Conclusions of the case study

Using dispersion models and solving the temporal variation of ozone measured at the Institute, we calculated approximately the presence of SO_2 and a H_2S pulse of at least 100 ppb during 4 hours. Moreover, the epidemiological studies of the affected population also concluded that the presence of SO_2 and H_2S, in the same order of magnitude as it was calculated, could produce the detected respiratory effects.

Several years later, during the trial process, personnel of the refinery declared that due to a wrong maneuver, sour gases were vented through a secondary torch tower and emitted to the atmosphere without burning it. The lapse time of the event and the amount of gases emitted were approximately the same as calculated.

As a consequence of this event, the refinery incorporated a Claus plant that treats more than 95% of the produced sour gases and only occasionally are they vented directly through the torch to the atmosphere.

5 Conclusions

The city of Mendoza has implemented a vast environmental program. This program includes various technical and legal aspects such as: a) environmental information; b) legal, and economic instruments for environmental protection; c) development of public environmental awareness; d) natural flora and fauna preservation; e) environmental control and sanitation; f) urban regulation and development; etc.

An important task of an air quality management system is to establish suitable air quality targets; and to determine how effective these targets are temporally and geographically complied. Another important aspect is to forecast the possibility to reduce pollution emissions. Such an air quality system has three important components: monitoring of the main pollutants; realisation of a comprehensive model, and development of satisfactory legal strategies. This articles emphasises the modelling efforts.

The implemented model is based on a Geographical Information System, which offers various advantages to handle spatial and temporal data. GIS combines data with algorithms, procedures and models that represent general scientific understanding.

In this article we showed the actual air quality situation in the city and we presented also possible future scenarios. Finally, we have presented a high-pollution episode occurred in the city, where many persons suffered respiratory effects. The modelling of this event was very useful to determine the legal responsibilities and to enforce the proper abatement strategies.

Acknowledgements

The authors would like to thank the authorities of the University of Mendoza for the support to the research activities. Also acknowledged are the engineers and technicians of our Institute IEMA for the permanent work in taking the daily measurements and calibrating the equipments. We thank as well the Ministry of Environment and Public Works of the Province of Mendoza, for permitting the use of the data from the monitoring stations. Our gratitude to the German colleagues from the Max Planck Institut für Aeronomie Lindau, in the person of Prof. Dr. Gerd Hartmann and the Centre for Environmental Research in Leipzig, represented by Prof. Dr. Olf Herbarth and Dr. Ludwig Weißflog for their friendly and fruitful cooperation. This study was partially funded by Argentine-German International Scientific Cooperation and the Argentine Agency for Science and Technology (FONCYT) under grant PICT97 N° 13-00000-01579. We are also grateful to WIT Press for the kind invitation to present this article.

References

[1] Schlink, U., Puliafito, E., Herbarth, O., Puliafito, J.L., Richter, M., Puliafito, C., Quero, J., Endlicher, W. & Zahnen, B., Ozone Air Pollution in Mendoza. *Air Pollution VI, Serie Advances in Air Pollution*, eds. C. Brebbia, C. Ratto & H. Power, WIT Press: Southampton, pp. 435–444, 1998.

[2] Schlink, U, Herbarth, O., Richter, M., Rhewagen, M., Puliafito, J.L., Puliafito, E., Puliafito, C., Guerreiro, P., Quero, J.L. & Behler, J.C., Ozone-Monitoring in Mendoza, Argentina: Initial results. *J. Air and Waste Management Assoc.*, **49**, pp. 82–87, 1999.

[3] Puliafito, E., Puliafito, J.L., Behler, J.C. & Alonso, P., La calidad del aire en Mendoza: Contaminación y efectos sobre la salud. *Mendoza Ambiental*, eds. E. Martinez & A. Dalmasso, pp. 207–242, Mendoza, Argentina. ISBN 950-43-7961-6, 1995.

[4] Weißflog, L., Paladini, E., Gantuz, M., Puliafito, J., Puliafito, S., Wenzel, K. & Schüürmann, G., Immission patterns of airborne pollutants in Argentina and Germany. I. First results of a heavy metal biomonitoring. *Fresenius Environm. Bullet.*, **3**, pp. 728-733.

[5] Wenzel, K., Weißflog, L., Paladini, E., Gantuz, M., Guerreiro, P., Puliafito, C. & Schüürmann, G., Immission patterns of airborne pollutants in Argentina and Germany. II. Biomonitoring of organo chlorine compounds and polyciclic aromatics. *Chemosphere*, **34**, pp. 2505–2518, 1997.

[6] Puliafito, J.L., A South American Country Example: Environmental legislation Enforcement in Mendoza; Experience and challenges. *Proc. of the Third International Conference on Environmental Enforcement*, **2**, pp. 51–62, Oxaca, México, April 25–28, J. Gerardu & C. Wesserman, USEPA and Ministry of Housing and Spatial Planning and Environment (VROM) the Nederlands; VROM8300/133, 1994.

[7] Herbarth, O., Fritz, G., Behler, J.C., Rehwagen, M., Puliafito, J.L., Richter, M, Schlink, U., Sernaglia, J., Puliafito, E., Puliafito, C., Schilde, M. & Wildführ, W., Epidemiologic Risk analysis of environmentally attributed exposure on airway diseases and allergies in children. *Centr. Eur. J. Publ. Health*, **7(2)**, pp. 72–76, 1999.

[8] Herbarth, O., Fritz, G., Wildfür, W. & Behler, J.C., An epidemiological study to the effects of traffic and domestic heating attributed emissions on respiratory illness and allergies among children, *Epidemiology*, **7(4)**, pp. 86, 1996.

[9] Zanetti, P., *Air pollution modelling*. Computational Mechanics Publications: Southampton, Boston, Van Nostrand Reinhold: New York, 1990.

[10] Seinfeld, J., *Air pollution, physical and chemical fundamentals*. McGraw-Hill: New York, 1975.

[11] CONCAWE Reports, Motor vehicle emission regulations fuel specifications, Part 1, summary and annual 1996 update. Reports, No. 5/97, 1997.

[12] Andrias, A., Samaras, Z., Zafiris, D. & Ziercok, K., CORNAIR Working Group on emission factors for calculating 1990 emission form road traffic. Final report, *Document of the European Commission ISBN 92-826-5572-X*, 1993.

[13] Hoydish, G. & Dabbert, F., Kinematics and dispersion characteristics of flows in asymmetric street canyons. *Atmospheric Environment*, **22**, pp. 2677–2689, 1988.

[14] Folinsbee, L., Human Health Effects of Air Pollution, *Env. Health Perspectives*, **100**, pp. 45–56, 1993.

[15] Pönkä, A. & Virtaten, A., Chronic Bronchitis, Emphysema, and Low-Level Air Pollution in Helsinki, 1987–1989. *Environmental Research*, **65**, pp. 207–217, 1994.

[16] Mustafa, M. & Tierney, D., Biochemical and Metabolic Changes in the Lung with Oxygen, Ozone, and Nitrogen Dioxide Toxicity. *American Review of Respiratory Disease*, **118**, 1978.

CHAPTER 5

Management of motor vehicle emissions in the United States

R. Baldauf[1], J. McDonald[1], R. Heck[2], R. Cook[1], J. Rege[1], L. Audette[1] & J. Armstrong[1]
[1] *National Vehicle and Fuel Emissions Laboratory, U.S. Environmental Protection Agency, United States of America*
[2] *Engelhard Corporation, United States of America*

Abstract

The management of motor vehicle emissions in the United States occurs on multiple levels. Federal statutes established in the United States' Clean Air Act and Transportation Equity Act of the 21st Century form the foundation of environmental regulations pertaining to motor vehicles. These regulations provide authority to federal, state, and local agencies in managing emissions. This chapter provides a background on motor vehicle emissions through the combustion process, the impact of these emissions on human health and the environment, and the role of government agencies and the public in managing these impacts. The chapter also provides a summary of engineering control methods used to reduce air emissions from current technology gasoline- and diesel-powered vehicles.

1 Introduction

Emissions from motor vehicles are a concern in almost all nations of the world. Since motor vehicles are present almost everywhere, these emissions can lead to adverse impacts at local, neighborhood, regional and global scales. Therefore, the control of vehicle emissions is important for the protection of public health and environmental resources. This chapter provides background on motor vehicle emissions and impacts, and provides legislative and engineering methods used in the United States to minimise these impacts.

2 Air pollution emissions from motor vehicles

In the United States, gasoline-powered vehicles predominate the passenger car segment of the national fleet, while diesel vehicles make up the majority of heavy-duty trucks and buses. Both gasoline and diesel combustion involve a complicated series of events, physical and chemical, which can be altered through engine design and operation. The sequence and rate at which these events occur are strongly related to the kinds and quantities of pollutants formed during combustion. Since the introduction and widespread use of electronic engine management systems, switching oxygen sensors, and three-way catalyst systems during the mid-1980s, control of emissions from spark-ignition, gasoline-powered engines has focused primarily on creating the conditions of temperature and chemistry in the exhaust that are most conducive to the catalytic reactivity of the exhaust emission control system. Conversely, diesel-emission control has focused almost entirely on control of combustion and other engine design parameters. These two differing approaches to emission control are beginning to converge as post-combustion catalytic exhaust control systems are developed for diesel-emission control in order to meet U.S. regulations. The fundamentals of gasoline- and diesel-powered combustion are provided in the following sections.

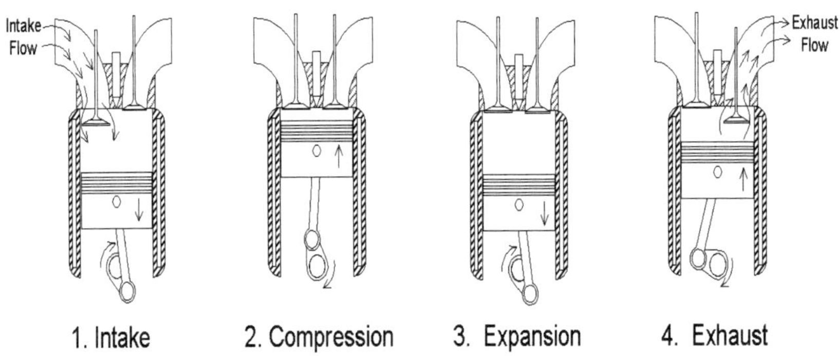

Figure 1: Four-stroke internal combustion engine operating cycle.

2.1 Gasoline combustion

The development of the spark-ignited combustion engine in the latter half of the 19th century permitted the controlled combustion of gasoline, which for the past century has provided the power plant to operate the automobile. These engines typically operate on the four-stroke principle of intake, compression, expansion or combustion, and exhaust (shown in Figure 1). On the first (downward) stroke of the piston, fuel and air are drawn into the cylinder. The following (upward) stroke compresses the fuel-air mixture, which is then ignited, expanding exhaust gases that force the piston downward for the third stroke, and the fourth and final (upward) stroke evacuates the spent exhaust gasses from the cylinder. The gasoline, which contains a mixture of paraffins and aromatic hydrocarbons, is combusted with controlled amounts of air producing complete combustion products of CO_2 and H_2O,

$$\text{Hydrocarbons in gasoline} + O_2 \text{ ------> } CO_2 + H_2O + \text{Heat} \quad (1)$$

Incomplete combustion products of CO and unburned hydrocarbons (UHC) are also formed in the combustion process. During combustion, very high temperatures are reached due to diffusion burning of the gasoline droplets, resulting in thermal fixation of the nitrogen in the air to form nitrogen oxides (NO_x) via the classic Zeldovich mechanism (Zeldovich 1946). The exhaust also contains H_2 from the combustion which usually occurs in a molar ratio of H_2 to CO of 0.3. The quantity of pollutants is influenced predominantly by the air-to-fuel (A/F) ratio in the combustion cylinder. Figure 2 shows the engine emissions from a spark-ignited gasoline engine as a function of the A/F ratio (Kummer 1980).

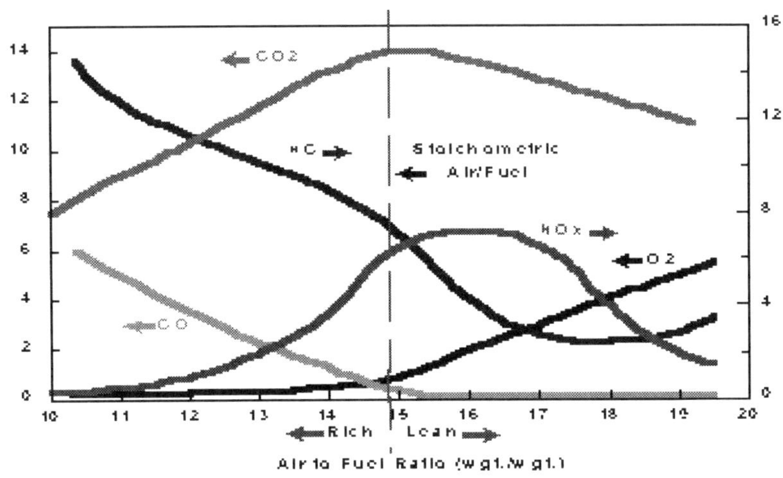

Figure 2: Spark-ignited gasoline engine emissions as a function of air-to-fuel ratio (Kummer 1980).

When the engine is operated rich of stoichiometric, the CO and HC emissions are highest while the NO_x emissions are depressed. This is due to incomplete burning of the gasoline caused by a deficiency in O_2. The level of NO_x is reduced because the adiabatic flame temperature is reduced. On the lean side of stoichiometric, the CO and HC are reduced since there is sufficient O_2 and nearly complete combustion dominates. Again, the NO_x is reduced since the operating temperature is decreased. Just lean of stoichiometric operation, the NO_x is a maximum, since the adiabatic flame temperature is the highest. At stoichiometric, the adiabatic flame temperature is lowered because of the heat of vaporisation of the liquid fuel gasoline. The actual operating region of combustion for the spark-ignited engine is defined by the lean and rich flame stability.

2.2 Diesel combustion

The typical diesel engine also operates on the four-stroke cycle. Combustion starts near the end of the compression stroke and continues through the first half of the expansion stroke. Near the end of the piston compression stroke, fuel is injected into the cylinder at high pressure, and mixes with the contents of the cylinder (air plus residual combustion gases). This period of premixing is referred to as ignition delay. Ignition delay ends when the premixed cylinder contents self-ignite due to the high temperature and pressure produced by the compression stroke in a relatively short, homogenous, premixed combustion event. Immediately following premixed combustion, diesel combustion becomes primarily non-homogeneous and diffusion-controlled (the rate of combustion is limited by the rate of fuel and oxygen mixing). During this phase of combustion, fuel injection continues creating a region that consists of fuel only. The fuel diffuses out of this region and air is entrained into this region, creating an area where the fuel to air ratio is balanced (i.e., near stoichiometric conditions) to support combustion. The fuel burns primarily in this region. The pollutant rate of formation in a diesel engine is largely defined by these combustion regions and how they evolve during the combustion process (Flynn, *et al.*, 2000).

NO_x is formed in diesel engines from molecular nitrogen (N_2) and oxygen (O_2) in the stoichiometric combustion regions of the diffusion-controlled and premixed diesel combustion phases described in the previous section. During the premixed phase of combustion, ignition and flame propagation occurs at high temperatures and at near stochiometric mixtures of fuel and air (Flynn, *et al.*, 2000). During diffusion-controlled combustion, the reaction zone is also near stoichiometric conditions. At the high temperatures present during premixed combustion or in the diffusion-controlled combustion reaction zone, a fraction of the nitrogen and oxygen can dissociate, forming radicals that can then combine through a series of reactions to form nitric oxide (NO), the primary NO_x constituent. Nitrogen dioxide (NO_2), the other NO_x constituent, is formed from NO in the flame region. NO_2 formed during combustion rapidly decomposes to NO and molecular oxygen unless the reaction is quenched by mixing with cooler cylinder contents. Engine-out emissions of NO are typically 80 percent or more of total NO_x from direct-injection diesel engines. The NO_x formation rate has a strong exponential relationship to temperature. Therefore, high temperatures result in high NO_x formation rates (Flynn, *et al.*, 2000). Any changes to engine design that can lower the peak temperature realised during combustion, the partial pressures of dissociated nitrogen and oxygen, or the duration of time at these peak temperatures can lower NO_x emissions. Reducing diesel NO_x emissions is most often accomplished by modifying combustion controls (e.g., injection timing) to reduce peak combustion temperatures.

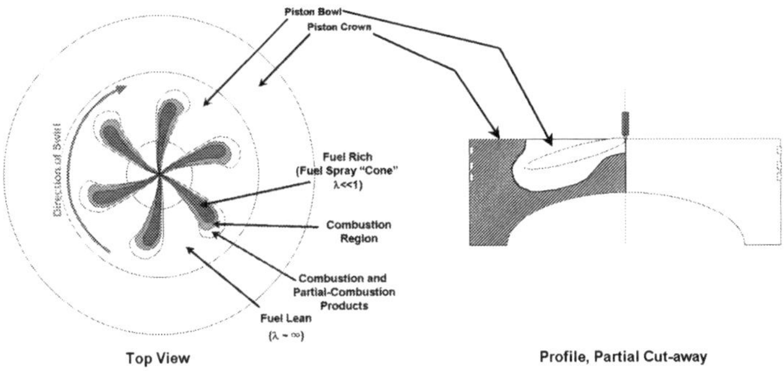

Figure 3: A top and cut-away view of direct diesel fuel injection into a common piston-bowl design showing fuel mixing and entrainment with air during the diffusion combustion phase. A value of "λ=1" corresponds to a stoichiometric air-fuel ratio.

PM emitted from diesel engines is a multi-component mixture composed mainly of elemental carbon (or soot), semi-volatile organic compounds, sulfate compounds (primarily sulfuric acid) with associated water, and trace quantities of metallic ash.

During diffusion-controlled combustion, fuel diffuses into a reaction zone and burns. Products of combustion and partial products of combustion diffuse away from the reaction zone where combustion occurs. At temperatures above 1,300 K, fuel compounds on the fuel-rich side of the reaction zone can be pyrolised to form elemental carbon particles (Flynn, *et al.*, 2000). Most of the elemental carbon formed (80 to 98 percent) is oxidised during later stages of combustion, most likely by hydroxyl radicals formed during combustion (Kittelson *et al.*, 1986; Foster and Tree, 1994).' The remaining elemental carbon agglomerates into complex chain-aggregate soot particles and exits the engine as a component of PM emissions.

From this description, the formation of elemental carbon particles during combustion and emission as PM following the combustion event can be summarised as being dependent upon three primary factors: 1) temperature, 2) residence time, and 3) availability of oxidants. Thus, in-cylinder control of elemental carbon PM is accomplished by varying engine parameters that affect these variables while balancing the resultant effects on NO_x emissions and fuel consumption.

The combination of organic compounds (volatile and semi-volatile) that contribute to PM are often referred to as the volatile organic fraction (VOF), or the soluble organic fraction (SOF), depending upon the analytical procedure used to measure the compounds. The test procedures used to demonstrate compliance with the PM emission standard in the U.S. do not differentiate between the various components of PM since emissions standards are based solely on total PM mass under specified test conditions (U.S. Code of Federal Regulations). Both VOF and sulfate PM are formed after cooling and air-dilution of the exhaust. Formation of sulfate PM is a function of fuel sulfur content. In the absence of post-combustion treatment of the exhaust (i.e., aftertreatment), approximately 1 to 3 percent of fuel sulfur is converted to sulfate, while the remainder is emitted as gaseous sulfur dioxide (SO_2). Post-combustion treatment of the exhaust using platinum catalysts can oxidise the VOF, thereby lowering PM emissions. VOF emissions are also reduced by the same in-cylinder emission control strategies used to reduce hydrocarbon (HC) emissions. Platinum catalysts can also oxidise up to 50 percent or more of the SO_2 to sulfate PM, depending on the exhaust temperature and the platinum content of the catalyst.

Hydrocarbon (HC) emissions from diesel engines primarily occur due to fuel and lubricant trapped in crevices (e.g., at the top ring land and the injector sac) that prevents sufficient mixing with air for complete combustion. Fuel-related HC can also be emitted due to "overmixing" during ignition delay, a condition where fuel in the induced swirl-flow has mixed beyond the lean flammability limit (Greeves *et al*., 1977). Higher molecular weight HC compounds (primarily lubricant-related compounds) adsorb to soot particles or nucleate and thus contribute to the VOF. Lower molecular weight HC compounds (mainly fuel-related compounds) are primarily emitted in the gas phase. During engine start-up under cold ambient conditions, fuel-related HC can be emitted as a concentrated, condensed aerosol (often seen as white smoke). Hydrocarbons can be controlled in-cylinder by reducing the size and number of crevices, and by reducing ignition delay. Post-combustion treatment of hydrocarbons can be accomplished via oxidation over precious-metal and base-metal catalysts.

Carbon monoxide emissions (CO) from diesel engines are generally insignificant compared to other mobile sources due to engine operation that, on a bulk-cylinder-content basis, is well fuel-lean of the stoichiometric air-to-fuel ratio.

3 Impacts from motor vehicle emissions

Exhaust emissions from motor vehicles result in impact to human health and the environment. The following sections describe the impacts of motor vehicle exhaust emissions on human health and the environment.

3.1 Human health effects

Over 700 compounds have been identified in exhaust and evaporative emissions from mobile sources (Sierra Research, 2001). Many of these compounds are associated with adverse health effects in humans and/or laboratory animals. The compounds associated with adverse health effects fall into two categories, based on how the emissions are controlled in the U.S.: criteria pollutants and air toxics.

The U.S. has National Ambient Air Quality Standards (NAAQS) to protect against adverse health effects from criteria pollutants. Air toxics are pollutants known to cause cancer or other serious human health effects or ecosystem damage, but the U.S. Environmental Protection Agency (EPA) has not developed a NAAQS. Examples of air toxic pollutants include gaseous hydrocarbons like benzene, formaldehyde, and 1,3-butadiene, metals such as chromium and nickel, and chemical mixtures from individual sources, such as diesel combustion exhaust. The following sections describe some of the health effects that have been associated with compounds found in motor vehicle exhaust.

3.1.1 Criteria pollutants

The EPA has NAAQS for six pollutants: carbon monoxide (CO), nitrogen dioxide (NO_2), ozone (O_3), particulate matter (PM), lead (Pb), and sulfur dioxide (SO_2). All of these pollutants are emitted from mobile sources.

3.1.1.1 Carbon monoxide CO is absorbed into the bloodstream through the lungs and reduces the amount of oxygen delivered to the body's organs and tissues. At lower levels of exposure, CO can increase the severity and duration of angina for individuals with cardiovascular disease. At higher levels of exposure, CO can cause fatigue, nausea, and impaired vigilance.

3.1.1.2 Nitrogen dioxide NO_2 is a reddish brown, highly reactive gas that is formed in the ambient air through the oxidation of nitric oxide (NO). Short-term exposures to low levels can result in decreased lung function in individuals with respiratory illnesses and in children. Long-term exposure can cause permanent lung impairment.

3.1.1.3 Ozone Ground level ozone is formed by a reaction of HCs and NO_x in the presence of heat and sunlight. The interaction between levels of HCs, NO_x, and weather is complex, as is the process of transport and accumulation. If NO_x levels are relatively high and HC levels are relatively low, NO_x contribute very little to ozone formation. Conversely, if HC levels are relatively high, small amounts of NO_x can lead to rapid formation of ozone. Thus, in determining which pollutants should be the focus of control efforts, it is critical to know the relative levels of the pollutants.

Increased hospital admissions and emergency room visits have been associated with ambient ozone exposures (EPA, 1996). Short-term and long-term exposures result in lung function impairment and exacerbation of asthma. Furthermore, long-term exposure can cause irreversible changes in lung structure, leading to chronic respiratory illness, such as emphysema and chronic bronchitis. Children and outdoor workers are most at risk from ozone exposure because they are typically active outside during summer when ozone levels are highest. Furthermore, children are more at risk than adults because their respiratory systems are still developing.

3.1.1.4 Particulate matter PM is emitted directly or formed secondarily in the atmosphere. Secondary particles are formed in the atmosphere from primary gaseous emissions. Examples of secondary particles include sulfate, formed from SO_2 emissions; and nitrates, formed from NO_x emissions. Particles are defined for regulatory purposes in the U.S. as either fine, with an aerodynamic diameter less than 2.5 microns ($PM_{2.5}$), or coarse, with an aerodynamic diameter in the range of 2.5 to 10 microns ($PM_{10-2.5}$). The chemical composition of particles depends on location, time of year, and weather.

Adverse health effects from exposures to PM result from penetration and deposition of particles in the various regions of the respiratory tract, and the biological response to these particles. Adverse health effects include decreased lung function, exacerbation of asthma symptoms, and changes to lung tissues and structure (EPA, 2001c). Furthermore, in recent years, results from numerous studies have provided evidence of an association between premature mortality and morbidity, and exposure to particulate matter (Dockery *et al.*, 1993; Pope *et al.*, 1995; Krewski *et al.*, 2000). This association appears to be attributable to fine rather than coarse PM (Schwartz *et al.*, 1996). In addition, source specific factor analysis shows that an increase in mobile source fine PM is associated with increased mortality (Laden *et al.*, 2000). The data also indicate an association between long-term exposure to fine PM, cardiopulmonary mortality, and lung cancer (Pope *et al.*, 2002).

3.1.1.5 Lead Exposure to lead occurs both through inhalation and ingestion. Excessive exposure results in neurological impairment, including mental retardation, attention deficit, anemia, and increased blood pressure in adults. Fortunately, lead emissions in the U.S. have decreased dramatically as a result of the phase-out of leaded gasoline.

3.1.1.6 Sulfur dioxide Short-term exposure to high concentrations of SO_2 can result in respiratory impairment to asthmatic children and adults who are active outdoors. Longer term exposure can cause lung impairment and aggravation of cardiovascular disease.

3.1.2 Air toxics

In a recent rule, EPA identified twenty-one pollutants as mobile source air toxics (shown in Table 1) (EPA, 2000). The discussion in this chapter focuses on mobile source air toxics that the U.S. has identified as having the greatest potential threat to human health, based on toxicity and the contribution from mobile sources (EPA, 2002a). These pollutants are diesel particulate matter and diesel exhaust organic gases, benzene, 1,3-butadiene, formaldehyde, acetaldehyde, and acrolein.

3.1.2.1 Diesel particulate matter and diesel exhaust organic gases The EPA has concluded that long term exposure to diesel exhaust likely poses a lung cancer hazard to humans (EPA, 2002b). Numerous other organisations have also classified diesel exhaust or diesel PM as a potential or probable carcinogen including the World Health Organisation and the International Agency for Research on Cancer. A large number of human epidemiology studies show increased lung cancer associated with diesel exhaust.

Table 1. List of Mobile Source Air Toxics (MSATs)

Acetaldehyde	Ethylbenzene	Naphthalene
Acrolein	Formaldehyde	Nickel
Arsenic	n-Hexane	Polycyclic Organic Matter[b]
Benzene	Lead	Styrene
1,3-Butadiene	Manganese	Toluene
Chromium	Mercury	Xylenes
Dioxin/Furans[a]	MTBE	Diesel PM + Exhaust Organic Gases

[a] This entry refers to two large groups of chlorinated compounds. In assessing their cancer risks, their quantitative potencies are usually derived from that of the most toxic, 2,3,7,8-tetrachlorodibenzodioxin.

[b] Polycyclic Organic Matter includes organic compounds with more than two benzene rings, and which have a boiling point greater than or equal to 100 degrees centigrade.

Furthermore, exposures in these epidemiology studies are in the same range as ambient exposures throughout the U.S. Although currently available science suggests an important role for the particulate phase component of diesel exhaust, the cancer and non-cancer health effects have been associated with diesel PM and the organic gas-phase components.

The EPA has not developed a cancer unit risk estimate, or risk value per unit of lifetime exposure, because of uncertainties related to discerning an exposure–response relationship for the observed lung cancer in epidemiological studies. Since the general population is exposed to levels close to or overlapping with apparent levels that result in increased cancer risk in epidemiologic studies, the inference of a public health concern is quite reasonable. The EPA has illustrated that ambient exposure to diesel exhaust could pose risk levels that are greater than one in a million and, therefore, are of interest to EPA's regulatory objectives to protect human health. The EPA showed that the "possible" risk to the general public could be greater than one in a hundred thousand (10^{-5}) to as high as one in a thousand (10^{-3}) over a lifetime of exposure. EPA was also clear to say that lower risks were possible and a zero risk could not be ruled out.

3.1.2.2 Benzene Benzene is an aromatic hydrocarbon present as a gas in both exhaust and evaporative emissions from motor vehicles. Benzene comprises 3 to 5 percent of mobile source exhaust total organic gases (TOG), which varies depending on control technology (e.g., type of catalyst) and the levels of benzene and aromatics in the fuel. The benzene fraction of evaporative TOG emissions is generally about one percent, but depends on control technology and fuel composition (e.g., benzene level and Reid Vapor Pressure) (EPA, 1993a).

The EPA has recently reconfirmed that benzene is a known human carcinogen by all routes of exposure, and a full cancer risk characterisation has been performed (EPA, 1998a). Benzene has been linked to acute (rapid and fatal) nonlymphocytic leukemia, chronic (lingering, lasting) lymphocytic leukemia and possibly multiple myeloma (primary malignant tumors in the bone marrow). Benzene exposures cannot be ruled out as a potential contributor to leukemia development in children (EPA, 1998a). Data from animal studies have shown benzene exposures result in damage to the hematopoietic system during development, and key changes related to the development of childhood leukemia occur in the developing fetus. An association between distance-weighted traffic density in proximity to homes and occurrence of childhood leukemia has also been identified (Pearson, 2000; Savitz *et al.*, 1989; Feychting *et al.*, 1998; Harrison *et al.*, 1999). Chronic exposure is also associated with a number of other blood disorders.

3.1.2.3 1,3-Butadiene 1,3-Butadiene is formed in vehicle exhaust by the incomplete combustion of gasoline and diesel fuel. 1,3-Butadiene accounts for 0.4 to 1.0 percent of total exhaust TOG, depending on control technology and fuel composition (EPA, 1993a).

The EPA recently designated 1,3-butadiene as a known human carcinogen (EPA, 2002c). Epidemiologic studies indicate that occupational exposure is associated with increased incidence of cancers of the lymphatic system and systems in the body that produce blood cells. In addition, 1,3-butadiene also causes a variety of reproductive and developmental effects in mice and rats exposed to long-term, low doses (EPA, 1998b).

3.1.2.4 Formaldehyde Formaldehyde is the most prevalent aldehyde in vehicle exhaust. It is formed from incomplete combustion of both gasoline and diesel fuel and accounts for 1 to 4 percent of total exhaust TOG emissions, depending on control technology and fuel composition (EPA, 1993a). Formaldehyde exhibits extremely complex atmospheric behavior. It is directly emitted and is formed by the atmospheric oxidation of virtually all organic species, including biogenic HCs.

EPA has classified formaldehyde as a probable human carcinogen, based on limited evidence for carcinogenicity in humans and sufficient evidence of carcinogenicity in animal studies (EPA, 1987). Epidemiological studies in occupationally exposed workers suggest that long-term inhalation of formaldehyde may be associated with tumors of the nasopharyngeal, nasal cavity, and sinus. Formaldehyde exposure can cause a range of noncancer health effects as well, including nasal, respiratory and eye irritation.

3.1.2.5 Acetaldehyde Acetaldehyde is a saturated aldehyde that is found in vehicle exhaust and is formed as a result of incomplete combustion of both gasoline and diesel fuel. It is not a component of evaporative emissions. Acetaldehyde comprises 0.4 to 1.0 percent of exhaust TOG, depending on control technology and fuel composition (EPA, 1993a). The atmospheric chemistry of acetaldehyde is similar in many respects to that of formaldehyde. Like formaldehyde, it can be both produced and destroyed by atmospheric chemical transformation.

Although human data on the carcinogenic potential of acetaldehyde are extremely limited, acetaldehyde is classified as a probable human carcinogen (EPA, 1999). This classification is based on a formal risk characterisation using a sufficient database of animal carcinogenicity studies. Specifically, increased incidences of nasal tumors in male and female rats and laryngeal tumors in male and female hamsters have been documented after inhalation exposure. The primary acute effects associated with exposure to acetaldehyde include irritation of the eyes, skin, and respiratory tract. Effects on the respiratory system have been reported from studies of animals exposed to long-term lower concentrations as well. The data from these studies was found to be sufficient for the EPA to develop an inhalation reference concentration (RfC) for acetaldehyde exposure. Although no information is available on the reproductive or developmental effects of acetaldehyde in humans, data from animal studies suggest that acetaldehyde may be a potential developmental toxicant.

3.1.2.6 Acrolein Acrolein is an exhaust emission formed as a result of incomplete combustion of both gasoline and diesel fuel. Acrolein comprises 0.05 to 0.4 percent of exhaust TOG, depending on control technology and fuel composition (EPA, 1993a). The atmospheric chemistry of acrolein is expected to be similar in many respects to that of formaldehyde and acetaldehyde.

The respiratory system is the primary target in humans and animals for acrolein toxicity resulting from inhalation exposure. Acute exposure results in upper respiratory tract irritation and congestion; whereas chronic exposures in animals indicated an increase in cell proliferation and in the numbers of white blood cells in the tissues lining the nasal passages. The data from these studies was found to be sufficient for the EPA to develop an RfC for acrolein exposure (EPA, 2002d). No information is available on the reproductive or developmental effects of acrolein in humans.

3.2 Ecological effects

The ubiquitous nature of mobile source emissions has resulted in a number of ecological effects that have local and global ramifications. This section focuses on two primary areas of ecological effects: global warming and atmospheric deposition. Other adverse environmental effects include impacts to visibility, increased corrosion and soiling of landmarks, and noise pollution.

3.2.1 Global warming

Human activities, such as fossil fuel burning, have altered the chemical composition of the atmosphere through the buildup of greenhouse gases (primarily carbon dioxide, methane, and nitrous oxide) and atmospheric PM. Mobile sources significantly contribute to the elevated concentrations of some of these gases and particles. Although uncertainty exists about how the earth's climate responds to increased concentrations of these constituents, the present scientific consensus maintains that global air temperatures and subsurface ocean temperatures are rising.

In the U.S., transportation sources account for approximately 27 percent of all US greenhouse gas emissions, which includes 31 percent of all CO_2 emissions, 1 percent of all CH_4 emissions, and 13 percent of all N_2O emissions (EPA, 2002d). In addition to the greenhouse gases, motor vehicles emit a significant amount of PM, which has also been attributed to global temperature changes.

3.2.2 Atmospheric deposition

Atmospheric deposition is the removal of contaminants from the air to soil, water, and other surfaces. Atmospheric deposition occurs through three major processes: wet deposition, dry deposition and air-surface exchange.

Atmospheric deposition of mobile-source pollutants can lead to a number of adverse health and environmental effects. Pollutants deposited into waterbodies may result in the contamination of the water and contamination of marine life residing in the affected area. Pollutant deposition onto vegetation may directly impact foliar surfaces or change the soil chemistry in the region.

A number of contaminants have been implicated in water and soil degradation. These contaminants may be present in the gas phase, particle phase, or a combination of both. Although other pollutants have been shown to adversely affect water and soil quality (for example, ozone and sulfur dioxide), the pollutants listed in the following sections have the potential for local and long-range transport.

3.2.2.1 Nitrogen Nitrogen is a beneficial nutrient to plant and animal life. The presence of nitrogen determines the nature, diversity, and productivity of all terrestrial and aquatic ecosystems. Diatomic nitrogen gas (N_2), an inert compound, comprises seventy-eight percent of the earth's atmosphere. However, elevated levels of nitrogen can lead to burdens that result in adverse effects to natural ecosystems and, ultimately, human health.

NO_x emissions occur mainly from combustion sources including motor vehicles. National emissions of NO_x have increased by four percent in the last twenty years, mainly due to emission increases in diesel vehicles. The conversion of NO_x to form nitrates is dependent on atmospheric conditions.

Therefore, NO_x emissions can contribute to both local impacts and long-range transport.

Nitrogen may also be supplied to water and soil surfaces by atmospheric deposition of ammonia. Since ammonia is very soluble, wet deposition will likely result in local effects. However, ammonia can also transform into the particle phase to ammonia-sulfate or ammonia-nitrate (Seinfeld and Pandis, 1998). Particle phase ammonia typically results in long-range transport. In urban areas, motor vehicles can be significant contributors to ammonia emissions, accounting for twenty to fifty percent of total ammonia emissions.

3.2.2.2 Heavy metals Heavy metals, including lead, cadmium, mercury, zinc, nickel, copper, and chromium have been implicated in soil and water quality degradation. Investigations of trace metals near roadways and industrial facilities indicate that a substantial burden of heavy metals can accumulate on vegetative surfaces. Elevated levels of zinc and lead in streambed sediments have been correlated with population density and motor vehicle use (Callender and Rice, 2000; Rice, 1999). In addition, platinum, palladium, and rhodium, metals found in the catalysts of modern motor vehicles, have been measured at elevated levels along roadsides (Ely *et al.*, 2001).

3.2.2.3 Polycyclic organic matter Polycyclic organic matter (POM) is a byproduct of incomplete combustion. Polycyclic aromatic hydrocarbons (PAHs) are a class of POM that are known or suspected carcinogens.

Since PAHs are insoluble, the compounds generally are particle reactive and accumulate in sediments. Atmospheric deposition of particles is believed to be the major source of PAHs to the sediments of Lake Michigan (Simcik *et al.*, 1996; Simcik *et al.*, 1999). Analyses of PAH deposition to Chesapeake and Galveston Bay indicate that dry deposition and gas exchange from the atmosphere to the surface water predominate (Arzayus *et al.*, 2001; Park *et al.*, 2001). PAHs that enter a waterbody through gas exchange likely partition into organic-rich particles and become biologically recycled, while dry deposition of aerosols containing PAHs tend to be more resistant to biological recycling (Arzayus *et al.*, 2001). Thus, dry deposition is likely the main pathway for PAH concentrations in sediments, while gas/water exchange at the surface may lead to PAH distribution into the food web, leading to increased health risk concerns.

4 Motor vehicle legislation

The following sections describe the role of the two primary legislative actions that regulate the management of motor vehicle emissions in the U.S.: the Clean Air Act (CAA) and the Transportation Equity Act for the 21st Century (TEA-21). This discussion focuses on the role of various agencies rather than specific legislative requirements due to the limited scope of this chapter.

4.1 The Clean Air Act

Over the years, the U.S. federal government's role in controlling air pollution from all sources, including motor vehicles, has become more defined and prominent through the progression of the Clean Air Act. The Air Pollution Control Act of 1955 was the nation's first federal air pollution legislation that identified air pollution as a national problem and gave Congress the right to control it. While an amendment to this Act requested the Surgeon General to research the health effects associated with motor vehicle exhaust, the Act did not include any regulations for vehicle emissions. Similarly, the Clean Air Act (CAA) of 1963 did not attempt to control mobile sources. An amendment to the CAA in 1965 authorised the Secretary of Health, Education, and Welfare to set vehicle emission standards.

The Clean Air Act of 1970 established new stringent standards for motor vehicle emissions, which were to be enforced by the federal and state governments. The EPA assumes a priority role for mobile sources in identifying and setting standards for pollutants, regulating motor vehicles, issuing waivers, collaborating with the states, enforcing standards, and providing information.

The timeline for meeting motor vehicle standards established in the CAA was extended in the amendments of 1977 in order to attain these standards more easily. The 1990 CAA takes a comprehensive approach to reducing pollution from motor vehicles, by cleaning up fuels, cars, trucks, buses and other motor vehicles. The Act includes auto inspection provisions to ensure cars are well maintained to help reduce air pollution (EPA, 1993b). In addition, the Act acknowledges the important link between transportation planning and air quality.

4.1.1 Role of the federal government

The EPA sets regulations applicable to the emission of any air pollutant from any class of new motor vehicle that may endanger public health or welfare. The agency is responsible for studying pollutants' health, welfare, and environmental effects in order to determine necessary levels of control for criteria pollutants. Furthermore, the EPA has authority to list and regulate air toxic pollutants. There are currently 188 air toxics listed, which were selected on the basis of potential health and environmental hazards.

In assessing new vehicle regulations, the EPA examines the availability of control technology and determines if a need exists for further emission reductions from motor vehicles. The need for air pollution control requires the consideration of the technology, lead time, safety, energy impacts, cost effectiveness of further reductions, alternative means of attaining or maintaining the NAAQS, and the feasibility and cost effectiveness of the alternative. Control techniques may include changes to vehicle design or fuel composition.

The EPA must also determine if motor vehicles conform to regulations by conducting tests, or requiring the manufacturer to conduct tests and provide information for any new motor vehicle, engine, or emission-control system. Tests are conducted under EPA prescribed procedures and conditions, and the results must be made public. EPA then issues either a certificate of conformity for vehicles or engines or a verification of emission-standard compliance for emission-control systems.

In order to carry out the CAA, the EPA plays the enforcement role and is responsible for overseeing and enforcing sources' compliance with air quality standards. Hence, designated EPA officers or employees are authorised to enter manufacturers' establishments in order to inspect or observe the manufacturers' records, files, papers, processes, controls, and facilities.

While the EPA has primary control over standards and regulations, the agency also works with the states in order to maintain those standards and assist the states by providing scientific research, expert studies, engineering designs and money to support clean-air programs. Moreover, the EPA is authorised to give grants to state agencies for the cost of developing and maintaining effective vehicle-emission devices, systems inspection, emission testing, and control programs or as reimbursement for state programs begun before the grant. Grants, however, are not given unless the program is certified as consistent with the Secretary of Transportation's highway safety program or unless the program includes provisions designed to insure that emission control devices and systems on vehicles in actual use have not been discontinued or rendered inoperative.

In general, the EPA's role is to protect public welfare, and in doing so, must keep the public informed about air quality issues. Thus, the EPA plays an important role in information distribution for the public and vehicle manufacturers. As part of this role, the 1990 CAA ordered the EPA to set up clearing houses to collect and make available all technical information, reports, and records (except those entitled to protection as trade secrets) to the public, states, and other air pollution control agencies.

Moreover, the EPA requires motor vehicle manufacturer's information to be promptly disclosed. For instance, the agency requires manufacturers to provide any needed information on emission control diagnostics systems and instructions for diagnosing or repairing related problems to motor vehicle repair people. The EPA must also disclose to the public in a non-technical manner manufacturer's test results for motor vehicles.

While states are responsible for attaining the national ambient air quality standards and must develop plans and adopt regulations to do so, EPA is responsible for approving those state implementation plans (SIPs) and imposing sanctions for failure to adopt or implement the SIPs. EPA therefore issues guidance to state and local agencies on preparing SIPs, including such motor-vehicle-related topics as inspection and maintenance programs, fuels programs, retrofits, transportation control measures, and other measures to reduce vehicle miles traveled.

4.1.2 Role of the state governments

The CAA allows states to take the lead in carrying out the provisions of the Act on the basis that pollution control problems often require special understanding of local industries, geography, housing patterns, and other local factors. Thus, state governments are responsible for establishing pollution mitigation plans in order to meet heath standards. While the CAA gives states more time to meet air quality standards, it also requires states to make constant formidable progress in reducing emissions. Therefore, the states' roles pertaining to motor vehicles include developing state implementation plans to attain air-quality standards, regulations, and conducting inspections.

4.1.2.1 State Implementation Plans State Implementation Plans (SIPs) describe how states plan to improve air quality in a given location. Often, SIPs include controls to reduce mobile source emissions. Motor vehicle SIP provisions must be developed in consultation with fleet operators, vehicle manufacturers, fuel producers or distributors, motor vehicle fuel suppliers, and other interested parties.

As part of motor vehicle emission control, states with areas not achieving the ozone or CO NAAQS must establish a clean-fuel vehicle fleets program in their SIPs. In addition, any state with a Serious-, Severe-, or Extreme-ozone nonattainment area may provide incentives in the SIP for the sale or use of clean-fuel vehicles and clean alternative fuels. A state may also prescribe and enforce the use of a fuel or fuel additive control or prohibition. In addition, states may adopt transportation planning measures for emissions reductions such as improved mass transit and high occupancy vehicle lanes.

4.1.2.2 Regulations For regulated vehicle or engine parts, states cannot adopt or attempt to enforce any standard or requirement of certification, inspection, or approval relating to motor-vehicle emissions. States are allowed to control, regulate, or restrict the use, operation, or movement of registered motor vehicles, but states cannot adopt or attempt to enforce any standard relating to the emission control from new motor vehicles or engines. The exception to this requirement is the state of California, which does have the authority to adopt standards. In addition, other states may adopt the standards set in California.

4.1.2.3 Inspection After the date of sale, states can implement or require motor vehicle testing, and thus motor vehicle or engine inspection is the states' responsibility. Additionally, SIPs containing motor vehicle inspection and maintenance programs must be amended to provide for inspection of new emission control regulations, such as onboard diagnostics systems and the maintenance of malfunctions or system deterioration identified by these diagnostics systems.

4.1.2.4 California Pilot Test Program Under the CAA, the state of California established a pilot program to demonstrate the effectiveness of clean-fuel, light-duty trucks and light-duty vehicles in controlling air pollution in ozone nonattainment areas, and revised its SIP to require the production, distribution, and availability of clean alternative fuels in California. California was also able to establish necessary specifications for these fuels so as to reduce any associated, unreasonable risk to public health, welfare, or safety and to ensure acceptable vehicle maintenance and performance characteristics. While California serves as the pilot test program, EPA also allows other states that submit SIP revisions to voluntarily opt-in to similar clean-fuel vehicles and clean alternative fuel programs.

4.1.3 Role of the cities

The CAA does not legally obligate cities to play a role in air quality management; however, many cities are involved in preparing their own SIPs. Some cities voluntarily participate in the SIP planning process while other cities are designated with the responsibility of preparing their SIPs by the state government. In addition, cities are actively involved in the TEA-21 regulations described below.

4.1.4 Role of the public

The public and the individual are allowed specific rights under the CAA. First, the public is allowed to attend hearings and provide comments on state and local cleanup plans. Secondly, the public is allowed to seek penalties against violators, including polluters or government agencies, that fail to carryout the CAA. Although the public can request the EPA or state to take action against violators, the government is not needed in order to create a suit. Thirdly, the public is entitled to reports required by the CAA, which contain information on amounts of released pollution. Lastly, individual motor vehicle owners are responsible, at their own expense, for properly maintaining their vehicle, and for replacing items such as spark plugs, points, condensers, and any other part, item, or device related to emissions that are not under manufacturer's warranty.

4.2 The Transportation Equity Act for the 21st Century

The Interstate Commerce Commission was originally responsible for administering highway projects, and the U.S. Congress funded highway construction projects; levied user-fee taxes on gasoline, tires, and truck manufacturers to raise these funds; and set up a permanent Highway Trust Fund to finance future projects. The Transportation Equity Act for the 21st Century (TEA-21) is the most recent motor vehicle legislation defining the government's role in highway planning. TEA-21 is an update of the Intermodal Surface Transportation Efficiency Act of 1991 (ISTEA), which establishes requirements for building and repairing highways and for encouraging states and localities to develop mass transportation systems and reduce urban highway congestion.

4.2.1 Role of the federal government

TEA-21 designates responsibilities to the U.S. Department of Transportation (DOT), the Bureau of Transportation Statistics, and the EPA. The DOT has the main responsibility for carrying out the tasks related to motor vehicle legislation in TEA-21 including a guaranteed level of Federal funds for surface transportation, improving safety, and advancing U.S. domestic and international economic growth and competitiveness.

A new transit program was established to assist in the early deployment of mass transportation services, management, operational practices, and applicable technology. Mass transit was identified as an important program to reduce the number of trips taken in personal passenger cars. Through TEA-21, DOT can form agreements with public or private research organisations, transit providers, and businesses for the study, design, and demonstration of mass transit options such as fixed guideway technology, bus technology, fuel-cell-powered transit buses, advanced propulsion control for rail transit, and low-speed magnetic levitation technology. TEA-21 also encourages mass transit and multi-modal transportation by offering transit and vanpool benefits to an employee in order to balance the differences between parking and transit/vanpool benefits.

4.2.2 Role of the state governments

TEA-21 streamlined many administration aspects of Federal highway programs, giving State transportation agencies additional authority. Under agreement with DOT, a State may assume DOT's responsibilities for contract awards, construction inspections, and approval of plans, specifications, and estimates. TEA-21 also streamlined the metropolitan and statewide transportation planning processes, thereby strengthening the local officials' role and increasing public involvement in the planning processes. This process gives states and localities greater flexibility in the use of federal funds and control of safety programs.

For each State, TEA-21 specifies a certain share of the aggregate funding for the following programs: Interstate Maintenance, National Highway System, Bridge, Congestion Mitigation and Air Quality (CMAQ) Improvement, Surface Transportation Program (STP), Metropolitan Planning, High Priority Projects, and Recreational Trails. The flexible funding provisions enable certain program funds to be used for either transit or highway projects.

The CMAQ program provides a flexible funding source to state and local governments to help meet the requirements of the CAA for transportation projects and programs, such as transit improvements, travel demand management strategies, traffic flow improvements, and public fleet conversions to cleaner fuels. Funding is available for areas that do not meet the NAAQS, as well as former nonattainment areas that are now in compliance.

The STP program provided flexible funding for States' and localities' use for any highway projects receiving federal funds. Ten percent of the funds are set aside for safety construction activities (i.e., hazard elimination, railway-highway crossing improvements, and bicycle safety), and ten percent are set aside for transportation enhancements. These enhancements encompass a broad range of surface transportation and environmental activities, (i.e. pedestrian and bicyclist safety education activities, transportation museums, and projects to reduce vehicle-caused wildlife mortality).

4.2.3 Role of the cities

Metropolitan planning organisations (MPOs) are responsible for helping city and state transportation needs and for adopting the metropolitan transportation plan. The core metropolitan and statewide transportation planning requirements emphasise the role of State and local officials, in cooperation with transit operators, to address provisions concerning fiscal constraint, planning timeframe, and public involvement. TEA-21 established seven broad 'areas' to be considered in the planning process, both at the metropolitan and statewide level. Transportation improvement plans, project or strategy plans, and certification actions are exempt from legal review for failure to consider any one of the 'areas'. The growing importance of operating and managing the transportation system is recognised as a focal point for transportation planning. Reflecting the importance of safety throughout all surface transportation programs, the safety and security of the transportation system for motorised and non-motorised users is one of the seven areas to be considered. States, local governments, and metropolitan planning organisations are eligible for discretionary grants to plan and implement strategies that improve the transportation system's efficiency; reduce environmental impacts; reduce costly future public infrastructure investments; ensure efficient access to jobs, services, and trade centres; and examine private sector development patterns and investments.

5 Controlling emissions from motor vehicles

Pollution-control techniques are required to minimise adverse impacts from motor vehicles and meet the regulatory requirements imposed by environmental legislation in the U.S. The following sections describe control techniques for gasoline- and diesel-powered vehicles, respectively.

5.1 Gasoline vehicles

The primary method of pollution control for gasoline vehicles is the use of catalytic converters. The automobile catalyst must perform the following reactions in the engine exhaust to oxidise CO and HC to CO_2 and H_2O,

$$C_yH_n + (1+n/4)O_2 \text{ ----> } yCO_2 + n/2H_2O \quad (2)$$

$$CO + 1/2O_2 \text{ ----> } CO_2 \quad (3)$$

$$CO + H_2O \text{ ----> } CO_2 + H_2 \quad (4)$$

and reduce NO/NO_2 to N_2

$$NO(\text{or } NO_2) + CO \text{ ---> } 1/2N_2 + CO_2 \quad (5)$$

$$NO(\text{or } NO_2) + H_2 \text{ ---> } 1/2N_2 + H_2O \quad (6)$$

$$(2+n/2)NO(\text{or } NO_2) + C_yH_n \text{ ---> } (1+n/4)N_2 + yCO_2 + n/2H_2O \quad (7)$$

When a driver first starts the automobile, both the engine and catalyst are cold. As the exhaust gradually warms, it reaches a temperature high enough to initiate the catalytic reactions. This is referred to as the light-off temperature. During this time period, the rate of reaction is kinetically controlled; that is, it depends on the chemistry of the catalyst since the transport reactions are fast. Typically, the CO and H_2 reactions begin first, followed by the HC and NO_x reaction. As the vehicle exhaust is hot, the chemical reaction rates become faster, and pore diffusion and/or bulk mass transfer begins to control the overall conversion of the exhaust pollutants.

Catalyst technology has progressed through a number of stages since the 1970 CAA, as depicted in Table 2. Also shown in this table are the enabling technologies that were developed concurrently to implement the next phase of catalyst technology.

Table 2. Technological progress in catalysts for auto emission control

Year	Technology	Enabling development
1976	HC and CO emissions, oxidation catalyst.	Monolith coating for forklift trucks, Pb removal from gasoline, alumina chemistry.
1979	HC, CO and NO_x emissions, three-way catalyst.	Oxygen sensor, Ce, stabilisation of alumina.
1986	High-temperature three-way catalyst.	Ce stabilisation, Precious metal segregation.
mid 1990's	All Pd three way catalyst.	Precious metal segregation, better engine controls, higher operating temperature.
1996	Low emission vehicles, high-temperature close-coupled catalyst.	Pd stabilisation.
2000+	Lean NO_x catalyst.	Lean-burn engines.

In order to accomplish the low emissions formulated through the California Pilot Program for Ultra-Low Emission Vehicles (ULEV) and Super Ultra-Low Emission Vehicles (SULEV), a number of approaches were investigated in the mid to late 1990s. The emphasis of the new regulations was the reduction of HCs in the exhaust. Analyses show that the majority of HC emissions (60–80 percent of the total emitted) are produced in the cold-start portion of the automobile, which includes the first two minutes of operation from the "key on" event.

During the 1990s, a number of technologies were investigated to resolve the cold start issues: 1) close-coupled catalyst, 2) electrically heated catalyzed metal monolith, 3) hydrocarbon trap, 4) chemically heated catalyst, 5) exhaust gas ignition, and 6) preheat burners

All of these approaches contain under-floor catalysts of various compositions. The leading technology for most low emission vehicle (LEV), ULEV, and SULEV applications has become the high-temperature, close-coupled catalyst (Heck and Farrauto 2002).

In combination with the close-coupled catalyst there have also been some unique system approaches such as: 1) cold-start spark retard or post-manifold combustion, 2) variable valve combustion chamber, and 3) double-walled exhaust pipe. A brief description of each of the various candidate technologies can be found in Heck (2002).

The modern three-way catalyst (TWC) uses a layered catalyst with wash-coat architectures to enhance the TWC reactions and the high-temperature durability of the catalyst. The major components in a modern TWC include: 1) an active component (precious metal), 2) oxygen storage component (OSC), 3) base metal oxide stabilisers (precious metal, alumina wash coat, oxygen-storage component), 4) moderator or scavenger for H_2S, 5) layered structure, and 6) segregated wash coat. Thus, the major components of Ce, Al, and precious metals are stabilised in the modern TWC.

The effect of sulfur continues to affect the modern catalyst technologies. Since the sulfur affects mostly the light-off characteristics of the TWC catalyst, low emission performance requires that sulfur be reduced in the gasoline. Sulfur is scheduled to be reduced to 30 ppm in the fuel by the year 2010 in the U.S. The P and Zn in the lubricating oil continue to be an issue. Since these materials deposit on the inlet section of the catalyst, light-off performance is affected.

The latter part of the 1990s was a time where close-couple catalyst technology was being utilised. In addition, the thermal stability of the NO_x function in the downstream catalyst was improved. Manufacturers learned how to make more sophisticated engine controls that enabled more complete system optimisation around the close coupled and under floor catalysts. This cooperation between the automobile manufacturers, the monolith suppliers, the exhaust system fabricators, the sensor manufacturers and the catalyst manufacturers intensified so that a complete engineered system approach was taken to achieve the goal of a "zero emission stoichiometric spark ignited (SI) vehicle". One of the major achievements of this effort was the introduction of SULEV vehicles into the 2001 model year (Honda Accord SULEV (2.3 litre, 4 cylinders); Toyota Prius SULEV (1.5 litre, 4 cylinders, 33 kW electric drive) gasoline hybrid electric; and Nissan Sentra-CA SULEV (1.8 litre, 4 cylinders) first that meets zero evaporative emissions and certified for 150,000 mile durability).

With the advent of a durable close-coupled catalyst, many of the engine modifications to provide rapid heat-up (e.g. spark retard on cold start) of the catalyst were now implemented as shown in Figure 4.

In this control strategy, the catalyst temperature is 190°C higher and the exhaust gas temperature 240°C higher at the 20 second time frame into the FTP. Of course temperature is just one of the key control parameters for the light-off of the oxidation reactions. Oxygen is required to achieve the oxidation reactions. The oxygen can be supplied via the engine operation or an external air pump. If oxygen is supplied via the engine, a special engine design (e.g. stratified charge or variable valve timing) is needed to extend the flame stability operating window for the lean air/fuel ratios. Figure 5 shows the air/fuel ratio during the 12 seconds for a ULEV and a SULEV control strategy. Note that stoichiometric A/F is around 14.65.

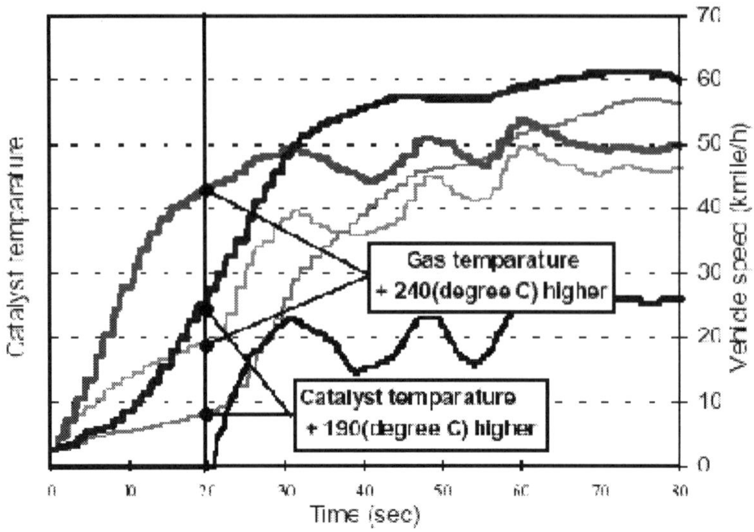

Figure 4: Quick warm-up systems solve low-emissions challenge (Kitagawa 2000).

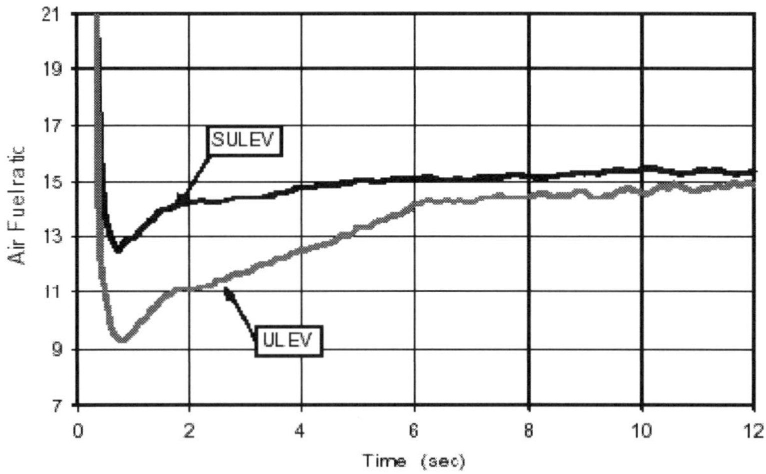

Figure 5: Leaner conditions required for earlier light-off of SULEV strategy (Kitagawa 2000).

The ULEV performance requirement for a four-cylinder vehicle, which may range from a hydrocarbon engine out emissions of 1.5 g/mile to 2.0 g/mile, is around 98 percent HC conversion. A SULEV vehicle attains greater than 99 percent HC conversion. Because of these high emission reduction efficiencies and a requirement for more geometric surface area, monolith suppliers made higher cell density substrates approaching 1200 cpsi (Kikuchi *et al.* 1999, Gulati 1999). The increased geometric surface of the higher density substrates is desirable but not at the expense of higher monolith bulk density, which is a key parameter affecting light-off due to thermal mass. The monolith suppliers responded by decreasing the cell wall thickness so that the 600 cpsi, 900 cpsi, and 1200 cpsi honeycombs have a lower bulk density than the 400 cpsi product, which was the standard until the year 2000.

The effect of the thinner wall monolith in going from 400 cpsi to 600 cpsi to 900 cpsi is shown in Figure 6. The thinner wall and higher cell density resulted in a cumulative 9-second decrease in light-off time.

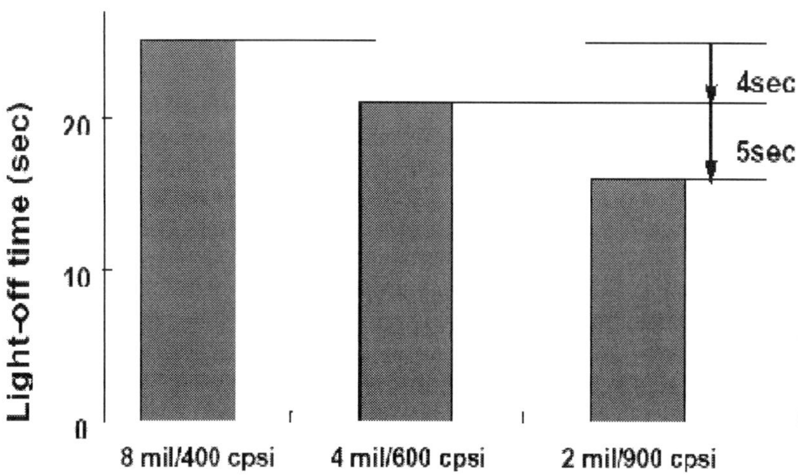

Figure 6: Decreasing the monolith wall thickness improves light-off. (Webster *et al* 2000).

A standard design addition for the ULEV and SULEV systems is a redesign of the manifold and exhaust piping to minimise heat loss during the critical cold start with fabrication of the low heat capacity piping (Kishi *et al.* 1998). The close-coupled catalyst can also be manufactured as a cascade design, which is a small diameter monolith followed by a larger diameter monolith. The smaller volume light-off catalyst has lower total thermal mass and can be heated very rapidly. In addition, some catalyst systems have a HC trap function incorporated in the catalyst for the first 10 seconds of the FTP to trap residual HCs leaving the close-coupled catalyst.

NO_x regulations are also a factor for the ULEV and SULEV regulations. The engine control must be better to minimise the perturbation effects on the TWC performance. To accomplish this a new sensor was developed based on the operating principles of the oxygen sensor but with more sophisticated design and electronics to give a gradual response curve to changes in A/F ratio or oxygen content in the engine exhaust (Anderson 1993, Bush *et al.* 1994).

With better control, the operating window for the TWC is narrowed for better overall HC, CO and NO_x conversion. In addition, the vehicles have control algorithms to maintain the set-point at stoichiometric operation for the engine and software that works with the dual oxygen sensors to predict the degree of oxygen saturation of the catalyst to prevent NO_x breakthrough.

All of the SULEV technologies were designed with reduced engine emissions by improving the combustion efficiency through better fuel atomisation (e.g. air atomisation) and higher energy ignition sources (e.g. improved spark plugs). This led to reduced HC and CO engine-out emissions but unfortunately increased the NO_x emissions. To counter these NO_x emissions, better exhaust gas recirculation (EGR) strategies were used that lowered the cylinder operating temperature and therefore reduced the thermal NO_x formation. This offset the effect of the improved combustion efficiencies. The engines were also designed to have a broader operating window for flame stability versus A/F ratio; thereby allowing leaner operation and more EGR dilution.

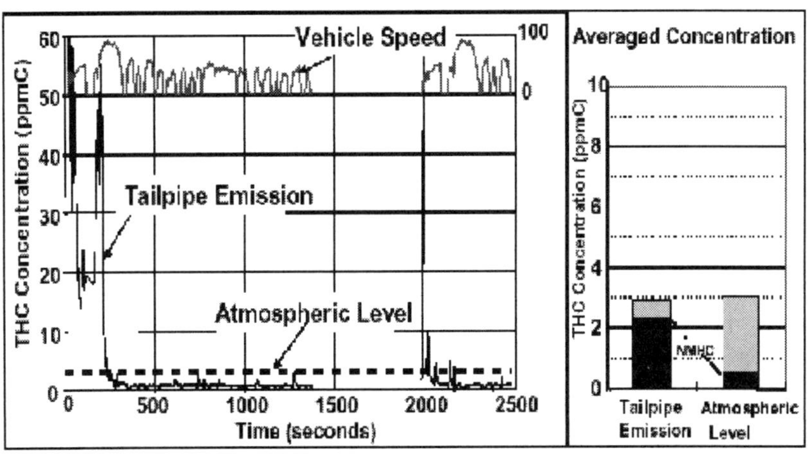

Figure 7: SULEV emissions demonstrated with Nissan Sentra CA. (Webster *et al.*, 2000).

As a result of the improved engine performance, engine control strategies, and catalyst technologies, the SULEV regulations could be obtained. Figure 7 shows the published emissions for a SULEV vehicle after 150,000 miles aging. This figure shows that HC emissions can be below the ambient background levels for HC during most of the operating cycle. Only during the cold and hot starts are HC emissions measurably above the ambient levels. Also shown in this figure are the total cumulative HC emissions during the FTP, which are comparable to the ambient or atmospheric level.

5.2 Diesel vehicles

The following sections describe some of the control technologies available to control emissions from modern diesel engines. The technologies described have all been applied in the U.S. in one form or another to highway diesel engines. The sections are divided into combustion controls and exhaust emission control devices.

5.2.1 Diesel emission control technologies – combustion controls

5.2.1.1 The NO_x vs. PM trade-off Diesel engine emission control technology performance is often characterised by its "NO_x vs. PM trade-off". This trade-off refers to the fact that, under many conditions, changes in engine control strategies designed to reduce one pollutant (e.g., NO_x) result in increased production of the other pollutant (e.g., PM). Diesel engine designers and environmental regulators must balance these trade-offs in order to accomplish both NO_x and PM control.

5.2.1.2 Charge air cooling Lowering the intake manifold temperature, and therefore the initial temperature of the gases entering the combustion chamber, lowers the peak temperature of combustion resulting in decreased NO_x emissions. The NO_x reduction realised from lowering the intake manifold temperature can vary depending upon the engine design, but generally, NO_x emissions can be reduced by five to seven percent with every 10°C decrease in intake manifold temperature. Typically, the intake manifold temperature is lowered by cooling the intake gases through a heat exchanger located between the turbocharger compressor outlet and the intake manifold. This type of heat exchanger is commonly called an aftercooler since it cools the gases after they are heated by the compression work done by the turbocharger compressor.

5.2.1.3 Fuel injection rate shaping and multiple injections Rate shaping or multiple injections vary the delivery of fuel over the course of a single combustion event. Igniting a small quantity of fuel initially limits the characteristic rapid increase in pressure and temperature that leads to high levels of NO_x formation. Injecting most of the fuel into an established flame then allows for a steady burn that limits NO_x emissions. Rate shaping may be done either mechanically or electronically. Rate shaping has been shown to reduce NO_x emissions by up to 20 percent (Ghaffarpour and Baranescu, 1996). These systems are primarily used on light and medium heavy-duty diesel trucks.

5.2.1.4 Injection timing retard Delaying the start of fuel injection and thus the start of combustion can significantly reduce NO_x emissions from a diesel engine. The effect of injection timing on emissions and performance is well established

(Herzog *et al.*, 1992; Uyehara, 1992; Durnholz *et al.*, 1992; Bazari and French, 1993). Delaying the start of combustion by retarding injection timing aligns the heat release from the fuel combustion with the portion of the power (or combustion) stroke of the engine cycle after the piston has begun to move down. This means that the cylinder volume is increasing and that work (and therefore heat) is being extracted from the hot gases. The removal of this heat through expansion lowers the temperature in the combustion gases. NO_x is reduced because the premixed burning phase is shortened and because cylinder temperature and pressure are lowered. Timing retard increases HC, CO, PM, and fuel consumption, because the end of injection comes later in the combustion stroke, where the time for extracting energy from fuel combustion is shortened and the cylinder temperature and pressure are too low for more complete oxidation of PM. Many of these tradeoffs with injection timing retard can be changed through the application of new technologies such as EGR.

5.2.1.5 Exhaust gas recirculation EGR reintroduces or retains a fraction of the exhaust gases into the cylinder. The use of EGR decreases NO_x formation in three different ways:

1) EGR can thermally reduce peak combustion temperature. Increasing the mass of the cylinder contents by increasing CO_2 and water vapor concentrations reduces peak cylinder temperatures during combustion (Ropke, 1995).

2) A fraction of the air within the cylinder is replaced with inert exhaust, primarily CO_2 and water vapor. This reduces the amount of molecular oxygen available for dissociation into atomic oxygen, an important step in NO_x formation via the Zeldovich mechanism (Dec and Espy, 1995).[1]

3) The high-temperature dissociation of CO_2 and water vapor is highly endothermic, and thus can reduce temperatures via absorption of thermal energy from the combustion process (Kreso *et al.*, 1998).

The use of EGR can increase the intake manifold temperature substantially. This reduces intake charge density and lowers the fresh A/F ratio for a given level of turbocharger. The result can be a large increase in PM emissions if the boost pressure cannot be increased to compensate for the lower intake charge density. For this reason, diesel EGR systems generally cool the exhaust gases using a heat exchanger in the exhaust recirculation loop. PM emissions can still be a challenge for EGR-equipped engines since the amount the EGR can be cooled is limited by the need to prevent acidic condensation of combustion products within the EGR system. The EGR rate must be electronically controlled to prevent temporary, overly fuel-rich conditions that can lead to high PM emissions during transient engine operation. Cooled EGR systems are used on all light-duty diesel engines built in 2000 or later.

5.2.1.6 Induced mixing/charge motion

Inducing turbulent mixing is one means of increasing the likelihood of soot particles interacting with oxidants within the cylinder. Turbulent mixing can be induced or increased by a number of means including increased (high) injection pressure, multiple/split injections, intake port/valve design and piston bowl design.

5.2.1.7 Control of air-to-fuel ratio Availability of oxidants (primarily oxygen) is highly dependent on the overall A/F ratio at which the engine is operated. Thus soot formation/oxidation and PM emissions are a strong function of the overall A/F ratio of the engine. Temporary conditions of overly fuel-rich A/F ratio can occur on turbocharged engines when the engine power demand increases sharply. This is because the fuel system can increase fueling immediately but the turbocharger takes some time to increase air flow. This effect is compounded on engines equipped with EGR. Black smoke emissions from older diesel vehicles (without electronic A/F control) during acceleration from a stoplight are a common example of overly fuel-rich operation. Electronic control of fuel injection can limit the injected quantity of fuel until sufficient air supply (i.e., boost pressure) can be obtained. Electronic wastegate control and electronically controlled, variable geometry turbochargers (VGT) can also be used to improve turbocharger response and improve PM emissions under such transient conditions. Electronic control of EGR is also necessary to prevent temporary conditions of overly fuel-rich A/F ratios.

5.2.1.8 Control of oil consumption Reducing oil consumption not only decreases maintenance costs, but also VOF and PM emissions. Reducing oil consumption has been one of the primary ways that highway diesel engines have complied with the PM standards. Piston rings can be designed to better "scrape" oil from the cylinder liner surface back into the crankcase, reducing the amount of oil consumed during combustion from the cylinder. Valve stem seals can be used to reduce oil leakage from the lubricated regions of the engines valvetrain into the intake and exhaust ports of the engine. Engine designs that incorporate these technologies have reduced VOF and PM emissions.

5.2.2 Diesel emission control technologies – post-combustion controls

5.2.2.1 Diesel oxidation catalysts The flow-through monolithic diesel oxidation catalyst (DOC) reduces HC and PM emissions by oxidising both gaseous (volatile) HCs and the VOF portion of PM forming CO_2 and H_2O. The VOF consists of hydrocarbons adsorbed to the carbonaceous solid particles and heavy HCs that, under atmospheric conditions, will condense to form liquid aerosols. The soot or elemental carbon portion of the PM remains largely unaffected by the catalyst. Although engine design improvements have reduced VOF emissions, VOF can still comprise a significant fraction of the total PM mass.

DOCs that rely heavily on platinum for catalysis can also oxidise sulfur species in the exhaust to form sulfate PM. At higher exhaust temperatures, platinum catalysts have a greater tendency to oxidise SO_2 to form sulfates, which contribute to total PM emissions. At exhaust temperatures between 250 °C and 400 °C, platinum catalysts can also oxidise a significant fraction (approximately

50 percent, depending on exhaust conditions and catalyst platinum loading) of NO to NO_2. Catalyst manufacturers have been successful in developing catalyst formulations that minimise sulfate formation at current U.S. highway diesel fuel sulfur levels by reducing the reliance on platinum (Voss *et al.*, 1994). Excess emissions of NO_2 are prevented via choosing appropriate platinum loadings and proper substrate sizing. Manufacturers have also adjusted the placement of the catalyst to a position where the needed VOF reduction is achieved, but sulfate and NO_2 formation is minimised (Johnson, *et al.*, 1994). DOCs are commonly used with light-duty vehicles and some light- and medium-heavy-duty trucks.

5.2.2.2 Catalyzed diesel particulate filters/continuously regenerating traps
One method of exhaust after-treatment for controlling diesel PM emissions is to pass diesel exhaust through a ceramic, monolithic filter or "PM trap" to collect the PM. The PM accumulated within the trap must be either continuously or periodically oxidised or cleaned from the trap. This process is often referred to as "trap regeneration". Traps have previously used catalyzed coatings, fuel additives, and electrical heating to assist with trap regeneration. Failure to consistently regenerate the trap can lead to plugging, excessive exhaust back-pressure, and eventually overheating and permanent damage to the trap. Inconsistent regeneration due to a high frequency of fairly low exhaust temperatures has been a particular problem in applying PM traps to light-duty diesel vehicles.

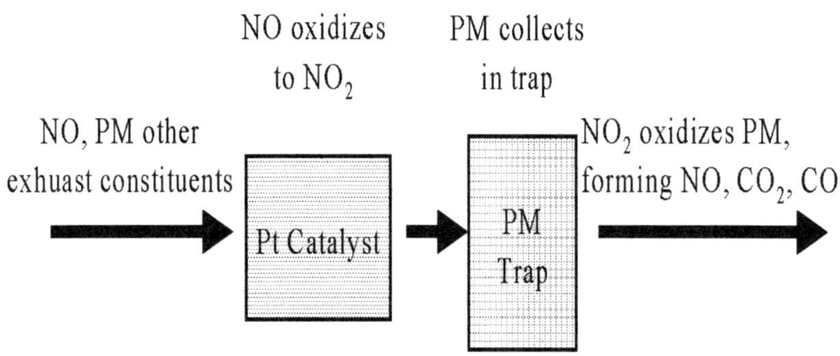

Figure 8: Functional schematic illustrating the operation of a CRT.

The continuously regenerating trap (CRT) has shown considerable promise for PM emission control in diesel applications due to its ability to consistently regenerate even at relatively low exhaust temperatures. The CRT uses a high-platinum-content DOC located upstream of the trap to oxidise NO in the exhaust to NO_2. This assists trap regeneration by lowering the initial temperature of PM oxidation from ~600 °C (PM oxidation via oxygen in the exhaust) to ~250 °C (PM oxidation by NO_2) (Hawker, *et al.*, 1997). This allows PM regeneration to occur nearly continuously for many light-duty diesel and most heavy-duty diesel vehicle and truck applications (Lüders, *et al.*). A general schematic of the CRT system is presented in Figure 8.

The catalyzed diesel particulate filter (CDPF) generally functions in a similar manner to the CRT with respect to NO_2 oxidation of PM, but the catalyst is applied directly onto the wall-flow monolith instead of using a separate, upstream DOC.

CRTs and CDPFs are capable of large reductions in PM emissions, typically exceeding 80–90 percent (Hawker *et al.* 1997, Hawker *et al.* 1998). SO_2 emissions from combustion of fuel sulfur components significantly impedes oxidation of NO to NO_2, and can eventually inhibit trap regeneration. SO_2 emissions can also be oxidised to sulfate and contribute to emissions of sulfuric acid PM, thus somewhat limiting the PM reduction ultimately achievable using CRTs and CDPFs. The reduction of diesel fuel sulfur in the U.S. to less than 15 ppm by 2006 will enable consistent regeneration of CRTs and CDPFs over a broad range of operating conditions, and to enable PM emission reductions of greater than 90 percent (U.S. Federal Register, 2000). PM emission reductions greater than 95 percent have been achieved from a modern heavy-duty highway diesel engine using a CDPF with a diesel fuel with less than 15 ppm sulfur content (Schenk *et al.*, 2001). Virtually all heavy-duty diesel engines sold in the U.S. after 2007 are likely to be equipped with CRTs, CDPFs, or other similar types of PM trap technology.

5.2.2.3 Lean NO_x catalysts Various types of active (i.e. require addition of a hydrocarbon fuel for reductant) and passive (no reductant addition) lean-NO_x catalysts have been investigated for NO_x reduction in lean exhaust environments such as those present in diesel exhaust. Lean-NO_x catalysts typically reduce NO_x efficiently over a fairly narrow range of exhaust temperatures. There are both "high" and "low" temperature varieties of lean-NO_x catalyst systems. Low temperature, precious-metal-based lean-NO_x catalysts are used in combination with zeolites support materials for catalyst promotion and adsorption of hydrocarbons and NO_x. High-temperature lean-NO_x catalyst formulations typically rely on base metals on zeolites (Cu-ZSM5, for example). For heavy-duty on-highway applications, a combination of both technologies would most likely be used to broaden the temperature range of effective NO_x reduction. The fuel economy penalty for hydrocarbon addition with active low-NO_x catalyst systems is approximately 3–7 percent for NO_x reductions of only 10 to 30 percent (Engler *et al.*, 1993; Kawanami *et al.*, 1995).[32]" Passive lean-NO_x catalyst systems generally provide only nominal NO_x reductions of 10 percent or less. Much of the apparent NO_x reduction with low-temperature, platinum-based lean-NO_x catalyst systems is due to conversion of NO_x to nitrous oxide (N_2O), which is a pollutant with severe climate change potential as discussed in Section 3 (Burch *et al.*, 1995).

5.2.2.4 NO_x adsorption catalysts The diesel NO_x adsorption catalyst system is an adaptation of technology originally developed for lean-NO_x control for gasoline direct injection (GDI) engines under development for the European and

Japanese automotive markets. GDI engines typically operate fuel lean in a manner similar to diesel engines, but also run near stoichiometric similar to standard gasoline spark-ignition engines. The NO_x adsorption catalyst was developed to provide storage of NO_x emissions during periods of fuel-lean operation when the three-way catalyst is not capable of effectively reducing NO_x. Under fuel-lean conditions, NO (the primary NO_x constituent) is oxidised to NO_2 using a platinum catalyst, and is stored as a nitrate using alkali or alkaline-earth adsorbant materials (barium and/or potassium compounds are most common). Periodic operation of the engine at conditions slightly fuel-rich of the stoichiometric point causes NO_x to be released from NO_x storage components on the catalyst. Because the NO_x release occurs under fuel rich (reducing) conditions, NO_x reduction to N_2 and O_2 can occur using platinum/rhodium catalysts in a manner similar to standard three-way catalytic converters used for spark-ignition automotive applications (Pott *et al.*, 1998). A functional schematic of this operation is presented in Figure 9

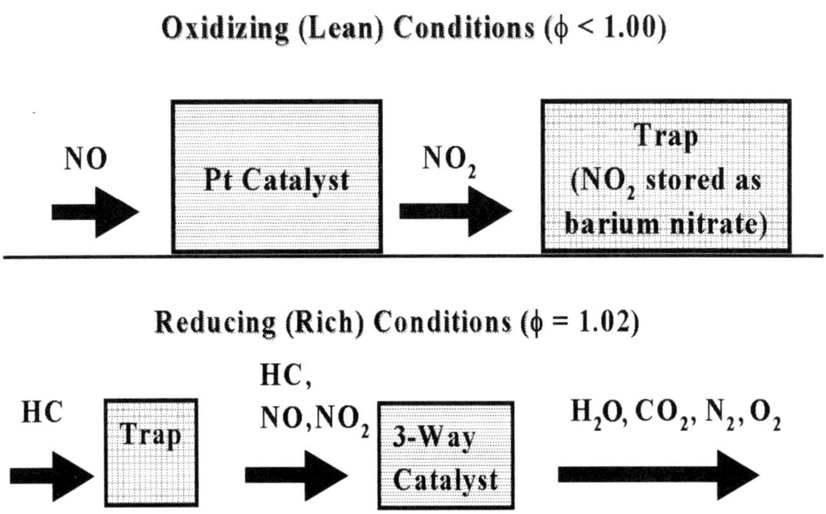

Figure 9: Functional schematic of barium-based NO_x adsorption catalyst operation. In actual systems the catalyst precious metals and storage components are typically washcoated together onto a single monolith, and the reactions shown in the schematic take place within the washcoat layers.

Diesel engines and vehicles equipped with NO_x adsorption catalyst systems have demonstrated very high NO_x reduction efficiencies (>90 percent) over a broad range of operating conditions with tailpipe NO_x emissions approaching that of three-way catalyst equipped low-emission SI engines and vehicles (U.S. Federal Register, 2001; Schenk *et al.*, 2001; McDonald and Bunker, 2002).

NO_x adsorption catalysts must be used with very low sulfur fuels. SO_2 can be oxidised and then stored as a sulfate on the storage components. The resultant sulfates are generally more stable than nitrates formed by NO_x adsorption. Higher temperatures and longer periods of reducing conditions are required for sulfate release. Repeated removal of adsorbed sulfate can significantly increase the amount of fuel reductant used, and prolonged high-temperature operation required for sulfur removal can damage NO_x adsorption catalyst components over time. The reduction of diesel fuel sulfur in the U.S. also facilitates the introduction of NO_x adsorption catalysts for highway diesel truck and bus applications.

5.2.2.5 Selective catalytic reduction (UREA SCR) Selective Catalytic Reduction (SCR) is a well-known process for reducing NO_x emissions from stationary sources and has been used industrially since the 1970s. A chemical agent (urea or ammonia) is metered into the exhaust stream and serves as a reducing agent for NO_x, even in the presence of high concentrations of oxygen. While vanadium-based SCR catalysts are largely insensitive to sulfur, systems incorporating platinum-based oxidation catalysts are only recommended for use in conjunction with low sulfur fuels (<50 ppm). Demonstrated NO_x reduction efficiencies typically range from 50 to 80 percent over transient cycles, while peak steady-state conversions of up to 90 percent are possible, depending upon NH_3/NO_x ratio and exhaust temperatures.

Urea is more typically used to supply ammonia reductant for SCR systems for mobile applications due to the safety concerns associated with the handling and dispensing of anhydrous ammonia. A number of issues are still associated with the application of urea-SCR technology on mobile diesel applications. For example, a distribution system for appropriately packaged urea solution or solution dispensers is not currently available. This limits practical application to commercial fleets with common fueling depots. Because ambient temperatures during the winter months are often below the freezing point of the aqueous urea solution, storage and on-board tanks must be adequately protected, typically through incorporation of an electric heating element. Without such provisions, practical application may again be limited. The application of stabilising additives to lower the freezing point of urea solution has so far been unsuccessful.

6 Voluntary motor vehicle emission reduction programs

In addition to legislative requirements to reduce motor vehicle emissions through the CAA and TEA-21, voluntary measures are encouraged through the EPA. For the purpose of this chapter, voluntary programs are defined as "emission reduction programs that relay on voluntary actions of individuals or other parties for achieving emission reductions."

Two basic types of voluntary programs exist in the U.S.: technology-based and activity-based. Examples of voluntary technology programs include, retrofitting existing heavy-duty diesel engines with exhaust emission control technologies, purchasing alternatively fueled vehicles, and changing to low-rolling resistance tires. Voluntary activity-based programs include Commuter Choice, Ozone Action Days, local and state-wide Smart Growth policies, reduced vehicle idling, and vehicle and lawnmower scrappage.

6.1 Creating the policy

The motivation for obtaining emission benefits by implementing "voluntary" programs was driven by passage of the 1990 CAA. Many areas not meeting NAAQS for ozone or CO were faced with not meeting SIP requirements even after applying all new federal regulations that increased the stringency of emission standards at the national level for a wide-range of mobile sources.

The Voluntary Emissions Reduction Policy (VMEP) was released in October 1997 and for the first time allowed emission credits in the SIP for voluntary mobile source reduction programs. The goal was to make it easier for States to implement voluntary activities and encourage innovation and investment in effective programs and actions. This policy provided new flexibility for State and local governments to experiment and create programs that addressed smaller or unconventional sources of mobile source air pollution, especially allowing a State to count emission benefits from programs that were implemented not by the State itself but by local governments or businesses.

Most recently the EPA published a national guidance entitled, "Improving Air Quality Through Land Use Activities" (EPA, 2001b) that serves as an additional incentive for state and local communities who wish to develop "smart growth" strategies and account for the air quality benefits associated with these various land use activities and policies. The guidance recognises air quality benefits in such land use activities as infill development, brownfield redevelopment, mixed use and transit-oriented development.

One key issue the EPA has tried to address, and which has caused the greatest hesitancy in allowing voluntary measures in SIPs, is the quantification of emission benefits. Over the past four years, EPA's Office of Transportation and Air Quality (OTAQ) has developed a series of peer-reviewed user-friendly tools that can quickly and cost-effectively generate data that can be used to reasonably quantify the emission benefits of voluntary programs. Specific tools developed include: Airport Ground Support Equipment Emissions Calculation Spreadsheet, Alternative Fueled Vehicles Light Day and Heavy Duty Spreadsheet, The COMMUTER Model, Retrofit Calculator, and Methodologies to Assess Transportation and Air Quality Impacts of Brownfields and Infill Development.

6.2 Creating national level voluntary transportation programs

Over the past two years, EPA has developed a new generation of national voluntary programs that aim to achieve measurable environmental results in a cost-effective and beneficial way without the need for regulation. These programs can reduce pollution and improve air quality by means of forming partnerships with small and large businesses, consumers, industry, manufacturers, trade associations, and state and local governments. Partners in these voluntary programs work together to improve environmental performance and in return receive such benefits as cost savings, public recognition, and the satisfaction of knowing that they are helping to clean the air.

6.2.1 Green Vehicle Guide

The Green Vehicle Guide was created by the EPA to help consumers choose the lowest emitting and most efficient vehicle that meets their needs. The Guide rates cars and trucks according to their emissions and fuel economy performance and provides consumers with information on how to make environmentally-informed choices when purchasing a vehicle.

6.2.2 Voluntary Diesel Retrofit Program

The Diesel Retrofit Program works to reduce pollution resulting from existing diesel vehicles and equipment by encouraging fleet owners to install pollution-reducing devices on the vehicles and to use cleaner-burning diesel fuel. This involves working with state, local, and industry partners to verify the effectiveness of pollution-reducing technology and to create retrofit projects around the country.

6.2.3 Best Workplaces for Commuters Program

Offering innovative solutions to commuting challenges faced by employers and employees, Best Workplaces for Commuters is a new business/government voluntary initiative promoting a new voluntary national standard of excellence for employee–commuter benefits.

Established by the EPA and the DOT, this initiative helps employers address limited or expensive parking, reduce traffic congestion, improve employee recruiting and retention, and minimise the environmental impacts associated with drive-alone commuting. Participating companies earn the designation "Best Workplace for Commuters" a mark of excellence for environmentally and employee-friendly organisations. In its first year of existence over 300 U.S. businesses, covering over 700,000 employees have joined this new partnership.

7 International programs

Environmental concerns associated with emissions from motor vehicles are not unique to the United States. Similar global, regional, and local scale concerns are prevalent throughout the world. As a result, an evaluation and integration of international programs in air quality management of motor vehicle emissions is extremely important. In addition, there is a responsibility to share information with other countries, especially those just beginning to experience adverse impacts from motor vehicle exhaust.

The EPA promotes coordinated research programs with other countries in an effort to harmonise standards among countries. Collaboration in the development of engine technologies, engine and ambient air testing procedures, and health effects information provides significant value for all countries. Within the United

Nations Economic Commission for Europe, the World Forum for Harmonisation of Motor Vehicle Regulations is working to develop the first global technical regulations under an agreement signed in 1998. There are also innovations in transportation planning that take place around the world that can potentially be incorporated into the strategies available in the U.S. such as congestion pricing, which is being tested in some of the large cities of Asia.

The demand for transportation is growing rapidly in developing countries, leading to growing levels of urban air pollution and congestion. The U.S. tries to encourage the development of partnerships with developing countries to assist in adoption of best practices for transportation planning including providing information, training, modeling tools, and support for sustainable urban design.

In developing countries, 2- and 3-wheeled vehicles and transit buses account for roughly half of urban transport energy use and 90 percent of passenger-vehicle-generated pollution. The use of lead in gasoline remains widespread, and public transit is often overcrowded and poorly maintained. The U.S. has joined with other countries and indisutry and NGO partners to promote the phase-out of leaded gasoline in order to facilitate exhaust emission control. The Partnership for Clean Fuels and Vehicles, launched at the World Summit on Sustainable Development in 2002, also promotes the reduction of sulfur in gasoline and diesel fuels concurrent with the adoption of cleaner vehicle technologies. The U.S. also encourages technology transfer such as the adaptation and demonstration of existing and proven technologies to clean up two-stroke scooters and cycles, and the support of urban bus retrofit strategies and alternative fuel conversions. The EPA promotes direct interaction with other countries as well as working with international development and health organisations including the World Bank and World Health Organisation.

References

[1] Anderson, M., A Feedback A/F Control System for Low Emission Vehicles. *Soc. Auto Eng.*, 930388, 1993.

[2] Bush, K., *et al.*, Automatic Control of Cylinder Air-Fuel Mixture Using a Proportional Exhaust Gas Sensor. *Soc. Auto Eng.*, 940149, 1994.

[3] Dec, J.E. & Espey, C., Ignition and early soot formation in a diesel engine using multiple 2-D imaging diagnostics. *Soc. Auto Eng.*, 950456, 1995.

[4] Dockery, D.W., *et al.*, An association between air pollution and mortality in six U.S. cities. *N. Engl. J. Med.*, **329**, pp. 1753–1759. 1993.

[5] Flynn, P., *et al.*, Minimum Engine Flame Temperature Impacts on Diesel and Spark-Ignition Engine NO_x Production. *Soc. Auto Eng.*, 2000-01-1177, 2000.

[6] Foster, D.E. & Tree, D.R., Optical measurements of soot particle size, number density and temperature in a direct injection diesel engine as a function of speed and load. *Soc. Auto Eng.*, 940270, 1994.

[7] Ghaffarpour, M. & Baranescu, R., NO_x Reduction Using Injection Rate Shaping and Intercooling in Diesel Engines. *Soc. Auto Eng.*, 960845, 1996.

[8] Greeves, G., *et al.*, Origins of Hydrocarbon Emissions from Diesel Engines. *Soc. Auto Eng.*, 770259, 1977.
[9] Gulati, S., Thin Wall Ceramic Catalyst Support. *Soc. Auto Eng.*, 1999-01-0269, 1999.
[10] Harrison, R.M., *et al.*, Analysis of incidence of childhood cancer in the West Midlands of the United Kingdom in relation to proximity of main roads and petrol stations. *Occupational and Environmental Medicine*, **56**, pp. 774–780, 1999.
[11] Heck, R. & Farrauto, R., Catalytic Air Pollution Control: Commercial Technology, 2nd edition, J. Wiley & Sons: New York, 2002.
[12] Jerrett, M., *et al.*, Reanalysis of the Harvard Six Cities Study and the American Can Society Study of Particulate Air Pollution and Mortality. Special Report of the Health Effects Institute, 2000.
[13] Johnson, J., *et al.*, A Review of Diesel Particulate Control Technology and Emissions Effects – 1992 Horning Memorial Award Lecture. *Soc. Auto Eng.*, 940233, 1994.
[14] Kishi, N., *et al.*, Development of the High Performance L4 ULEV System. *Soc. Auto Eng.*, 980937, 1998.
[15] Kitagawa, H., L4-Engine Development for a Super-Ultra-Low Emissions Vehicle. *Soc. Auto Eng.*, 2000-01-0887, 2000.
[16] Kittelson, *et al.*, Particle concentrations in a diesel cylinder: comparison of theory and experiment. *Soc. Auto Eng.*, 861569, 1986.
[17] Kreso, A.M., *et al.*, A Study of the Effects of Exhaust Gas Recirculation on Heavy-Duty Diesel Engine Emissions. *Soc. Auto Eng.*, 981422, 1998.
[18] Krewski, D., *et al.*, High Cell Density and Thin Wall Substrate for Higher Conversion Ratio Catalyst. *Soc. Auto Eng.*, 1999-01-0268, 1999.
[19] Laden, F., *et al.*, Association of fine particulate matter from different sources with daily mortality in six U. S. cities. *Environ. Health Perspect*, **108**, pp. 941–947. 2000.
[20] McDonald, J. & Bunker, B., Testing of the Toyota Avensis DPNR at U.S. EPA-NVFEL. *Soc. Auto Eng.*, 2002-01-2877.
[21] Pearson, R.L., *et al.*, Distance-Weighted Traffic Density in Proximity to a Home is a Risk Factor for Leukemia and Other Childhood Cancers. *J. Air Waste Manag. Assoc.*, **50**, pp. 175–180. 2000.
[22] Pope, C.A., *et al.*, Particulate air pollution as a predictor of mortality in a prospective study of U. S. adults. *Am. J. Respir. Crit. Care Med.*, **151**, pp. 669–674. 1995.
[23] Röpke, S., *et al.*, NO_x Formation in Diesel Engines for Various Fuels and Intake Gases. Soc. Auto Eng. 950213. 1995.
[24] Savitz, D. A., and L. Feingold. Association of childhood cancer with residential traffic density. Scandinavian Journal of Work, Environment and Health 15: 360–363, 1989.
[25] Schenk, C., *et al.*, High Efficiency NO_x and PM Exhaust Emission Control for Heavy-Duty On-Highway Engines. Soc. Auto Eng. 2001-01-1351, 2001.

[26] Schwartz, J., *et al.* Is daily mortality associated specifically with fine particles? J. Air Waste Manag. Assoc. 46: 927–939. 1996.
[27] Seinfeld, J.H. & Pandis, S.N., *Atmospheric Chemistry and Physics*, John Wiley and Sons, Inc.: New York, 1998.
[28] Sierra Research, A List of Compounds Emitted from On-Road and Non-Road Mobile Sources. Report No. SR01-02-01, *U.S. Environmental Protection Agency*, 2001.
[29] U.S. Environmental Protection Agency, Air Quality Criteria for Ozone and Related Photochemical Oxidants. EPA/600/P-93/004a-cF, 1996.
[30] U.S. Environmental Protection Agency, Air Quality Criteria for Particulate Matter: Second External Review Draft. EPA 600/P-99/002aB. 2001.
[31] U.S. Environmental Protection Agency, Improving Air Quality Through Land Use Activities. EPA420-R-01-001. 2001b.
[32] U.S. Environmental Protection Agency, Control of emissions of hazardous air pollutants from mobile sources. *Federal Register*, **65**, pp. 48058–48105. 2000.
[33] U.S. Environmental Protection Agency, National-Scale Air Toxics Assessment for 1996. Office of Air Quality Planning and Standards. 2002a.
[34] U.S. Environmental Protection Agency, Health Assessment Document for Diesel Engine Exhaust. Office of Research and Development. Report No. EPA/600/8-90/057F, May, 2002b.
[35] U.S. Environmental Protection Agency, Health Assessment of 1,3-Butadiene. Office of Research and Development, National Center for Environmental Assessment, Washington Office, Washington, DC., Report No. EPA600-P-98-001F, 2002c.
[36] U.S. Environmental Protection Agency, Motor Vehicle-Related Air Toxics Study, U.S. Environmental Protection Agency, Office of Mobile Sources, Ann Arbor, MI, EPA Report No. EPA 420-R-93-005, April 1993a.
[37] U.S. Environmental Protection Agency, The Plain English Guide To The Clean Air Act. Office of Air Quality Planning and Standards. EPA-400-K-93-001. 1993b.
[38] U.S. Environmental Protection Agency, Carcinogenic Effects of Benzene: An Update, National Center for Environmental Assessment, Washington, DC. Report No. EPA600-P-97-001F. 1998a.
[39] U.S. Environmental Protection Agency, Health Risk Assessment of 1,3-Butadiene. EPA/600/P-98/001A, February 1998b.
[40] U.S. Environmental Protection Agency, Environmental Protection Agency, Assessment of health risks to garment workers and certain home residents from exposure to formaldehyde, Office of Pesticides and Toxic Substances, 1987.
[41] U.S. Environmental Protection Agency, Environmental Protection Agency, Integrated Risk Information System (IRIS), Office of Health and Environmental Assessment, Environmental Criteria and Assessment Office, Cincinnati, OH, 1999.

[42] U.S. Environmental Protection Agency, Toxicological Review of Acrolein and IRIS Summary. External Review Draft. Office of Research and Development, Research Triangle Park, NC. 2002c.
[43] U.S. Environmental Protection Agency, The U.S. Greenhouse Gas Inventory. EPA430/F-02-008. Office of Air and Radiation, Washington, D.C. 2002d.
[44] U.S. Code of Federal Regulations, Title 40 § 86.1310–2007. Exhaust gas sampling and analytical system for gaseous emissions from heavy-duty diesel-fueled engines and particulate emissions from all engines.
[45] U.S. Federal Register, January 18, 2001, Volume 66, Number 12, pages 5001-5050 40 CFR Parts 69, 80, and 86 Control of Air Pollution From New Motor Vehicles: Heavy-Duty Engine and Vehicle Standards and Highway Diesel Fuel Sulfur Control Requirements; Final Rule. 2001.
[46] Voss, K.Y., *et al.*, Performance Characteristics of a Novel Diesel Oxidation Catalyst. *Soc. Auto Eng.*, 940239, 1994.
[47] Webster, L., *et al.*, Nissan's Gasoline SULEV Technology, *Soc. Auto Eng.*, 2000-01-1583, 2000.
[48] Zeldovich, J., The oxidation of nitrogen in combustion and explosions, *Acta Physiochimica U.R.S.S.*, **XXI(4)**, 1946.

CHAPTER 6

Air quality management in Bogota

A. Valencia & J. Huertas
Department of Mechanical Engineering,
Los Andes University, Colombia.

Abstract

This chapter describes briefly the programs that the government of the city of Bogota has implemented to assure the air quality that the natives of Bogota breathe. Initially a short description of the city appears. It is emphasised that Bogota is a city of about 7.5 million inhabitants located in the tropical zone at a high altitude above sea level (2700 masl). Then the present state of the air quality of the city is described. It is shown that particulate matter and ozone are the pollutants that exceed the limit fixed by the local government.

It is continued by presenting the environmental regulations that the local government has adopted in order to control the air pollution of the city. It is emphasised that these environmental regulations have been strongly influenced by the regulations stipulated by the US-EPA. Later a description of the main sources of emission of the city appears. It is concluded that the industries dedicated to food production and vehicles constitute the main source of pollutant emissions in the city.

Finally, the programs that have been implemented and those that will be implemented to control the air pollution are shown. It is shown that the fuel substitution by less-polluting fuels and the reorganisation of the public transportation have been the more successful air pollution control programs that have been implemented in the city.

1 Introduction

Bogota, the capital of Colombia, is at present the main geographic, financial, political, industrial and cultural centre of the country. It is seat of the national government and the most extensive city of Colombia. With more than 7.5 million inhabitants, Bogota concentrates 17% of the total population of the nation and has a population density of approximately 4800 inhabitants/km^2. It generates almost 22% of the Gross Internal Product (GIP) of Colombia and concentrates 35% of the industrial employment of the country.

Bogota is located at 4 degrees of North latitude, on the Eastern Cordillera of the Andes, at 2,650 metres of altitude above sea level. It presents an annual average temperature of 14°C, an annual average precipitation of 1013 mm and an annual average relative humidity of 52%. In the city, two rainy periods and two dry periods occur in the year. The dry periods occur between January and February and between June and September. Nevertheless, the climatologic phenomena that have occurred lately have caused variations in these periods.

Due to its climatologic characteristics, the pollution problem of Bogota has been increased in recent years. Due to this, the environmental authorities have designed and implemented diverse strategies to reduce and to prevent the polluting emissions in the city. This chapter describes these plans.

Initially, a description of the air quality of Bogota is shown. Later, it is shown the legislation adopted by the environmental authorities with the purpose of controlling the air pollution of the city. It continues with a description of the main polluting sources of the city. Finally, the programs that the environmental authorities have implemented and those that are going to be implemented to assure the air quality of the city are described.

2 State of the air quality in Bogota

The state of the air quality of Bogota is described below. Initially, the regulation of air quality adopted by the environmental authorities of the city is shown as a comparison point. Later the infrastructure used to monitor the air quality of the city is described. Finally, the average concentration values of the pollutants, measured at different points in the city are presented.

2.1 Air quality regulation

The Department of the Environment is the National environmental authority and therefore it is the governmental organisation ordered to establish the environmental regulation that prevails over the country. Nevertheless, each region relies on an environmental authority of a local type. These organisations can promulgate environmental regulations if they are more restrictive than the current are at the national level. In the case of Bogota, the Technical and Administrative Environmental Department (DAMA) is the environmental authority.

In 2000, DAMA updated the environmental regulation and established the air quality regulation that is shown in Table 1. Additionally, this regulation establishes the maximum permissible values for the inorganic substances contained in the particulate matter, in steam or gases and in carcinogenic substances.

This regulation also specifies the sampling and analysis methods that are used to determine the concentration of the pollutants regulated. Tables 2 and 3 show the specified methods. The regulation accepts new technologies as long as they count on the approval of the US–EPA. For the determination of the concentrations of lead, cadmium, mercury, arsenic and compounds of these inorganic substances contained in the particulate matter high volume particle sampling is used, continuously during 24 hours, taking four samples per month, and using the atomic absorption spectrophotometry for its analysis.

Table 1: Air quality regulation for Bogota.

Substance	***Period***	***Description***	***Unit***	***Year of application***			
				2001	***2003***	***2006***	***2010***
Carbon monoxide (CO)	8 hours	Hour concentration average in 8 hours	μg/m^3	13	12	11	10
	1 hour	Hour concentration average	μg/m^3	45	40	35	30
Sulfur oxides (SO_2)	Annual	Arithmetic average of the daily average concentrations in 365 days	μg/m^3	90	80	70	60
	24 hours	Hour concentration average in 24 hours	μg/m^3	375	350	325	300
	3 hours	Hour concentration average in 3 hours	μg/m^3	1450	1400	1350	1300
Nitrogen oxides (NO_2)	Annual	Arithmetic average of the daily average concentrations in 365 days	μg/m^3	100			
	24 hours	Hour concentration average in 24 hours	μg/m^3	230	220	210	200
	1 hour	Hour concentration average	μg/m^3	320			
Ozono (O_3)	8 hours	Hour concentration average in 8 hours	μg/m^3	130	120	110	100
	1 hour	Hour concentration average	μg/m^3	165	160	155	150
Total Suspended Particulate (TSP)	Annual	Geometric average of the daily average concentrations in 365 days	μg/m^3	95	90	85	80
	24 hours	Hour concentration average in 24 hours	μg/m^3	340	300	280	260
Particulate Matter lower than 10 microns (PM10)	Annual	Geometric average of the daily average concentrations in 365 days	μg/m^3	65	60	55	50
	24 hours	Hour concentration average in 24 hours	μg/m^3	170	160	155	150

Table 2: Semiautomatic methods used for the taking of samples and analysis of the pollutants.

Pollutant	***Sample taking***	***Analysis method***
Suspended Particulate	High volume	Gravimetric
Sulfur dioxide	Gas impinger	Colorimetric (ParaRosanilina)
Nitrogen dioxide	Gas impinger	Colorimetric (NEDA)
PM_{10}	High volume classifier	Gravimetric

Table 3: Automatic methods used for the taking of samples and analysis of pollutants.

Pollutant	*Sample taking*	*Analysis Method*
Suspended Particulate	Analyzer	Beta rays absorption
Sulfur dioxide	Analyzer	Conductimetry
Nitrogen dioxide	Analyzer	Quimioluminescence
Ozone	Analyzer	Quimioluminescence
Carbon monoxide	Analyzer	Non-disperse infrared
Hydrocarbons	Analyzer	Flame ionisation

2.2 Air quality network of Bogota

Bogota has had air quality networks since 1983. At the beginning it used manual equipments for the measurement of the pollutants. Only in 1996 was the present air quality network of Bogota (RMCAB) installed, which is composed of 14 automatic stations. Figure 1 shows the location of the stations and the equipment whereupon it counts. Figure 2 shows the meteorological equipment installed in each one of the stations.

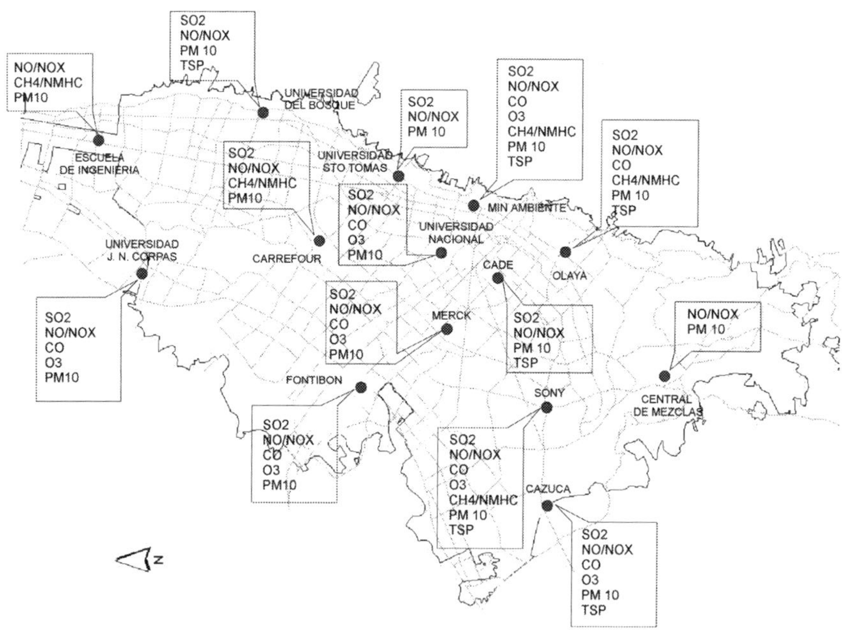

Figure 1: Location of the RMCAB stations in the city of Bogota. Shown is the air quality equipments of each station.

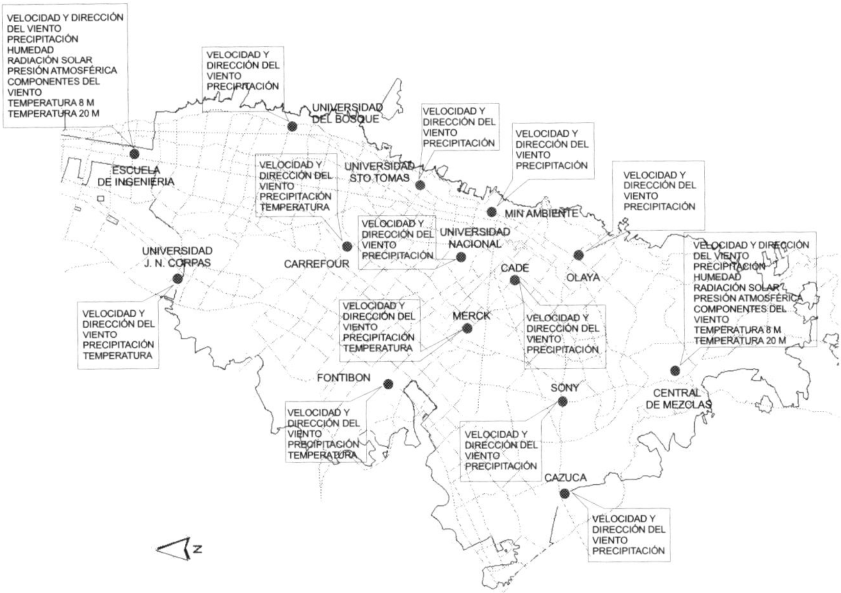

Figure 2: Meteorological equipments installed in the stations of the RMCAB.

Table 4 presents a comparison between the air quality networks of Bogota, Santiago de Chile and Mexico City. This table and the guidelines established by the US-EPA about the number of stations required to evaluate the exposure levels of urban populations [1], show that the number of stations of the RMCAB is adequate to evaluate the general exposure of the people in Bogota to the air contamination levels presented in the city.

Table 4: Comparison between the number of stations of the air quality networks of Bogota, Santiago de Chile and Mexico City.

City	*Stations*	*Surface area (Km^2)*	*Population (Million inhabitants)*
Bogota	14	1732	10 [2]
Santiago de Chile [3]	8	1400	6
Mexico City [4]	32	4607	25

Table 5 describes the technology used by the air quality analyzing equipments used in the stations of the RMCAB. It is shown that the EPA acknowledges the equipments used for the criteria pollutants (SO_2, NOx, O_3, CO and PM_{10}), except the particulate matter meters. Nevertheless, these meters use the Beta attenuation method, which has been homologated by the EPA in equipments similar to the equipments used by the network.

Table 5: Description of the technology used by the air quality equipments of the RMCAB.

Criteria pollutants	***Detection Method***	***Automatic***	***Acknowledged by EPA***
CO	IR Correlation	Yes	Yes
NO_2/NO_X	Quimioluminescence	Yes	Yes
SO_2	UV fluorescence	Yes	Yes
PM_{10} and TSP	Beta attenuation	Yes	No
O_3	UV fluorescence	Yes	Yes

Similarly, Table 6 describes the detection methods employed by the RMCAB meteorological sensors. It is observed that these are the sensors commonly used for these purposes.

Table 6: Description of the principle of operation used by the meteorological equipments that are used in the RMCAB.

Sensor	***Detection Method***	***Type***
Temperature	Electrical resistance	Analog
Humidity	Capacitor	Digital
Wind direction	Vane	Analog
Wind speed	Rotating casserole anemometer	Digital
Precipitation	Tilted deposit	Digital
Solar radiation	Thermo-electrical	Digital
Atmospheric Pressure	Transducer	Digital
Wind component	Helix	Analog

The above information concludes that the RMCAB has a technological infrastructure adapted to evaluate the fulfillment of the air quality standards and the progress obtained in this sense; to activate control procedures in order to prevent or to alleviate contamination episodes; to provide databases for research and evaluation of the urban effects, the land and transportation use and to provide bases for planning, development and evaluation of control strategies, development and validation of diffusion models.

2.3 Air quality in Bogota

Annually, DAMA presents a report with the average criteria pollutants concentration and the meteorological variables measured by the RMCAB. Tables 7 and 8 show these values. The shaded values correspond to the values that surpass the actual regulation.

Table 7: Annual average concentration of sulfur and nitrogen oxides in Bogota [5].

Sector	Station	SO_2 (µg/m³)				NO_2 (µg/m³)			
		1998	1999	2000	2001	1998	1999	2000	2001
South west	Sony	77	51	72	43	36	31	29	94
	Cazuca	48	37	27	45	34	27		
Centre west	Cade	77	69	77	67	34	15	15	17
	Merck		52	69	35		51	38	36
South	Olaya	29	59	27		31	55	36	
Centre	MMA	13	29	13	11	27	54	40	38
	UNAL	27	24	27	24	36	40	46	44
	U. Tomas	21	16	16	21	44	38		57
North	Bosque	11	8	8	27	25	15	17	
North West	Corpas	18	10	11	16		12		
	Carrefour			4	24			7	31
	Fontibon		29	58	24		37	29	27

Table 8: Annual average concentration of particulate matter smaller than 10 microns in Bogota [5].

Sector	Station	PM_{10} (µg/m³)			
		1998	1999	2000	2001
South west	Sony	**70**	**68**	58	**86**
	Cazuca	**69**	57	52	62
Centre West	Cade	62	57	44	54
	Merck			**87**	**93**
South	Olaya	54	47	44	
Centre	MMA	39	45	39	49
	UNAL				
	U. Tomas	21	23	21	37
North	Bosque	26	24	24	29
North West	Corpas	46	34	42	**81**
	Carrefour				**78**
	Fontibon			**82**	**91**

As shown in Tables 7 and 8, the air quality levels do not present a clear tendency to decrease or increase in previous years. This is caused by the interaction between the measures adopted by the local authority to improve the air quality, the climatic phenomena that occurred in previous years and the economic recession that has affected the country.

In Bogota there are no clearly defined industrial zones. Generally, around the industrial zones there are residential zones. That is why it is necessary to take a strict control of the different pollutants concentration. Figures 3 to 7 show the distribution of the contamination levels of the different criteria pollutants in Bogota for the year 2001 [5]. It is shown that in the zones of greater industrial activity (North-western and South-western) there are high levels of particulate matter and ozone.

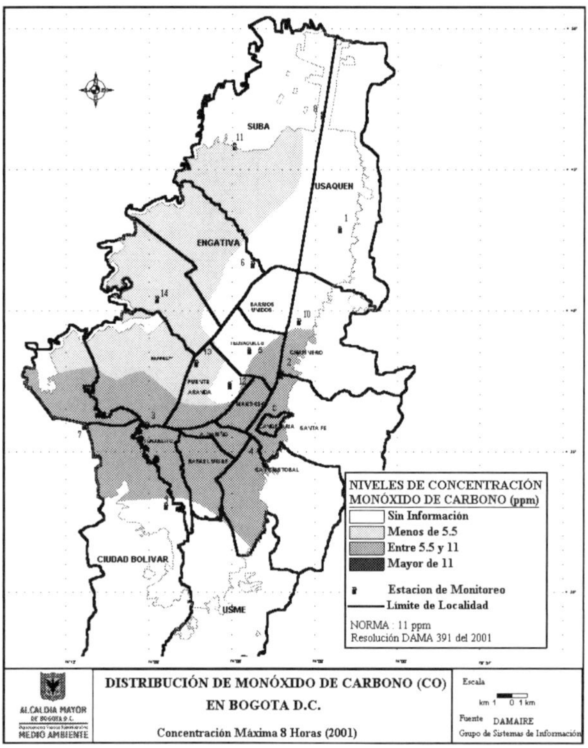

Figure 3: Spatial distribution of the average concentration for 8 hours of the carbon monoxide (CO) in 2001 [5].

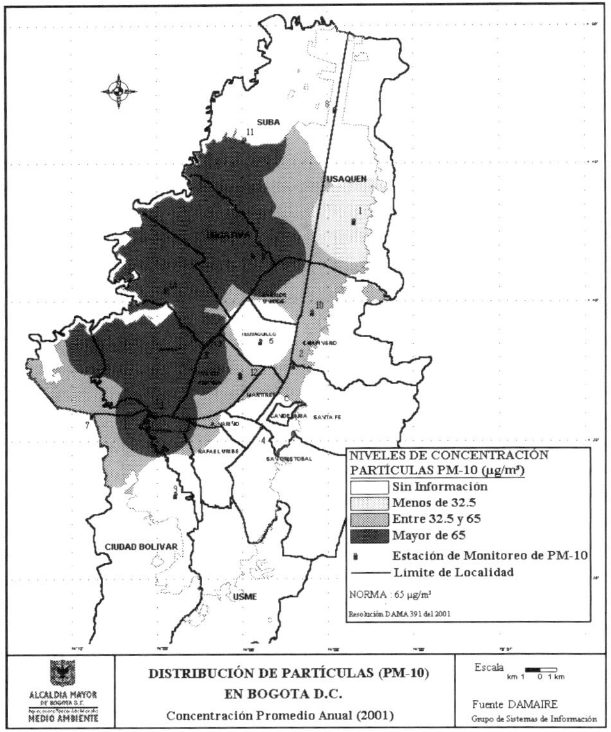

Figure 4: Spatial distribution of the annual average concentration of particulate matter smaller than ten microns (PM_{10}) in the year 2001 [5].

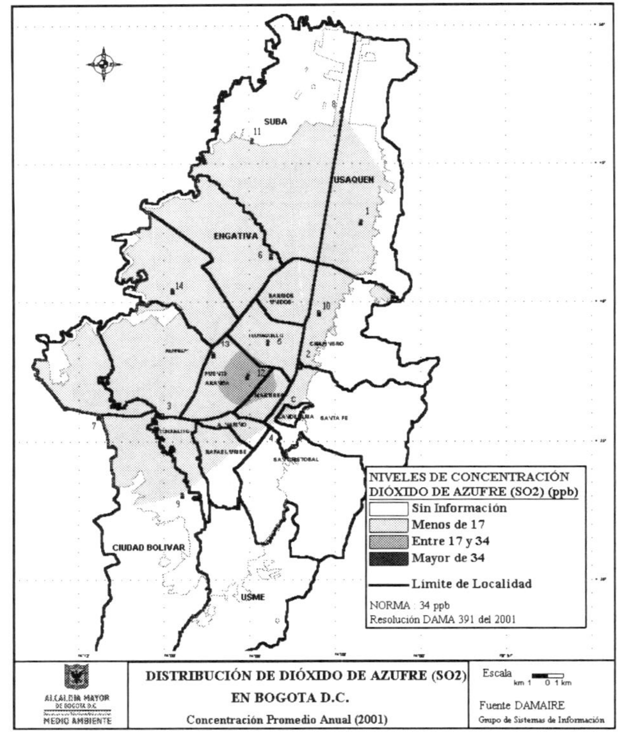

Figure 5: Spatial distribution of the annual average concentration of sulfur oxides (SO_2) in the year 2001 [5].

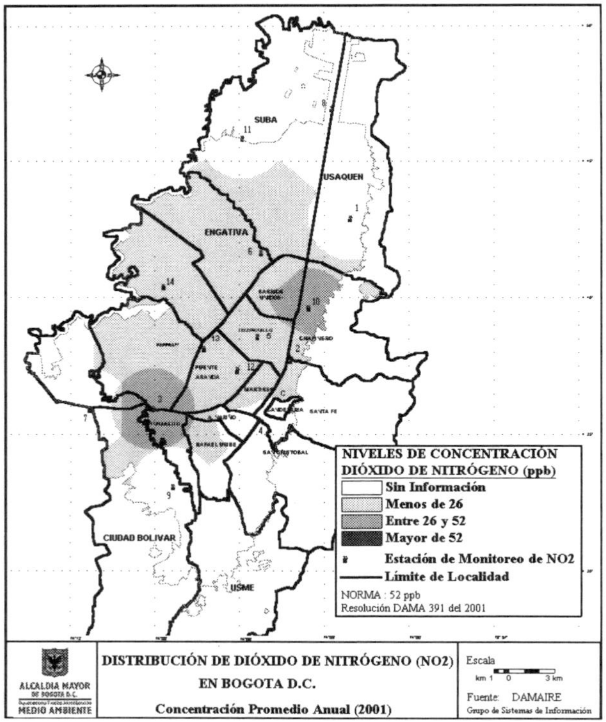

Figure 6: Spatial distribution of the annual average concentration of nitrogen oxides (NO_2) in the year 2001 [5].

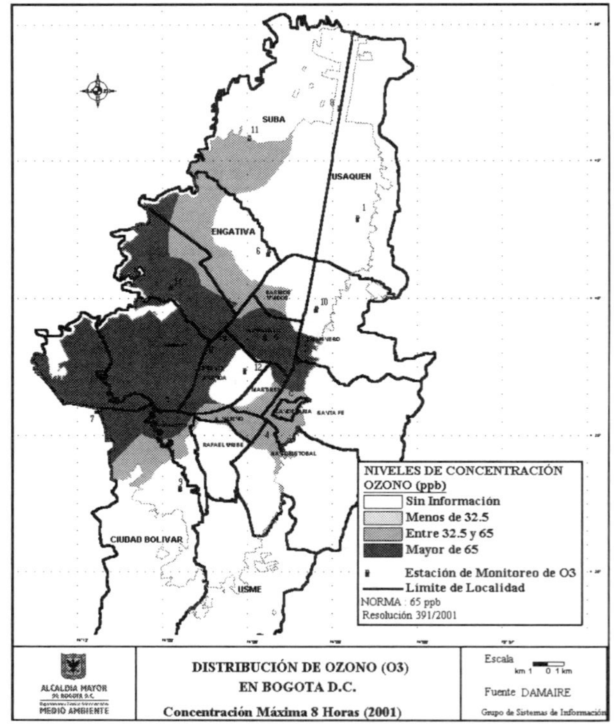

Figure 7: Spatial distribution of the maximum concentration for 8 hours of ozone (O_3) in the year 2001 [5].

2.4 Air Quality Index for Bogota

With the purpose of informing to the community, in a simple way, about the air quality of the city, DAMA designed the air quality index of Bogota (IBOCA). This index evaluates qualitatively the air quality according to the concentration of the criteria pollutants measured by the RMCAB. It has five categories that range from Good to Dangerous, depending on the proximity of the measured concentration to the limits established in the air quality regulation.

The city is divided into five sectors as is shown in Fig. 8. For each sector a daily IBOCA is calculated. Figure 9 presents the predominant pollutants in the different sectors of the city during year 2001. Figure 10 shows the categories in which IBOCA stood in the different sectors of the city during the year 2001.

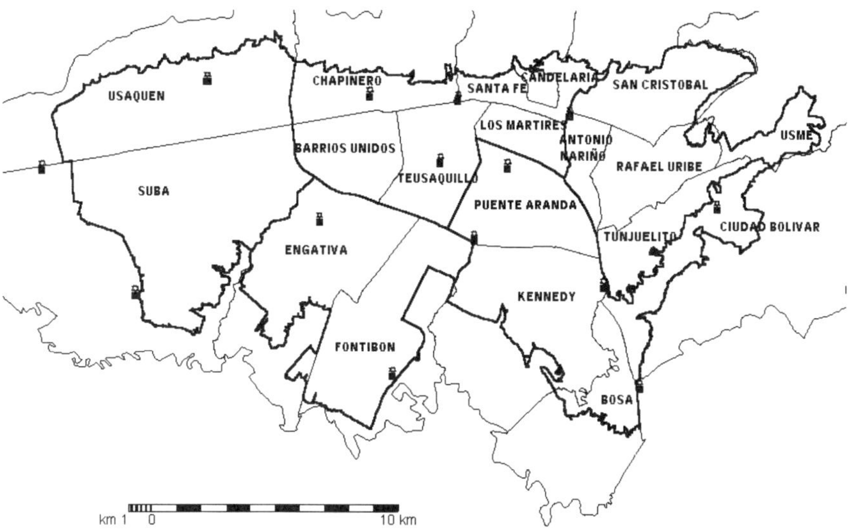

Figure 8: Division of the city by sectors and monitoring stations used to calculate IBOCA.

From Fig. 9, it is possible to observe that during the year 2001 the critical pollutant was the particulate matter smaller than ten microns (PM_{10}), with exception of the centre zone where there is a combination of pollutants that include ozone, nitrogen oxides and particulate matter. The high vehicular circulation in the zone explains this occurrence. In general, the air quality in all of the zones of Bogota remains more than a 50% of the time in the Unhealthy qualification of IBOCA. This corresponds with a rank between 25% and 75% of the air quality regulation for Bogota.

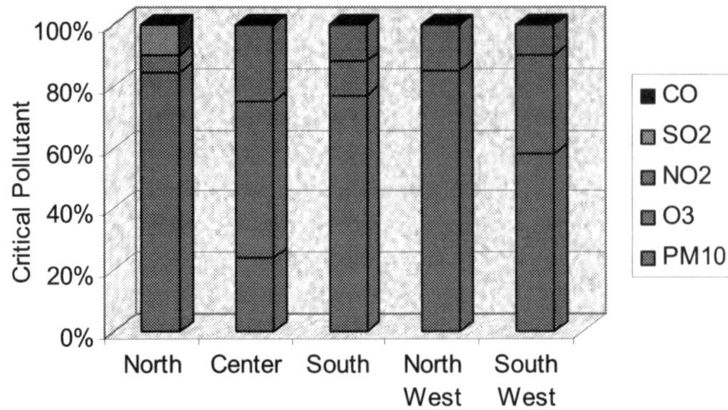

Figure 9: Pollutants that predominated in the different sectors of the city during the year 2001 according to IBOCA classification [6].

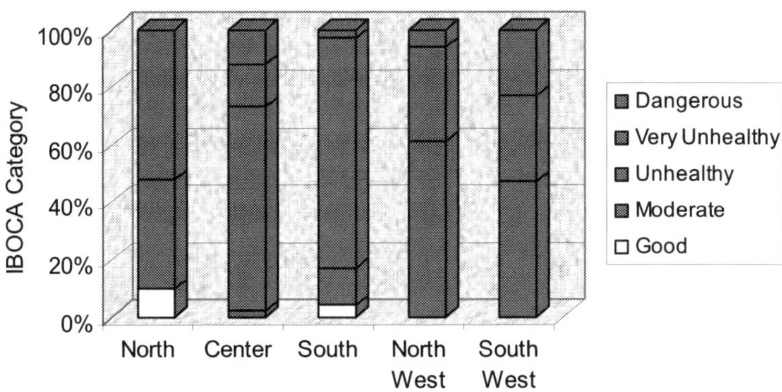

Figure 10: Categories in which IBOCA stood in the different sectors of the city during the year 2001 [6]

From Fig. 10 it is possible to observe that the North-western and South-western sectors of the city (where there is a concentration of industry) remained, an important percentage of the year (6% and 23% respectively), exposed at levels of concentration over the established maximum levels in the air quality regulation.

Similarly, the centre zone was exposed for 12% of the year at levels of concentration of pollutants over the established in the regulation. Nevertheless, in this case the contamination is associated to the movable sources since this zone corresponds to a zone of little industry and high vehicular flow.

3 Regulatory framework

In order to prevent air pollution of the city, the environmental authorities have imposed restrictions on the type of fuel that can be used in industry and automotive vehicles. They have also imposed restrictions for the maximum levels of pollutants concentration that the industrial sector and the automotive vehicles can emit to the atmosphere. These restrictions have been imposed by implementing environmental regulations. These regulations have been influenced by similar regulations established by the US-EPA. The current environmental regulations for the city of Bogota are synthesised below.

3.1 Fuels

3.1.1 Fuels for vehicles

The fuels used in Colombia for automotive vehicles are gasoline, diesel and lately natural gas. Most of these fuels are extracted and/or refined within the country.

Table 9 presents the specifications of the Colombian gasoline. Two types of gasoline with different octane number are specified. It is compulsory for use of dispersing additives, detergents and deposits controller in the admission system, including the valves. The use of additives containing heavy metals is prohibited as well as additives containing polyaromatic hydrocarbons and hydrocarbons with chains of more than fourteen (14) carbons as diluents. The imported gasoline to be distributed within the national territory, without previous preparations, such as mixtures or other that modify their characteristics, must fulfill the quality requirements specified in Table 9.

Table 9: Specifications of the Colombian gasoline [7]

Parameter		***Unit***	***Date of use***		
			1996	***2001***	***2006***
Minimum Octane Number	Regular Gasoline	(RON+MON)/2 Index	81	81	81
		(RON) Index	86	86	86
	Extra Gasoline	(RON+MON)/2 Index	86	86	86
		(RON) Index	94	94	94
RVP, maximum		Psia	8.5	8.1	8.1
Sulfur, maximum		% Weight	0.10	0.05	0.03
Oxygen, minimum		% Weight	-	2.0	2.0
Aromatics, maximum		% Volume	28.0	25.0	25.0
Benzene, maximum		% Volume	1.1	1.0	1.0

Table 10 presents the specifications of the diesel fuel that is distributed in the country. The additives containing heavy metals are also prohibited. As is shown in this table, the sulfur content of the diesel used in Bogota requires higher specifications than the diesel used in the rest of the country.

Table 10: Specifications of the Colombian diesel [7].

Property	***Units***	***Bogota***		***Rest of the country***	
		2001	***2005***	***2001***	***2005***
Sulfur, maximum	% Mass	0.12	0.05	0.45	0.05
Aromatics, maximum	% Volume	35	35	35	35
Cetane index, minimum		45	45	45	45
Copper corrosion, 3h @50°C, maximum	Classification	2	2	2	2
ASTM Color, maximum		2.0	2.0	3.0	3.0
Micro coal remainders, maximum (10% bottom)		0.20	0.20	0.20	0.20
Viscosity @ 40 °C, Minimum – maximum	mm^2/s	1.9–4.1	1.9–4.1	1.9–5.0	1.9–5.0
Water and sediment, maximum	% Volume	0.05	0.05	0.05	0.05
Fluidity point, maximum	°C	4	4	4	4
Flash point, minimum	°C	52	52	52	52
Ash, maximum	% Mass	0.01	0.01	0.01	0.01

The use of natural gas in automotive vehicles in the city began recently (~5 years ago). Therefore, the local government has not even established a regulation on the specifications that must fulfill this fuel. The Colombian organisation for normalisation (ICONTEC) has established a technical normative with the requirements on the natural gas quality that must provide the service station to the vehicles in order to be operated successfully [8]. Table 11 shows the typical characteristics of the natural gas used for automotive vehicles.

3.1.2 Fuels for industrial use

The fuels mainly used in Colombia for industrial use are fuel oil, diesel, coal, natural gas and LPG.

The environmental legislation regulates the maximum sulfur content for the fuel oil that is distributed in the country for consumption in boilers and furnaces of industrial and commercial use, as well as, in heat or energy generation as shown in Table 12. Additionally, the sulfur content of the emulsions or suspensions of heavy oils in water or any other liquid, or those of mineral coal in water, hydrocarbon, or in any other liquid, should not be greater than 1.7% in weight, when they are used in furnaces or boilers within the country.

It would be possible to employ the used oils as unique fuel or mixed with other types of fuels, in any proportion, in furnaces or boilers with an installed thermal power equal to or greater than 10 megawatts. For boilers or furnaces with a thermal power smaller than 10 megawatts, the used oil will be able to be used whenever it is mixed with other fuels, in a proportion smaller or equal to 5% in volume of used oil.

Table 13 shows the composition of the fuels typically used by the industries in Bogota and its physical properties.

Table 11: Specifications of the natural gas used for automotive vehicles [9].

Substance	***Volumetric composition of the natural gas [%]***								
	Guajira	***Apiay***	***Cusiana***	***Opón***	***Payoa***	***Güepajé***	***Neiva 540***	***Monta-ñuelo***	***Cerrito 1***
Methane	97.76	78.46	76.55	91.86	95.68	96.94	91.78	96.17	93.98
Ethane	0.38	14.06	10.86	5.52	4.20	0.55	2.77	0.70	0.63
Propane	0.20	1.96	5.36	1.32	0.12	0.15	2.28	0.24	0.02
i-Butane	0.00	0.48	0.68	0.12	0.00	0.07	0.54	0.05	0.01
n-Butane	0.00	0.45	0.78	0.05	0.00	0.03	0.90	0.00	0.00
i-Pentane	0.00	0.41	0.13	0.00	0.00	0.02	0.45	0.00	0.00
n-Pentane	0.00	0.09	0.08	0.00	0.00	0.01	0.21	0.00	0.00
Hexane	0.00	0.00	0.05	0.00	0.00	0.01	0.00	0.00	0.00
Heptanes	0.00	0.00	0.00	0.01	0.00	0.07	0.20	0.00	0.00
Nitrogen	1.29	1.69	0.44	0.15	0.00	2.12	0.63	0.42	0.18
Carbon dioxide	0.37	2.40	5.07	0.96	0.00	0.03	0.24	2.42	5.18
Water	0.00	0.00	0.00	0.01	0.00	0.00	0.00	0.00	0.00
Physical properties									
High heating value [kJ/kg]	55368	51513	49281	55141	56637	55257	55838	52851	49777
Molecular weight	16.41	20.12	21.31	17.56	16.67	16.55	18.36	16.96	17.61
Density [kg/m^3]	0.67	0.83	0.88	0.72	0.69	0.67	0.75	0.70	0.72

** Composition and properties could change with time*

Table 12: Sulfur content of the fuel oil.

Parameter	***Units***	***Date of use***		
		1997	***2001***	***2006***
Sulfur, maximum	% Peso	1.7	1.5	1.0

Table 13: Composition, physical and chemical properties of the fuels used in Colombia [10–14].

Element	***Ultimate analysis of the fuels % Weight***				
	Natural Gas*	***Fuel Oil****	***Diesel****	***Coal*****	***LPG****
C	73.07	83.88	87.3	79.89	82.09
H2	23.65	11.19	12.5	5.45	17.71
O_2	0.00	0.16	-	3.85	0.00
N_2	3.28	0.54	-	1.36	0.00
S	0.00	3.71	0.21	0.32	0.00
H_2O	0.00	0.50	-	4.24	0.00
Ash	0.00	0.00	0.0025	4.88	0.00
Others	0.00	0.02	-	0.01	0.20
Physical properties					
Net Heating Value	55291 [kJ/kg]	48619 [kJ/kg]	44729 [kJ/kg]	25532 [kJ/kg]	47438 [kJ/kg]
	35315 [BTU/m^3]	150000 [BTU/gallon]	138000 [BTU/gallon]	24200 [BTU/Kg]	92000 [BTU/gallon]
Density [kg/m^3]	0.7	860	860	N/D	2.1

* *Composition of the fuels according to the National Commission of Energy Saving of Mexico*

** *Composition of the coal found in the Arrayan mine in the municipality of Lenguazaque (Cundinamrca). The composition of the coal can vary considerably depending on the extraction zone.*

3.2 Stationary sources

The environmental regulation for Bogota controls the maximum permissible emission levels of pollutants in stationary sources by type of fuel used. Tables 14, 15 and 16, respectively, show the emission levels for sources of external burning from solid fuels (firewood, lignites, soft coals, anthracite, mineral coal, coke), liquid fuels (diesel, fuel oil) and gaseous fuels (methane, ethane, propane, butane, natural gas, gas of refinery, gas of high furnace, biogas or mixtures of these). Table 17 shows the regulation for asphalt plants emission. Additionally, the regulation contemplates emission levels for incinerators and crematoria furnaces.

Tables 14 to 17 show that the regulation looks for a decrease in the emission levels of pollutants in a horizon of 10 years. Indirectly, it seeks that the companies use cleaner fuels instead of investing in technology of emissions control. In the opinion of the authors, the strategy must be more aggressive and impose the same maximum allowed emission levels for all type of fuels after 5 years.

Table 14: Emission levels for sources of external burning from solid fuels.

Pollutant	***Concentration [mg/m^3 N]***			
	2001	***2005***	***2008***	***2011***
Particulate matter (TSP)	300	200	150	100
Carbon monoxide (CO)	300	280	260	250
Nitrogen oxides (NO_2)	400	350	300	250
Sulfur oxides (SO_2)	600	500	450	400
Heavy metals	0.8	0.7	0.6	0.5
HCl	300	270	240	200
HF	30	25	20	15

Table 15: Emission levels for sources of external burning from liquid fuels.

Pollutant	***Concentration [mg/m^3 N]***			
	2001	***2005***	***2008***	***2011***
Particulate matter (TSP)	300	200	150	100
Carbon monoxide (CO)	200	190	180	170
Nitrogen oxides (NO_2)	400	350	300	250
Sulfur oxides (SO_2)	600	500	450	400
Heavy metals	4.0	3.0	2.5	2.0
HCl	50	40	35	30
HF	8	7	6	5

Table 16: Emission levels for sources of external burning from gaseous fuels.

Pollutant	***Concentration [mg/m^3 N]***
	2009
Particulate matter (TSP)	100
Carbon monoxide (CO)	100
Nitrogen oxides (NO_2)	350
Sulfur oxides (SO_2)	35

Table 17: Emission levels for asphalt plants.

Pollutant	***Concentration [mg/m^3 N]***			
	2001	***2005***	***2008***	***2011***
Particulate matter (TSP)	1000	800	500	100
Carbon monoxide (CO)	200	190	180	170
Nitrogen oxides (NO_2)	400	350	300	250
Sulfur oxides (SO_2)	600	500	450	400

3.3 Mobile sources

With the purpose of controlling the air pollution originated for mobile sources, the environmental authorities have promulgated two types of regulations.

The first one is directed to new vehicles. In this regulation, the maximum emission values measured under dynamic tests, following the established protocols for such aim by the US-EPA or the European Union, are specified. Tables 18 and 19 show the allowed maximum values. However, the city does not count on the infrastructure necessary to make these tests and to verify the fulfillment of the regulation. That is why, it is only demanded that the manufacturer (assembler or importer) demonstrates the fulfillment of the regulation by means of a written certification from a recognised laboratory that has made the test to an engine or prototype vehicle of the model that is imported or assembled in the country.

The second type of regulation is directed to used vehicles and seeks that the owner maintains his vehicle in good state*. For the gasoline vehicles, there are maximum emissions levels of CO and HC measured on idle, depending on the model of the vehicle. Table 20 shows these values. For diesel vehicles, maximum levels of opacity measured under free acceleration conditions are specified, based on the model of the vehicle. Table 21 shows these values.

Table 18: Maximum permissible emission levels of pollutants for vehicles when they are evaluated following cycles FTP–75 or USA–13.

<table>
<tr><th rowspan="2">Model</th><th rowspan="2">Category</th><th colspan="3">Permissible emissions (g/km)</th></tr>
<tr><th>CO</th><th>HC</th><th>NO_x</th></tr>
<tr><td rowspan="3">>1998</td><td>Light</td><td>2.10</td><td>0.25</td><td>0.62</td></tr>
<tr><td>Medium</td><td>11.2</td><td>1.05</td><td>1.43</td></tr>
<tr><td>Heavy</td><td>25.0</td><td colspan="2">10.0 g/HP-h</td></tr>
</table>

Light vehicle: *Passengers <12*
Medium vehicle: *GVW < 3.860 kg, passengers >12*
Heavy vehicle: *GVW > 3.860 kg*

Table 19: Maximum permissible emission levels of pollutants for vehicles when they are evaluated following cycles ECE-15+EUDC.

<table>
<tr><th rowspan="2">Category</th><th rowspan="2">Weight (kg)</th><th colspan="3">Emissions (g/km)</th></tr>
<tr><th>CO</th><th>HC</th><th>NO_X</th></tr>
<tr><td>M1</td><td>-</td><td>2.72</td><td colspan="2">0.97</td></tr>
<tr><td rowspan="3">M1, N1</td><td><1250</td><td>2.72</td><td colspan="2">0.97</td></tr>
<tr><td>>1250
<1700</td><td>5.17</td><td colspan="2">1.40</td></tr>
<tr><td>>1700</td><td>6.90</td><td colspan="2">1.70</td></tr>
<tr><td>N2, N3, M2, M3</td><td>-</td><td>11.20*</td><td>2.40*</td><td>14.40*</td></tr>
</table>

**g/kW-h measured by following cycle ECE-15*

Table 20: Maximum permissible emission levels of pollutants for gasoline vehicles, evaluated according to the vehicle model.

Model year	*CO [%]*	*HC [ppm]*
2001 and later	1.0	200
1998 – 2000	2.5	300
1996 – 1997	3.5	450
1991 – 1995	4.0	500
1981 – 1990	5.0	650
1975 – 1980	5.5	800
Previous to 1974	7.0	1000

Table 21: Maximum permissible emission levels of pollutants for diesel vehicles under free acceleration conditions, evaluated according to the vehicle model

Model year	*Opacity [%]*		
	Light vehicle	*Medium vehicle*	*Heavy vehicle*
2001 and later	40	40	40
1996 – 2000	50	50	50
1991 – 1995	50	50	50
1986 – 1990	55	55	55
1981 – 1985	60	60	60
Previous to 1980	65	65	65

In order to verify the fulfillment of the second type of regulation, DAMA decided that the vehicles must go once a year to a Recognised Diagnosis Centre (CDR) with the purpose of obtaining their Emissions Certificate. These centres are private automotive laboratories that the environmental authority recognises as suitable for issuing the emissions certificates as they have the infrastructure and personnel with the qualification necessary to do this work suitably.

4 Emission sources

The main air pollution sources of Bogota are the industries and the vehicles. These emission sources are described below.

4.1 Stationary sources

4.1.1 Characterisation of the stationary sources

In the 2001, DAMA made a characterisation study of the stationary sources [15] in Bogota. In this study, around 4500 industries with more than 9400 emissions points were identified.

These emission points are divided into point sources (chimneys), fugitive emissions that imply the presence of particulate matter or gases, evaporative sources that involve volatile organic compounds and area sources that refer to the storage of materials outdoors.

The industries were classified in 12 economic activities as shown in Table 22. This table shows the economic activity, the number of industries and the participation of each economic activity.

Table 22: Classification of the industries in Bogota by economic activity.

Economic activity	*Number of industries*	*Participation [%]*
Extraction of non metallic minerals	3	0.1
Manufacture of dress articles	50	1.1
Manufacture of textile products	63	1.4
Manufacture of metallurgical products	71	1.6
Manufacture of non metallic products	91	2.0
Manufacture of chemical products	184	4.1
Manufacture of footwear	194	4.3
Manufacture of rubber and plastic products	213	4.7
Manufacture of furniture and others	305	6.8
Manufacture of food	663	14.8
Hotels, bars, restaurants and similar	1125	25.1
Other: laundry services, health, education, printing, engineering activities	1528	34.0
Total	**4490**	**100.0**

From Table 22 it is possible to observe that the more predominant economic activity corresponds to hotels, bars and restaurants, followed by food manufacture and the manufacture of furniture and others.

From this study it was concluded that the most used process in the industries of Bogota was baking, followed by steam production and water and oil heating, welding and painting, polishing and cutting. The most used equipment in the industrial processes includes furnaces, boilers, equipment for the food baking, welding equipment and equipment to cut and to polish. Table 23 shows that the fuel mainly used by the equipments previously mentioned is natural gas, followed by LPG, diesel, coal and wood.

Table 23 show that the equipments that use heavy liquid fuels, as fuel oil, participates with only 2%. Nevertheless, these equipments generally work 24 hours a day, every day of the week. Therefore, the distribution by fuel consumption is quite different from the presented in Table 23.

Table 23: Fuels used by the stationary emission sources in Bogota.

Fuel	*Number of equipments*	*Participation [%]*
Natural Gas	1265	32.1%
LPG	1019	25.9%
Diesel	992	25.2%
Coal and wood	570	14.5%
Heavy liquid fuels	80	2.0%
Gasoline	6	0.2%
Used oil	3	0.1%
Total	**3935**	**100.0%**

4.1.2 Polluting emissions by stationary sources

Using this study, reference [16] estimated the atmospheric emissions caused by the stationary sources of the city. This estimation was made using the emission factors of the US-EPA. Table 24 shows the results obtained. This table shows that the heavy liquid fuels, like fuel oil, cause the greater contamination by sulfur oxides, nitrogen oxides and particulate matter.

Table 24: Polluting emissions of the industries in Bogota.

Fuel	*Polluting emissions (Tons/year)**						
	SO_2	SO_3	NO_x	*MP*	*CO*	CO_2	*COT*
Natural Gas	0	0	37	2	9	0	0
LPG	0	0	1	0	0	514	0
Diesel	2184	80	1176	175	88	0	0
Coal	715	0	619	1489	7843	0	0
Wood	0	0	2	8	15	2368	0.25
Heavy liquid fuels	27580	1008	5467	3145	407	0	0
Used oil	0	0	0	20	0	0	0
Total	**30480**	**1088**	**7301**	**4839**	**8363**	**2882**	**0.25**

**Zero values refer to very small quantities*

4.1.3 Typical emission values

On the other hand, the authors developed an experimental study oriented to determine the typical levels of pollutant emissions by chimneys of the most used thermal equipments in the industries of Bogota [17].

In this study a representative sample of thermal equipments by type of fuel was selected. Steam boilers, thermal oil boilers, water heaters and furnaces were evaluated. Coal, fuel oil No. 2, fuel oil No. 6, Natural gas and LPG were included. Altogether, 26 companies were visited and 36 thermal equipments were evaluated.

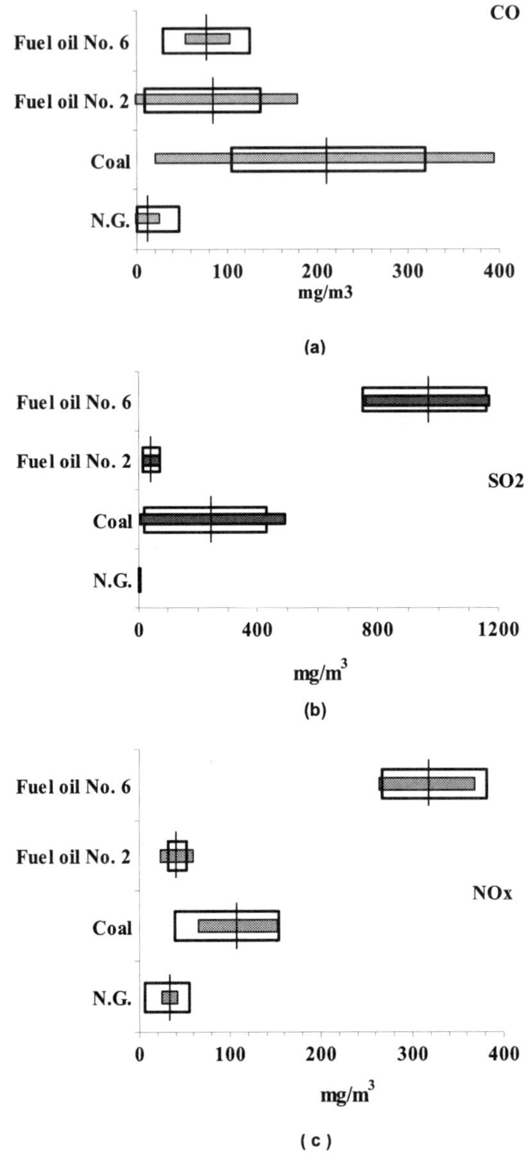

Figure 11: Comparison of concentrations of pollutants in the stack of fire tube boilers of different heating capacities. The box in the graphs indicates the maximum and the minimum values measured. The vertical line shows the average value obtained and the stuffed bar shows the rank where, with a confidence of 90%, it is expected to find the average of the population.

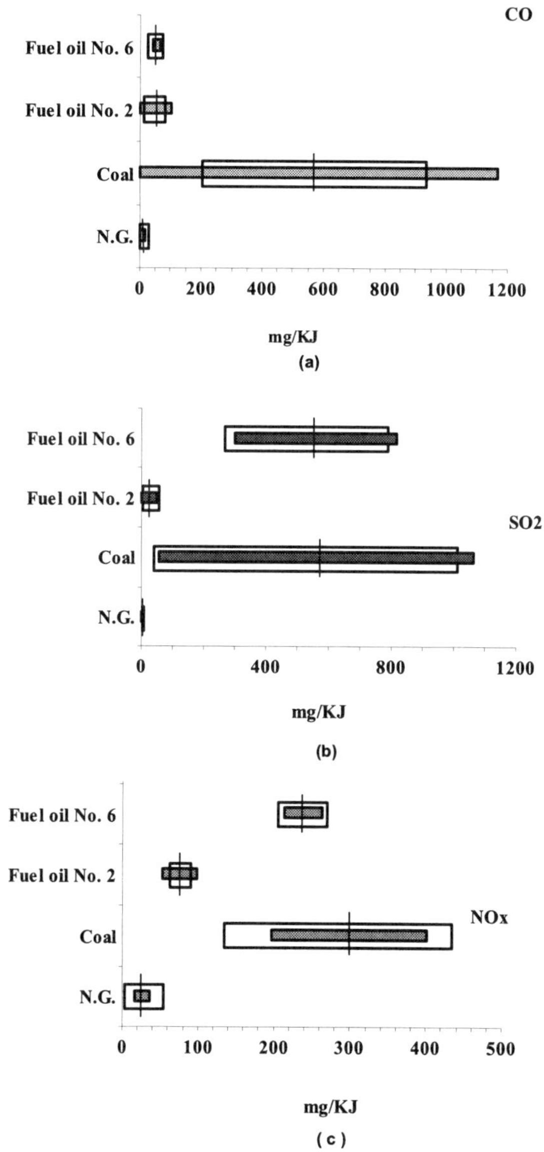

Figure 12: Mass emissions of pollutants by unit of energy used in the stack of fire tube boilers of different heating capacities. The box in the graphs indicates the maximum and the minimum values measured. The vertical line shows the average value obtained and the filled bar shows the rank where, with a confidence of 90%, it is expected to find the average of the population.

It was determined that considering the fuel consumption, the steam boilers are the thermal equipment most used in industry.

Most thermal equipments fulfill the actual environmental regulation. Nevertheless, the thermal equipments that work with heavy solid or liquid fuels show emission levels that are close to the maximum value allowed.

Figure 11 show the results obtained for the case of steam boilers in terms of the pollutants concentration as measured in the stack. It is observed that coal and fuel oil are the fuels that on average have the higher emissions of contaminants.

However, in order to compare pollutants emissions of different fuels on the same basis, the emissions have to be expressed in terms of mass per unit of energy released. Figure 12 shows the average values of mass emissions of pollutants found in fire tube boilers of different power in the city of Bogotá. It shows that, when the steam boilers operate with natural gas, the mass emissions of CO, SO_2 and NOx by unit of energy used are lower than when they operate with fuel oil No. 2, fuel oil No. 6 and coal in that order, respectively.

From these results it is possible to conclude that the thermal equipments that use heavy liquid and solid fuels in intensive form are the main contamination sources of the Bogota city. Nevertheless, industry continues using these fuels. The study of reference [17] showed that for the conditions established in 2002, it is cheaper to use coal in the thermal equipments than any other type of fuel and that the costs of operating thermal equipments with fuel oil No. 6 and natural gas are similar. Nevertheless, the industry will not make a massive decision by less polluting fuels while there are not greater economic incentives or a more restrictive environmental legislation.

4.2 Contamination originated in the ways of the city

In the 2001, DAMA estimated the particulate matter emission generated in the streets of the city, caused by the Aeolian erosion, precipitations, the sweeping of streets, the decomposition of the pavement, depositions caused by the automotive vehicles, dust fall, mud transport and spillage of substances. To determine the polluting load in the paved ways (good conditions) and not paved ways (regular or bad conditions), the methodology and the emission factors of the US-EPA were considered. Table 25 presents the obtained results.

Table 25: Estimation of the particulate matter emission in the streets of Bogota city, using the emission factors of the US–E PA.

Ways	***Particulate matter (Tons/year)***
Good conditions	132745
Regular conditions	133616
Bad conditions	206470
Total	**472831**

4.3 Mobile sources

DAMA also estimated the pollutants emission generated by the vehicles of Bogota [16]. The emission factors established by the Japanese International Cooperation Agency (JICA) [17] and those from the US–EPA were considered.

4.3.1 Characterisation of the vehicles in the city

For the estimation, the composition of the vehicles that circulate around the city was used. According to data of the Secretary of Transit and Transport of Bogota, in 2001 there were more than 650,000 vehicles registered in the city, of which more than 79,000 were used for public transportation [18]. In addition to the vehicles registered in Bogota, an additional 35% circulate daily in the city (~230,000 vehicles) that corresponds to the vehicles registered in the neighboring municipalities and other cities of the country [19, 20]. The composition of the vehicles registered in the city is shown in Table 26.

Table 26: Composition of the vehicles registered in Bogota.

Type of Vehicle	*Number of private vehicles*	*Number of vehicles used for public transportation*
Automobile	551236	51770
Bus, shuttle and van	9061	23755
Truck and light truck	13047	3725
TOTAL	**573344**	**79250**

Table 27. Estimation of pollutant emissions by mobile sources in Bogota

Type of vehicle	*Number of vehicles*	*Pollutant emissions [Tons /year]*					
		SO_x	**NO_x**	**CO**	**PM**	**HC**	**Total**
Automobile	549350	1623	14470	523475	43138	672	583378
Large size bus	22525	701	3786	165735	9630	339	180191
Medium size bus	15364	446	2408	105394	6124	216	114587
Trucks	41923	1471	7364	295103	15805	397	320141
4x4	102669	257	2292	82922	6833	107	92411
Pick up	124743	326	2908	105194	8669	439	117536
Small size bus	8713	186	1034	38365	2204	41	41830
Truck tractor	2762	96	481	19261	1032	26	20895
Dump truck	4771	79	429	18759	1090	38	20395
Others	228	10	81	3349	285	12	3737
TOTAL	**873048**	**5195**	**35252**	**1357557**	**94809**	**2289**	**1495102**

4.3.2 Pollutant emissions by mobile sources

Additionally, in the emission estimation by mobile sources, it was assumed for the vehicles an average speed ranging from 15 to 25 km/h and an annual traveled distance of 20,000 kilometres for private vehicles and 60,000 kilometres for vehicles used for public transportation. Results are shown in Table 27.

Table 27 shows that the main pollutant emitted by the mobile sources is the carbon monoxide with a participation of 91% in the total emission of the mobile sources. On the other hand, the automobile is the type of vehicle that, by its quantity, produces the greater amount of pollutants (39%). Nevertheless, the citizenship perceives that buses are the type of vehicles that produce more contamination. This observation agrees with the reported in table 27.

4.4 Participation of the stationary and mobile sources in the air pollution of Bogota

Based on the results shown in the previous sections, Fig. 13 shows the relative contribution of each pollutant in the air contamination of Bogota, considering only the stationary and mobile sources. It is observed that CO constitutes the 88.3% of the air pollution of Bogota. Figure 14 shows the participation of the stationary and mobile sources in the pollutant emissions in Bogota. The value in parenthesis corresponds to the total emitted tons. It is shown that the mobile sources formed by ~800,000 vehicles, generate 99% of the carbon monoxide emissions, 100% of hydrocarbons and 83% of nitrogen oxides, whereas the stationary sources formed by ~4000 sources contribute 86% of sulfur oxides and 67% of particulate matter.

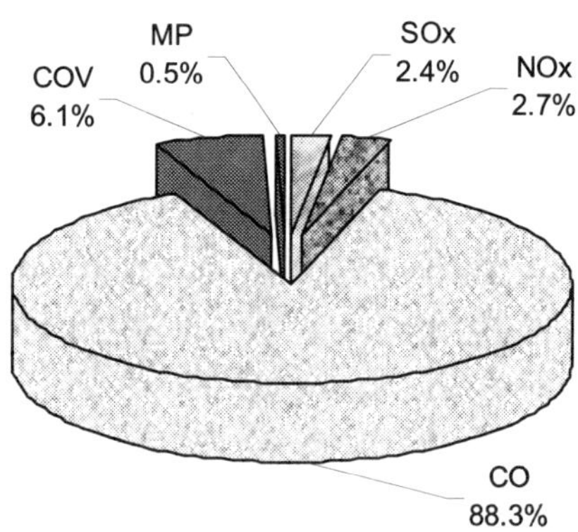

Figure 13: Relative contribution of each pollutant in the air pollution of Bogota, considering only the stationary and mobile sources.

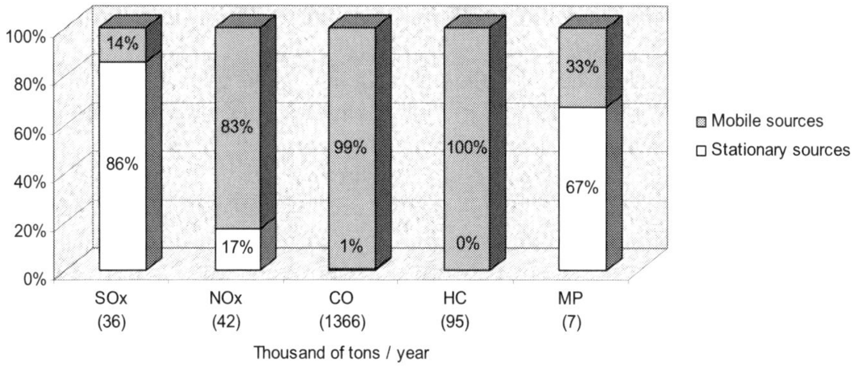

Figure 14: Participation of the stationary and mobile sources in the pollutant emissions of Bogota city.

The previous results allow us to conclude that the control programs of air pollution in Bogota must focus on primarily in controlling the vehicle emissions, beginning with the private automobiles since they are the main contamination source in the city. Additionally, it is necessary to develop experimental studies considering the local conditions and verifying the results obtained using the emission factors of the US–EPA.

5 Actions being taken to improve the air quality of Bogota

Trying to assure the air quality for the Bogota citizens, DAMA and the local government have established a series of actions directed to control the quality of the fuels used in the city and to control the pollutant emissions in vehicles and the industry. The most relevant actions are described below.

5.1 Fuels

Since 1996 the addition of lead tetraethyl in the gasoline, produced and commercialised in the country, was eliminated. Lead tetraethyl was used to increase the octane number in the gasoline, improving the anti-knocking characteristics of the fuel used in internal combustion engines.

Since 1995, the Department of Environment regulated the quality of fuels used in boilers, furnaces and internal combustion engines. For the gasoline it established that from year 2001, the maximum sulfur content would be 500 ppm and 300 ppm from year 2005. In addition, it was required to include dispersing additives, detergents and deposit controllers in the admission system, including the valves. These specifications were determined not only for the gasoline produced in the country but for the imported one. For diesel it was decided that for the year 2001, the sulfur content could not be higher to 1200 ppm and 500 ppm for the year 2005. The fuel oil No. 6 used in boilers and furnaces of industrial and commercial use, must have a maximum sulfur content of 1,5% for the year 2001 and 1% for 2005.

Since 1998, the Department of Environment limited the use of dielectric oil or fluids of power transformers (PCB´s) to a maximum concentration of 50 ppm in used oils that are going to be used in combustion processes. These substances are recognised as a threat to the health and the environment.

In the year 2000, DAMA prohibited the use of Castile crude as a fuel for thermal equipments that operate in the city, due to its high sulfur content.

For more than 5 years, natural gas has been promoted in the city as an alternative fuel to replace the use of traditional fuels use in industrial, residential and vehicular sectors. Natural gas has entered the residential sector in massive form, replacing the consumption of gasoline and firewood in this sector. Also, it has entered strongly the industrial sector, replacing the use of liquid fuels like fuel oil No. 2 and fuel oil No. 6.

However, in the transportation sector, the penetration of natural gas has just began in recent years and its penetration has been limited by the effect of the height of the city on the mechanical performance of the vehicle and by the lack of fuel stations. At the moment, more than 12,000 vehicles converted to natural gas circulate in the city. These vehicles are mainly taxis, small-size buses and fleets of vehicles used in the distribution of goods in the city. There are only 8 fuel stations distributed all over the city.

In recent years, DAMA created the *Management to promote the use of cleaner fuels* [21] with the purpose of boosting the use of natural gas, especially in the industrial and vehicular sectors, which has allowed stimulation of its use. As part of the work in this Management, the authors have promoted the use of vehicles with dedicated and converted motors to natural gas for public transportation (passengers transportation, dump trucks, police vehicles) [22–30], solving the adaptation problems of the existing technology to the conditions of Bogota (high altitude and natural gas with low methane concentration and variation in its composition). The studies carried out show that for the fuel price conditions in the 2002, the gasoline substitution by natural gas is highly profitable. Nevertheless, even the diesel substitution by natural gas is slightly favorable from the economic point of view; the investors will not decide to invest in a new technology until the economic benefit is more attractive.

The aspects that require greater research are the determination of an objective methodology to synchronise the converted engines, alternatives to decrease the loss of power and the adaptation of natural gas engine technology to heavy work under city conditions.

5.2 Stationary sources

DAMA designed the *Unit of Atmospheric Contamination (UCA)* for the industrial sector. This unit weighs the criteria pollutant emissions produced by the factories and classifies them using a qualitative scale that goes from *little polluting* to *very highly polluting*. In this way, DAMA prioritise its control emissions work in the more-polluting factories. The emission measurements frequency of a factory is proportional to its classification in UCA terms.

According to the reference [25], industries that use fuel oil No 6. and coal in their combustion processes are classified as *very highly polluting* by the UCA, whereas the industries that use natural gas are classified as *little polluting*. Depending on this rank the companies will have to make a stricter pursuit of its emissions.

Another aspect regulated by DAMA is the height of the emission point of stationary sources, which is determined using the volumetric and mass flow of the pollutants, the speed of the gases, the stack diameter and the average wind speed.

With the purpose of motivating the use of cleaner technologies, the local environmental authority began the operation of the fourth phase of the *Unit of Environmental and Technical Attendance for the Small and Medium Industry* (ACERCAR). This unit offers environmental information guide, integral diagnoses and technical attendance for the implementation of strategies that allows the small and medium industries to improve their competitiveness and environmental management. In addition, it serves as a financing source for the implementation of these projects.

5.3 Mobile sources

The effort of the local government has been centred in two aspects to control the air pollution by mobile sources.

The first one is directed to assure the good condition of the vehicles that circulate in the city. As described in Section 3.3, since 1996 DAMA regulated the maximum allowed emission levels for automotive vehicles according to the vehicle manufacture year. In order to verify the fulfillment of the measurement, DAMA forces the vehicles to go to the Recognise Diagnosis Centre (CDR) at least once a year to obtain an *Emissions Certificate*. This certificate must be shown to the police when required. Not to carry this certificate is penalised with a fine of one minimum wage (approximately US$ 100).

An automotive laboratory can emit the *Emissions Certificate* after DAMA recognise it as authorised laboratory. To do this, DAMA visits the laboratory and verifies the suitable technology and personnel to measure the vehicle pollutant emissions. The recognised centres are forced to report the results of the measurements to DAMA. The certificate is made of security paper provided by DAMA. This allows DAMA to take a control on the emissions certification process and to finance the program.

Since year 2001, DAMA exempted the vehicles that use natural gas as fuel from the fulfillment of pollutant emissions regulation. This measurement was promoted with the purpose of stimulating the use of natural gas in the vehicles.

Figure 15: Massive transportation system implemented in Bogota.

Figure 16: Bicycle routes constructed in Bogota to substitute the vehicle use.

On the other hand, the effort of the local government has been centreed in improving the mobility of the city. The city has neither metro nor subway and then public transportation was an increasing problem. A large number of private and public vehicles resulted in traffic jams and deioration of the air quality. To address this problem, during recent years, local government has implemented several important measures to organise traffic. The actions taken include:

- Restriction to the circulation of private vehicles during four hours a day, two days a week, according to the last digit of the license plate.
- Restriction to the circulation of vehicles, used for public transportation, 16 hours a day, once a week, according to the last digit of the license plate.
- Implementation of a massive transportation system (Transmilenio) based on the use of exclusive lanes with fixed stops for brand new buses with large transportation capacity (160 passengers). Feeding routes that take the passengers from different places to the main lines. Figure 15 shows a picture of the system. At the beginning of 2003 there are 42 km constructed, 470

articulated buses and 241 feeding buses in operation. The system has been very successful. It reduced the transportation time up to one fourth of the usual one for passengers.

- Construction of 269 km of exclusive routes for bicycles all over the city, trying to stimulate the use of the bicycle to go to work. Figure 16 shows this system.

Educative campaigns for pedestrians and drivers. The Natives of Bogota learned to respect the pedestrian routes and the transit signal.

5.4 Other actions

Since year 2002, DAMA managed the inclusion of Bogota in the Clean Air Initiative of the World Bank, whose mission is to improve the air quality in the cities of Latin America to protect the health of its inhabitants and to mitigate the global contamination. The Eighth Directive Committee of the Initiative approved the inclusion of Bogota in Sao Paulo.

The objectives of the Initiative include to facilitate the public regulation decisions and investment associated to the air decontamination in the partner cities; to facilitate the private investment decisions that stimulate the use of technologies, fuels and operation practices associated to less air emissions and to provide elements that guarantee the sustainability of the decontamination actions undertaken in the city.

The Initiative also wants to fortify the Atmospheric Pollution Prevention and Control Program in Bogota, based on the participation of all the actors, to protect the health of the inhabitants, interchange experiences and knowledge with the other participants of the Environment National Commission of the Metropolitan Region of Santiago de Chile (CONAMA), the Ecology Secretariat of Mexico State, the Metropolitan Municipality of Lima, the Environment Municipal Secretariat of Rio de Janeiro and the Environment Secretariat of Sao Paulo [31].

6 Actions to be taken to improve the air quality of Bogota

In [16] the environmental authority of Bogota defined the guidelines to follow during the following years to improve the air quality in Bogota. Some of the measures that are going to be taken are described below.

6.1 Stationary sources

6.1.1 Pollutant emissions control

In order to control the pollutant emissions, DAMA will continue making periodic audits to the commercial and industrial establishments, prioritising those classified as highly polluting according to the UCA, to evaluate the stationary sources emissions and to guide them towards an effective emissions control.

Additionally, the emissions inventory will be updated for stationary and mobile sources. In this sense, the industry must present an annual emission evaluation that will be unified in a database.

The verification of the stacks height will be continued in order to guarantee the fulfillment of the new legislation.

6.1.2 Fuel substitution incentives

It is planned to continue with the incentive program to the traditional fuel substitution by natural gas in the industries of the city. Initially, it is considered them exempt from the emission permissions to the stationary sources that use this type of fuel.

In the industrial sector it is planned to design an *"Ecologic Tax"* that penalises the polluting companies and subsidises the conversion to natural gas of industries that use coal in their processes. In addition, it is planned to assist the companies that use diesel or fuel oil since in most of the cases it is more profitable to use natural gas that these fuels.

Finally, it was agreed with the companies involved in the provision of natural gas, to maintain a fuel price policy that guarantees the use of natural gas as an economic alternative.

6.1.3 Other actions

Epidemiological studies will be made in the city to determine the atmospheric contamination effects in the health of children less than five years old in different localities of the city.

Additionally, the definition of the industrial sectors will continue through the land use plan, as well as the definition of the air quality model for the city.

The formation of combustion technicians, boiler-inspection technicians and smoke supervisors will continue by means of the elaboration of a combustion manual and the accomplishment of combustion courses for the furnace and boiler operators in the industry.

It is planned to execute tree cultivation programs in different sectors of the city.

6.2 Mobile sources

6.2.1 Pollutant emissions control

The annual emissions certificate program and the campaigns to assure the synchronisation of the vehicles that circulate in the city will continue. The vehicle-emission audits will continue by using the seven mobile monitoring units existing in the city.

The restriction program on private vehicles and vehicles used for public transportation will continue.

Suitable measures for the disposition of used tires generated by the mobile sources will be implemented, as well as alternatives to dispose of the used batteries.

Qualification programs to mechanical workers in the diagnosis, synchronisation and gas analysis of the diesel and gasoline vehicles will continue.

It is planned to continue with the *Management to promote the use of cleaner fuels* in the vehicular sector of the city, with the purpose of assuring the conversion of the public gasoline vehicles to natural gas and the economically profitable private vehicles.

The national government is gradually eliminating the diesel fuel subsidy. This will promote the use of natural gas in automotive vehicles.

6.2.2 Reorganisation of the public transportation

It is planned to continue with the implementation and expansion of the mass passenger transportation system (Transmilenio) existing in the city. At the moment, the local government is preparing a study to reorganise the rest of the public passenger transportation in such way that harmonises with the Transmilenio transportation system. The study includes relocating some of the public transportation terminals, decreasing the number of existing routes and creating shorter circular routes. Also it is planned to continue with the destruction of three or more vehicles used for public transportation by each new bus that enters the mass transportation system.

The presentation of an environmental management plan will be required to urban transportation and school service companies.

Execution of the bicycle routes program will be continued, which today has almost 270 kilometres of exclusive routes for bicycles, in such a way that allows the bicycle mobilisation to any point of the city.

6.2.3 Natural gas vehicle stimulation

The local authority plans to implement measures that stimulate the use of natural gas in public service vehicles. Some of the proposed measures for the implementation are: to define exclusive or preferential routes in the centre and concurred places of the city for vehicles that operate with natural gas, force the public service fleet owners (passengers transportation, garbage collection) to incorporate natural gas vehicles within its vehicles fleet, allow independent waiting lines for natural gas vehicles in the airport and transportation terminals and turn the vehicles of the government to natural gas.

In the private sector the possibility of designating preferential parking lots for private natural gas vehicles and the elimination of the restriction in the circulation for this type of vehicles is under study.

Mechanisms that assure the development of suitable infrastructure to provide the fuel and qualified technical service are being implemented.

The reduction of taxes in the purchase of vehicles, equipment and accessories related to the use of the natural gas is being implemented

References

[1] Environmental Protection Agency. Air Quality Planning and Standards Office. Quality Assurance Handbook for Air Pollution Measurement Systems Volume II: Part 1 Ambient Air Quality Monitoring Program Quality System Development. August 1998.

[2] National Administrative Department of Statistics – DANE. DANE projection for 2002, http://www.dane.gov.co
[3] Alvarado, G., Environmental National Center - CENMA. Informal communication, 2002.
[4] Statistics synthesis of the Mexico city metropolitan zone - SEMARNAP. Environmental statistics, Mexico, 1999.
[5] Technical and Administrative Environmental Department – DAMA. Air Quality Monitoring Network of Bogota. Annual information for 1998, 1999 and 2000.
[6] Institute for Hydrology, Meteorology and Environmental Studies – IDEAM. Colombian Environmental Information System – SIAC. Perfil del Estado de los Recursos Naturales y del Medio Ambiente en Colombia 2001, Volume 3, 2002
[7] www.ecopetrol.com.co, Consulted on April, 2003.
[8] Normalization Colombian Organism – ICONTEC. Colombian Technical Normative – NTC 4826. Quality of compressed natural gas for vehicular use.
[9] www.ecopetrol.com.co/comercio/tipicas.htm, Consulted on April, 2001.
[10] Interlabco Ltda. Ultimate analysis of a coal sample in the Arrayán coal mine. Cundinamarca, June 3, 1999.
[11] www.conae.gob.mx/programas/vapor/anexovap5.html, Consulted on June, 2001.
[12] Sonntag, R. & Van Wylen, G., Fundamentals of thermodynamics. Fifth edition, USA, 1998.
[13] Huertas, J., Beltrán, R. & Valencia, A., Evaluación preliminar de combustibles diesel utilizados actualmente para vehículos en el área urbana del Distrito. Universidad de los Andes – DAMA. Bogota, 1999.
[14] Gas Natural S.A. E.S.P. Energía para la industria del nuevo milenio. Bogota, 2000.
[15] INAMCO & Technical and Administrative Environmental Department – DAMA. Inventario de fuentes fijas de emisión de contaminantes a la atmósfera en la ciudad de Bogota D.C. Bogota. November, 2001.
[16] Technical and Administrative Environmental Department – DAMA. Plan de Gestión del Aire para el Distrito Capital 2000 – 2009, Bogota, February, 2001.
[17] Huertas, J. & Valencia, A., Beneficio económico y ambiental para las empresas que se convierten a gas natural. Los Andes University, Bogotá, 2001.
[18] Secretary of Transit and Transport. Database of the vehicles registered in Bogota. Bogota, 2001.
[19] Japanese International Cooperation Agency – JICA. Estudio del Plan Maestreo de Transporte para Bogotá. Bogota, 1996.
[20] PROEZA & Technical and Administrative Environmental Department – DAMA. Diseño y Ejecución de un Programa de Seguimiento a los Centros de Diagnostico Reconocidos por el DAMA. Bogota, 2001.

[21] Huertas, J. & Valencia A., Management to promote the use of cleaner fuels in the vehicle sector of Bogota. Technical report. Universidad de los Andes. Bogota, 2002.

[22] Huertas, J., Valencia, A. & Giraldo, N., Conversion to natural gas and technical evaluation of the police vehicles in Bogota. To be published. Universidad de los Andes. Bogota, 2003.

[23] Huertas, J. & Arias, D., Dynamic Tests of CNG and Diesel Fueled Buses at High Altitude. http://www.ngvc.org/ngv/ngvc.nsf. The gas technology institute's first annual conference and exhibition on natural gas technologies: what's new & what's next. Florida, USA. September, 2002.

[24] Huertas, J & De Lemos, L., Conclusiones del seminario combustibles, vehículos y polución del aire. Gas natural vehicular No. 10. Orvisa comunicaciones Ltda. Bogota. August, 2002.

[25] Huertas J. & Valencia A., Economical and environmental benefits for the industries that convert its thermal equipment to natural gas. Tenth international conference on modeling, monitoring and management of air pollution. Air pollution X. Wessex Institute of Technology, WIT press, Segovia, July, 2002.

[26] Valencia, A. & Huertas, J., Reducing air pollution of a large city by organizing public transportation. Tenth international conference on modeling, monitoring and management of air pollution. Air pollution X. Wessex Institute of Technology. WIT Press. Segovia. July, 2002.

[27] Huertas, J. & Arias, D., Evaluación de flota de buses a GNCV en el mundo y su posible desempeño en Bogotá. Gas natural vehicular No 9. Orvisa comunicaciones Ltda. Bogota. May, 2002.

[28] Huertas, J. & Valencia, A., Guía ambiental para distribución de GNCV Gas natural vehicular No 8. Orvisa comunicaciones Ltda. Bogota, March, 2002.

[29] Huertas, J. & Arias, D., Emisiones de NO_x en vehículos a GNCV. Gas natural vehicular No 6. Orvisa comunicaciones Ltda. Bogota. July, 2001.

[30] Huertas J., Báez, F. & Moreno, F., Diseño y ejecución de un conjunto de pruebas sobre los buses Diesel y Gas Natural, para determinar su adaptabilidad a las condiciones de Santafé de Bogotá y conocer en detalle sus costos de operación. Universidad de los Andes. Bogota, 1999.

[31] www.dama.gov.co. Consulted on August 2002.

CHAPTER 7

Sectoral analysis of air pollution control in Delhi

M. Khare[1] & A. Kansal[2]
[1]*Civil Engineering Department, Indian Institute of Technology Delhi.*
[2]*School of Environmental Management, G G S Indraprastha University Delhi.*

Abstract

The chapter discusses the present air quality status and strategies for air pollution control in Delhi, which is amongst the most polluted cities in the world. In recent years, the on-going strategies adopted by the government have been able to curb the rise in air pollution levels of Delhi. Strategies include introduction of compressed natural gas (CNG) powered engines in all commercial and public transport, implementation of a metro rail project, re-siting of polluting industries, enforcement of stringent air pollution emission norms and introduction of cleaner fuels for domestic fuel burning. However, the continued increasing pressures from growing human numbers, vehicular traffic, industries and domestic fuel burning may offset such success in future. Hence, it is desirable to complement the 'supply side interventions' (technological fix) with 'demand management measures' (alternatives) to secure improved levels of air quality.

1 Introduction

Urban areas represent complex environmental problems. In particular, the quality of the ambient air in such areas of developing countries has been deteriorating rapidly over the past few decades. In an 'uncontrolled' scenario, the higher a region is in the energy use ladder, the more are the pollutant emissions, whereas, the concentration of pollutants in the ambient air depends not only on the local meteorological and anthropogenic factors but also on the policies and strategies that are in force to minimize release, and dissipate whatever is released. Comparison of the air quality status of two regions having equivalent energy-use patterns and meteorological conditions reflects the efficacy or success of environmental policies of one region over the other. Similar regions elsewhere in the world can thus adopt the successful policies implemented in other regions.

Delhi, the national capital of India, has the distinction of being not only the most urbanized and densely populated city but also one of the most polluted cities in the world. An investigation into the present status, causal factors and past policies brings out important clues for future actions for an effective air quality improvement programme.

The chapter presents case studies of air pollution control strategies in Delhi. A large number of local initiatives for combating air pollution emissions have their genesis in the regional or country's environmental programme. The first section of the chapter describes the evolution of environmental policy in India, especially in the context of air pollution. A discussion on India's environmental organizational set-up for air policy, follows. The next section describes the current status of ambient air quality in Delhi. An analysis of causal factors, current strategies and future policy options are discussed in the context of the most significant activities, viz. transport sector, manufacturing industries and thermal power plants and domestic fuel combustion. There is a final summing up of all the issues in the concluding section.

2 India's legislative framework for air pollution control

India's attempts to operationalize the concepts of environmental protection were initiated in the early 1970s in the wake of the UN conference on human environment, held in Stockholm in 1972. The Water (Prevention and Control of Pollution) Act 1974, the Water (Prevention and Control of Pollution) Cess Act 1977 and the Air (Prevention and Control of Pollution) Act 1981 followed. Based on the experience gained over a decade, these laws as well as those relating to forests conservation and motor vehicles were amended in the 1980s and an umbrella type of legislation in the form of the Environment Protection Act, 1986 and the Public Liability Insurance Act, 1991 (in the wake of the Bhopal tragedy) were enacted. Table 1 lists the major legislations/policies formulated for the prevention of environmental degradation, especially that of air (MoEF [1–4]).

By and large these enactments seek to create and sustain a 'command and control' regime of environment protection. They enable the central and state governments to control the air pollution, and to regulate the activities that cause such pollution. Government thus formulated the 'National Minimum Standards' for the ambient air; standards for emissions from polluting activities; and compelled the existing and new industries to conform to these standards, either by adopting end-of-the-pipe solutions (i.e., installing pollution control devices) or by changing over to new technologies or fuels. The government regulated the entry and existence of polluting activities, both temporally and spatially, using a system of issuing "consent to establish" and "consent to operate".

Ambient air quality standards (both short-term, i.e. *24-hourly*, and long-term, i.e., *annual*) have been laid down for industrial, residential/rural/other, and sensitive areas with respect to various pollutants like sulfur dioxide (SO_2), oxides of nitrogen (NO_x), suspended particulate matter (SPM), respirable particulate matter (RPM), lead (Pb), and carbon monoxide (CO) (Table 2). In addition, source-specific emission standards have also been laid down as 'minimal national standards' (MINAS).

Table 1: Environmental legislation/policy in India.

Year	Environmental legislation	Year	Policy statement
1981	The Air (Prevention and Control of Pollution) Act	1992	The policy statement on Abatement of Pollution
1986	The Environment (Protection) Act	1992	The National Conservation Strategy and Policy Statement on Environment and Development
1991	The Public Liability Insurance Act	1993	The Environment Action Programme
1995	The National Environment Tribunal Act		
1997	The National Environment Appellate Authority Act		

Table 2: National ambient air quality standards (CPCB [5]).

Pollutant	Time-weighted average	Concentration in ambient air ($\mu g/m^3$)		
		Sensitive area	Industrial area	Residential rural & other areas
Sulfur dioxide (SO_2)	Annual*	15	80	60
	24-hour**	30	120	80
Oxides of nitrogen as NO_2	Annual*	15	80	60
	24-hour**	30	120	80
SPM	Annual*	70	360	140
	24-hour**	100	500	200
RPM (< 10 μm)	Annual*	50	120	60
	24-hour**	75	150	100
Lead (Pb)	Annual*	0.50	1.0	0.75
	24-hour**	0.75	1.5	1.00
Carbon monoxide (CO)	8-hour**	1000	5000	2000
	1-hour	2000	10000	4000

* Annual arithmetic mean of minimum 104 measurements in a year taken twice a week, 24-hourly, at uniform intervals.

** 24-hourly/8-hourly values should be met 98% of the time in a year. However, 2% of the time, it may exceed, but not on two consecutive days.

For the assessment and monitoring of air quality, India has established a National Ambient Air Quality Monitoring (NAAQM) network, covering 290 stations, spread over 90 cities/towns. Under the NAAQM programme, monitoring of pollutants like, SO_2, NO_2, and SPM is carried out. In selected stations, specific parameters like ammonia (NH_3), hydrogen sulfide (H_2S), RPM, and polycyclic aromatic hydrocarbons (PAH) are also monitored (CPCB [6]). Surveys are also conducted at traffic intersections, and source-specific monitoring is carried out in industrial and thermal power plants.

In order to reduce vehicular pollution, vehicular emission norms for both petrol- and diesel-driven vehicles at the manufacturing stage were introduced in 1990 and made tighter to EURO-II norms in April 2000. Another small but significant step taken in the intermediate phase was the installation of catalytic converters in new petrol-driven four wheelers in four metro cities (Delhi, Mumbai, Chennai, and Kolkotta). Steps have also been taken for improving the fuel quality. The lead content in gasoline was progressively reduced and now unleaded petrol is available throughout the country. Similarly, sulfur content in diesel and the benzene content in gasoline were also reduced in a phased manner.

For siting of industries, a project has been initiated to prepare a *district-wise zoning atlas* to delineate the areas that are suitable for industrial activities. Also, after the intervention of the courts, polluting industries have been shifted away from urban centers and restrictions have been imposed on siting of polluting industries in identified areas.

Among other measures, environmental impact assessments (EIAs) were made mandatory for 29 categories of projects, both in the public and private sectors. To ensure public participation in the EIA process, public hearings were made mandatory. Submission of environmental statement by the polluting units to the concerned state pollution control board was also made mandatory in 1992 (MoEF [4]).

3 Institutional framework

The Ministry of Environment and Forests (MoEF) in the central government and similarly named ministries in the state governments are the principal nodal agencies for all environment-related matters. Other ministries such as Road Transport & Highways, Power, Commerce & Industry, Urban Affairs and Employment, and Petroleum and Natural Gas also share responsibility in air-quality-related problems. The Planning Commission of India plays a key role in environmental decision making.

MoEF is responsible for proposing government's environmental legislation, maintaining public information system on environmental matters, development of environmental standards and developing a programme for meeting these standards. It also co-ordinates international collaborations and finances research projects, aimed at environmental protection. The ministry is also responsible for providing environmental clearance for all major projects in the public and private sectors, on the basis of the review of the environmental impact assessment reports and through its regional offices, for monitoring compliance with conditions attached to such clearances.

The basic executive authority under the environmental laws is vested in another organ of MoEF, viz., the Central Pollution Control Board (CPCB). The CPCB co-ordinates the activities of the State Pollution Control Boards (SPCBs) and Pollution Control Committees (PCCs) of the Union Territories (regions or territories directly administered under central government authority). These are autonomous statutory bodies created under the acts relating to Water and Air pollution control. CPCB also advises the MoEF and other central government ministries on all matters concerning the control of pollution.

CPCB determines the minimal national standards for ambient air quality and emissions of gaseous pollutants into air. SPCBs in the states enforce these standards within their jurisdiction, punish transgressions by imposing fines and order closure of activities that continue to violate the standards. These organizations can make the standards more stringent for their jurisdictions or parts thereof, depending on the special circumstances prevailing there, but cannot relax them or lower the standards. Appellate bodies have been established in each state and Union Territories to hear appeals against the decisions of SPCBs or PCCs.

The key institutions at local levels are the 'corporations' (*in large cities*), 'municipalities' (*in smaller cities and towns*) and 'panchayats' (*in villages*). These institutions have a key role to play in activities that reduce pollution and improve environment such as land use planning, health and sanitation, traffic planning and regulation, enforcement of building codes, and creation and maintenance of green belts and parks. The 74th Amendment to the constitution of India seeks to put them on a solid foundation based on devolution of political authority and financial resources from the state governments. A large number of non-governmental organizations (NGOs) are also active in respect of the environment- and relocation-related issues.

4 Study area

Geographically, the national capital territory (NCT) of Delhi (Fig. 1) is 28°35' N latitude and 77°12'E longitude. It is 160 km south of the Himalayas, at an elevation of 216 m above mean sea level. It has a semi-arid climate with extremely hot summers, heavy rainfall in monsoon and very cold winters. The annual mean temperature is 25.3°C and the annual mean rainfall is 715 mm (IMD [7]). Northwesterly winds normally prevail here; while in June and July, south-easterlies predominate. Wind speeds are typically higher in summer and monsoon; in winters, 'calms' are frequent. Pre-monsoon dust storms are westerly from the great Indian desert, depositing large concentrations of SPM. Inversion conditions mostly prevail in winters, increasing the pollution load (CPCB [5]).

Delhi is facing tremendous stress in providing basic urban services for safe and wholesome living. Delhi, with an area of 1485 km^2, is inhabited by about 13.78 million people (population density 9294 per km^2), of which 12.82 million are in urban areas (Census of India [8]). The decadal growth rate of the region is

46.31% (urban growth is 51.33%). Such a high growth rate is primarily attributable to large and uncontrolled in-migration from the surrounding states for reasons of better employment opportunities and higher quality of life. In order to restrict the growth of the city, the national capital region (NCR), covering about 30,000 km^2 has been delineated. However, the runaway growth of Delhi continues unrestricted as it continues to expand. It is no longer just the national capital but a nerve centre of national trade, politics, culture and education. Due to space constraints, the population started living in the slum areas under extremely deplorable conditions.

Figure 1: Geographical location of Delhi.

Besides, various other urban pressures, such as industrial activity, transport infrastructure, economic status, etc., also confer a continued high growth trend to Delhi. Industrial units have grown from 26000 in 1971 to 129000 in 1998 (DoES [9]), a nearly 5-fold increase within 27 years. The number of registered vehicles has grown from 0.2 million (1970/71) to 1.8 million (1990/91), reflecting about a 9-fold increase in the last three decades (DUEIIP [10]). The economic status in the form of per capita income has also increased from Rs 17,355 in 1994 to Rs 29,623 in 2000, which is double the national per capita income of Rs 15841 at current prices (GNCTD [11]). Open spaces are mainly uncultivated agricultural lands. Mixed urbanization pattern (residential and commercial) led to the haphazard growth of Delhi. Thus, the built-up area of Delhi has increased to 750 km^2, comprising over 50% of the geographical area of Delhi (NCRPB [12]). This indicates the pace of urban sprawl in Delhi causing an immense resultant pressure on the air and land environments.

5 Status of ambient air quality

Ambient air quality monitoring is being carried out regularly at nine stations in Delhi by the CPCB; these includes five stations in the industrial area (Nizamudin, Ashok Vihar, Shahzada Bagh, Nazafgarh, and Shadara) and four stations in the residential area (Janakpuri, Siri Fort, Sarojini Nagar and Town Hall) (Fig. 2). The annual average concentrations of SO_2, SPM and NO_2 are monitored in these areas. In addition, CPCB, in February 1997, undertook detailed monitoring of air quality at 19 traffic intersections recognized as 'microenvironments', experiencing high pollution loads (Fig. 2). Besides SPM, SO_2 and NO_2, other pollutants monitored at traffic intersections, include RPM, lead, PAH and ozone (O_3). Monitoring for these parameters is done for 24-hr, except for O_3, for which hourly observations are recorded between 9:00 AM and 5:00 PM. Figures 3 to 11 describe the air quality status for all major pollutants (CPCB [5], [13–15]).

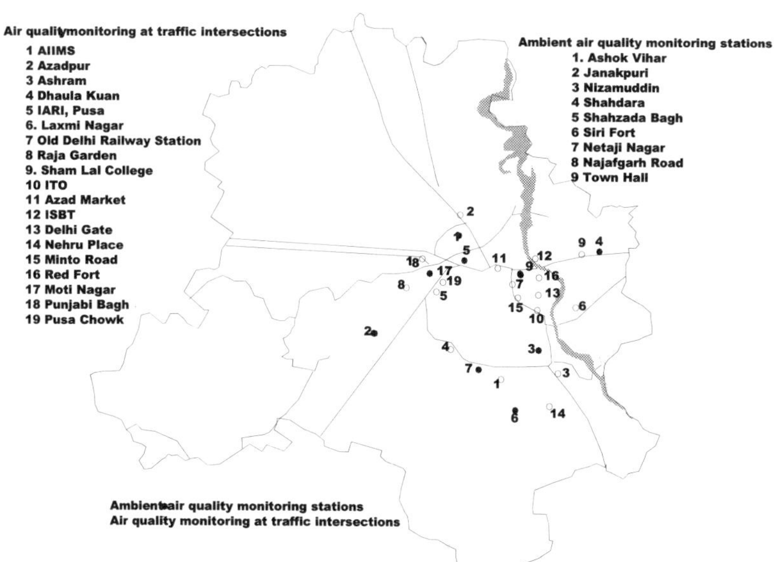

Figure 2: Location of air quality monitoring stations (CPCB [13]).

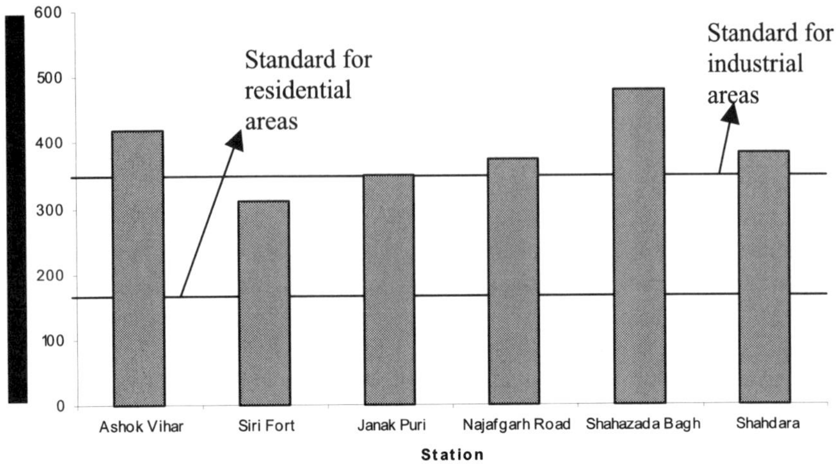

Figure 3: Annual average concentration of SPM during 2000 (CPCB [13]).

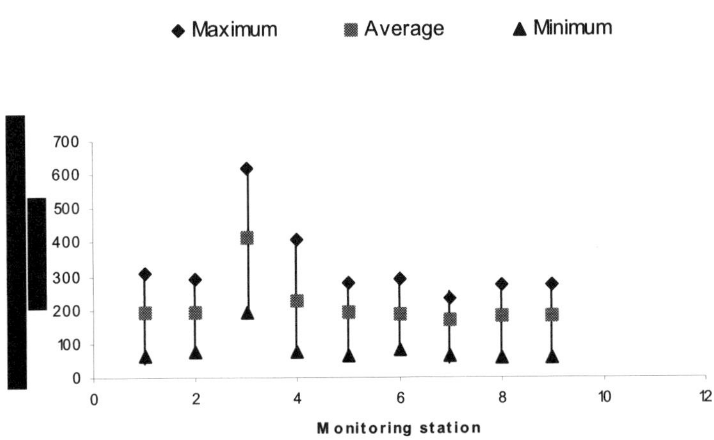

Figure 4: Ambient concentration of RPM at various monitoring sites during July 2000–January 2001 (TERI [16]).

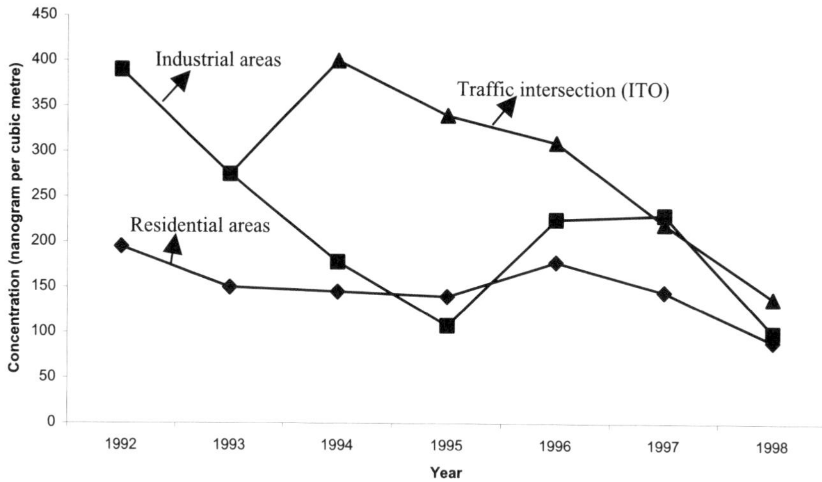

Figure 5: Trends of annual average particulate lead in ambient air in Delhi (CPCB [5]).

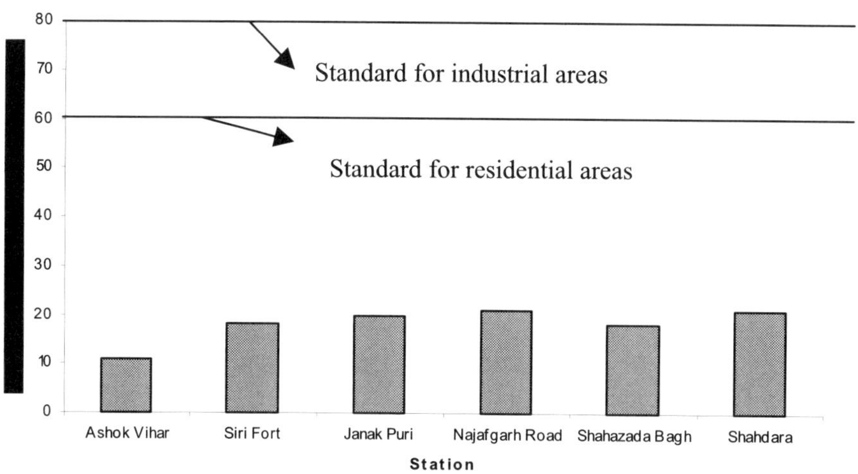

Figure 6: Annual average concentration of SO_2 during the year 2000 (CPCB [13]).

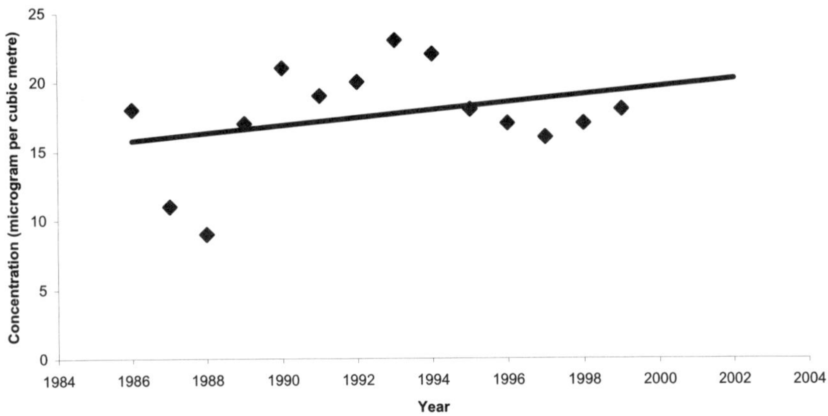

Figure 7: Annual average concentration of SO_2 in Delhi during 1987–2000 (CPCB [5]).

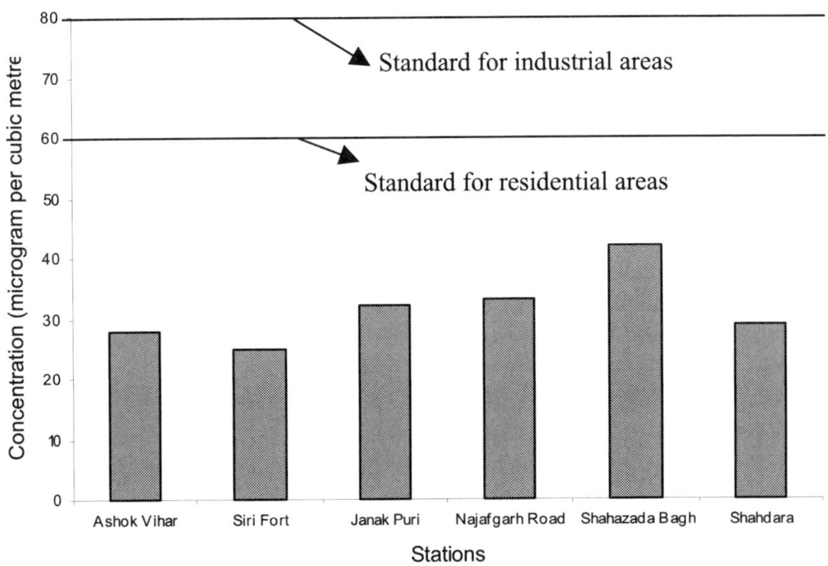

Figure 8: Annual average concentration of NO_x during 2000 (CPCB [13]).

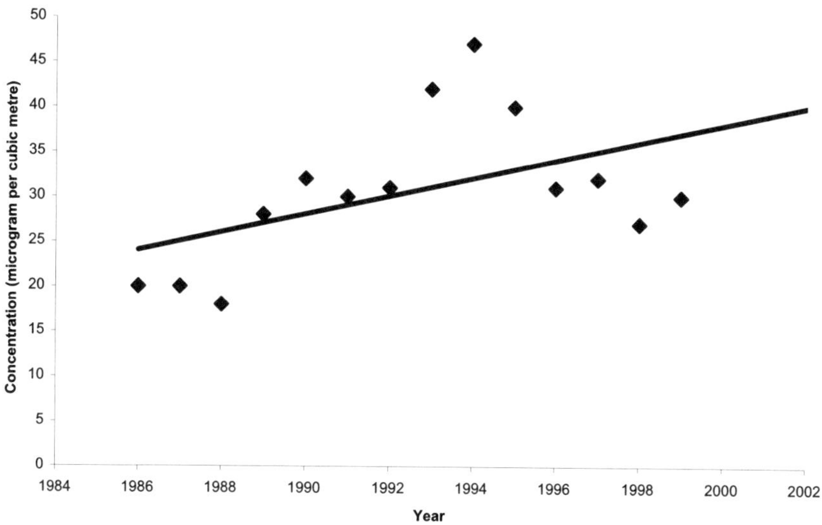

Figure 9: Annual average concentration of NO_x in Delhi during the period 1987–2000 (CPCB [5]).

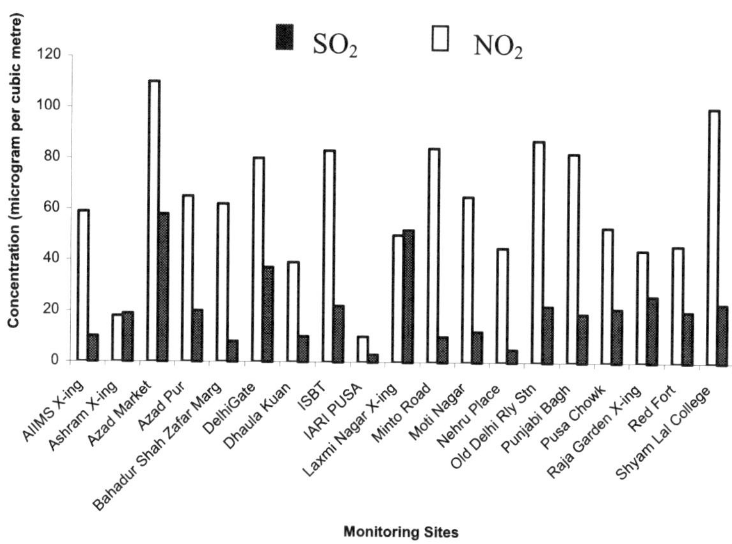

Figure 10: Concentrations of NO_2 and SO_2 at major traffic intersections of Delhi (CPCB [14]).

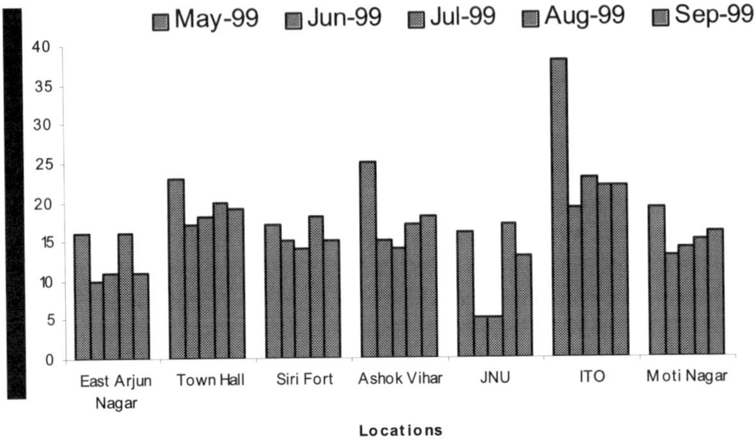

Figure 11: Benzene concentrations in the ambient air of Delhi (CPCB [15]).

5.1 Particulate matter

Annual average concentration of SPM at all stations is much higher than the CPCB standard for residential areas (140 $\mu g/m^3$) (Fig. 3). The highest SPM level (475 $\mu g/m^3$) is recorded at Shahzada Bagh (representing industrial zones). The case for RPM (PM_{10}) (Fig. 4) is similar. At the Income Tax Office (ITO), the annual average concentration during 2000 was 190 ($\mu g/m^3$), which is more than three times the standard prescribed for residential areas. The ratio of RPM to SPM at this site is 0.39. This reflects the possibility of high contribution from various man-made sources such as roadside dust, auto exhaust and coal combustion (though, further long-term studies adopting the receptor modeling are required to corroborate this).

The lead emissions in ambient air are mostly from vehicular exhaust. The trend analysis of particulate lead from 1992 to 1998 indicates a fluctuating pattern of particulate lead concentrations up to 1996 (Fig. 5). However, subsequently, a downward trend is observed both at residential and industrial areas that is well below the national standards. This may be due to the introduction of catalytic converters in 1995 and gradual introduction of unleaded petrol (CPCB [5]).

Table 3 presents the results of 15 PAH fractions in particulates that are being analyzed routinely during the past few years (CPCB [5]). Owing to the carcinogenic character, the WHO and the United States Environmental Protection Agency (USEPA) have recommended no safe limit of PAH. Significantly, high levels of PAH in Delhi's atmosphere may be due to the fairly large percentage of two/three wheelers on the road and significant use of coal and kerosene as domestic fuels.

Table 3: Range of concentration of PAH in various seasons in Delhi during 1998.

Season	Range of PAH (ng/m^3)
Winter	30.3–60.9
Summer	16.0–29.3
Monsoon	9.4–27.8

5.2 Gaseous pollutants

The annual concentration of SO_2 at all the ambient air quality stations is well within the CPCB standard, even for residential areas (60 $\mu g/m^3$) (Fig. 6). Long-term annual trends of SO_2 concentrations have also not shown appreciable increase over the years (Fig. 7). This is despite the fact that there has been a manifold rise in industrial, vehicular, and domestic activities over this period. The plausible reasons for this could be the combined impacts of implementation of various control measures (such as restriction of SO_2 emissions from various industrial sources and stack height norms applicable to all major industrial furnaces) and introduction of low 'sulfur' content auto fuels. Also, the possibility of high SO_2 conversion to sulphates in the polluted atmosphere cannot be ruled out due to high levels of soot particles in Delhi's atmosphere.

The sources of NO_x in Delhi are same as that of SO_2, namely industrial processes, vehicular and domestic combustions. The annual average concentration of NO_x in Delhi, at all the ambient air quality stations, is well within the CPCB standard (Fig. 8). However, unlike SO_2, long-term trends of NO_x, reflect a significant rise in ambient concentrations over the years (Fig. 9). Furthermore, at major traffic intersections, the 24-hour standard for NO_2 (80 $\mu g/m^3$) is not exceeded at any of the locations (Fig. 10); whereas, some locations do exceed the prescribed limit of NO_2. This may be due to the rapid increase in total vehicular activity (resulting in higher conversion rate of atmospheric N_2 to NO at high combustion temperatures prevailing in combustion chambers), and also due to the rising percentage of diesel-powered vehicles.

Benzene concentrations have, however, increased in the ambient air after the introduction of unleaded petrol. Table 4 describes benzene concentration during August 1998 (prior to phasing out of leaded petrol) and subsequently in November 1998 (after phasing out of leaded petrol). The CPCB has also been monitoring benzene concentration at seven residential, industrial, and commercial locations since May 1999 (Fig. 11). The overall mean benzene concentration is 17 $\mu g/m^3$ for seven locations, monitored during May–September 1999, which is higher when compared with the United Kingdom's recommended annual average limit of 16 $\mu g/m^3$.

Table 4: Average concentration of benzene ($\mu g/m^3$) in ambient air in Delhi (CPCB [13]).

Location	August 1998	November 1998
Residential area	Not available	88
Traffic intersection	42	195
Vicinity of petrol pump	164	182

CO concentration is being regularly monitored by CPCB at one of the busiest traffic intersections, the ITO bridge, since 1992. The annual average concentration during 1992-98 ranged between 2.69 and 5.59 $\mu g/m^3$ in comparison to the national air quality standard of 2.0 $\mu g/m^3$ (8-hourly) at residential areas (CPCB [14]). The frequent violations of CO values have been because of the large number of motor vehicles operating on Delhi's road.

6 Health implications of air quality

Air pollutants cause responses ranging from reversible changes in respiratory systems and lung functions, changes in airway reactivity and inflammation, structural remodeling of pulmonary airways, and impairment of pulmonary host defenses, to increased respiratory morbidity and mortality. Women and infants are the most vulnerable group due to poor nutrition and lack of health care.

Current research trends showed that in developing countries air pollution levels and associated negative health impacts are increasing. According to one estimate, out of a total of 8.2 million deaths in India in 1989, anywhere between 0.9 and 3.6 million could be attributed to particulate air pollution (Banerjee [17]). Respiratory diseases accounted for about 20% of the annual registered deaths in rural India during 1985–1991 (VHAI [18]). It is also estimated that 40–60% of pre-school children and 25–30% of women in the reproductive age, suffer from chronic lung diseases (VHAI [18]). Figure 12 shows the trend of public health in Delhi (TERI [19]). In addition, a total of 22.6% non-trauma deaths during 1991–94 were directly attributable to air-pollution-related diseases such as pneumonia, bronchitis and asthama (CPCB [5]).

Furthermore, an epidemiological study done in 1999, revealed that over 40% of the emergency hospital admissions of patients with breathing and heart problems were due to air pollution. According to another study conducted in 10 schools of Delhi in 1996, the cumulative prevalence of Asthma in children was 15.3% (GHK International Ltd [20]). Table 5 shows the estimated annual episodes of illness and premature deaths due to ambient SPM levels in Delhi and the corresponding monetary costs (GHK International Ltd [20]). Yet another study done by the World Bank, put the cost of health due to air pollution in the range of $100–400 million (Brandon and Hommann [21]).

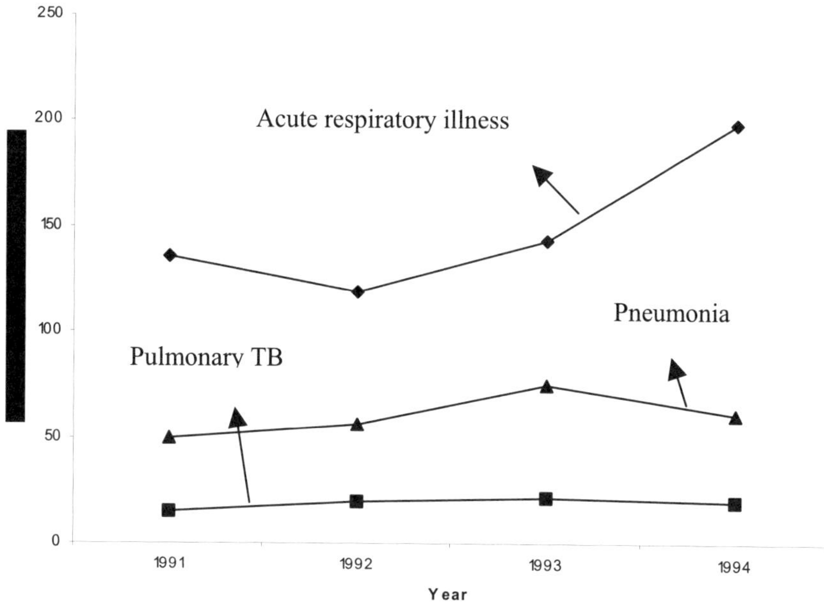

Figure 12: Trends in increase in air pollution related diseases in Delhi.

Table 5: Estimates of annual episodes of illness and premature deaths and corresponding monetary costs (Rs million) due to ambient SPM.

Illness/death	1991/92	1995
Premature deaths (no.)	7491	9859
Annual episodes of illness (no.)	3948923	5197018
Cost of premature deaths (Rs million)	6510	8567
Costs of medical treatment (Rs million)	246	324

7 Sectoral analysis

The majority of the pollutant emissions are mainly derived from energy sources. Emissions depend upon the fuel used and the conditions of combustion. Urban transport, manufacturing industries, thermal power plants, and domestic fuel combustion have emerged as the major sources of human-induced pollution (CPCB [22]). Figure 13 gives estimated daily air pollution loads in Delhi. Vehicular sources contribute about 67% of the total pollutants emitted, followed by thermal power and industries (25%) and domestic sector (8%) (MoEF [23]).

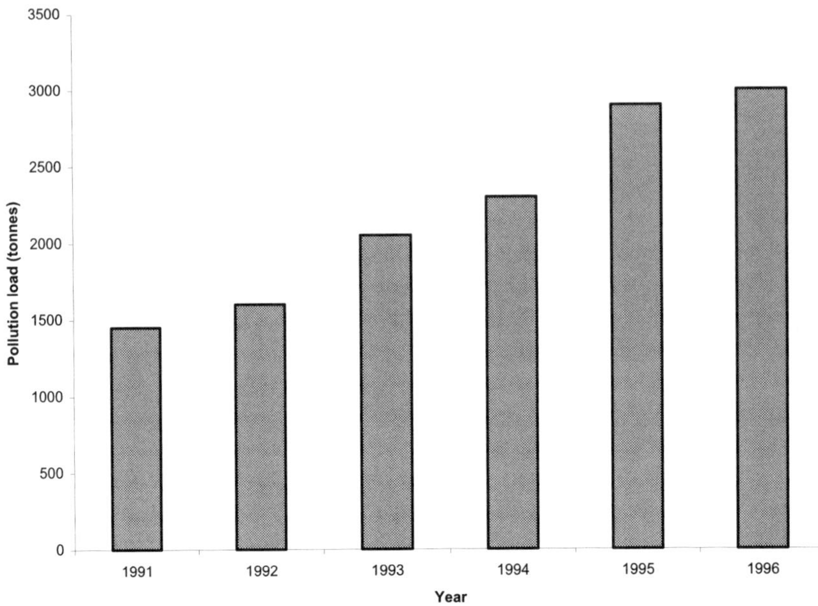

Figure 13: Estimated daily air pollution loads in Delhi (MoEF [23]).

The emission accounts are the logical starting point for the evaluation of abatement options. The following sections discuss each of the main sectors responsible for air pollution in Delhi and elaborate on their particular features, current strategies, and suggest future options.

7.1 Transport sector

Delhi has experienced an exponential growth in personalized vehicles in the last two decades. Figure 14 shows the growth in the number of motor vehicles in the four most populated Indian cities. It is evident that Delhi has registered a phenomenal increase in the number of vehicles. Delhi itself accounts for about 8% of the total registered vehicles (MoST [24]). Furthermore, 95% of the total registered vehicles are personal transport vehicles with the share of two-wheeled vehicles (scooters, motorcycles, mopeds, etc.) being about 67% (Mahajan [25]). Figure 15 describes the number of motor vehicles of different types (two-wheeled vehicles, 3-wheelers, cars, taxis, and buses) in use in Delhi for a given mode in the year 2000 (Mahajan [25]).

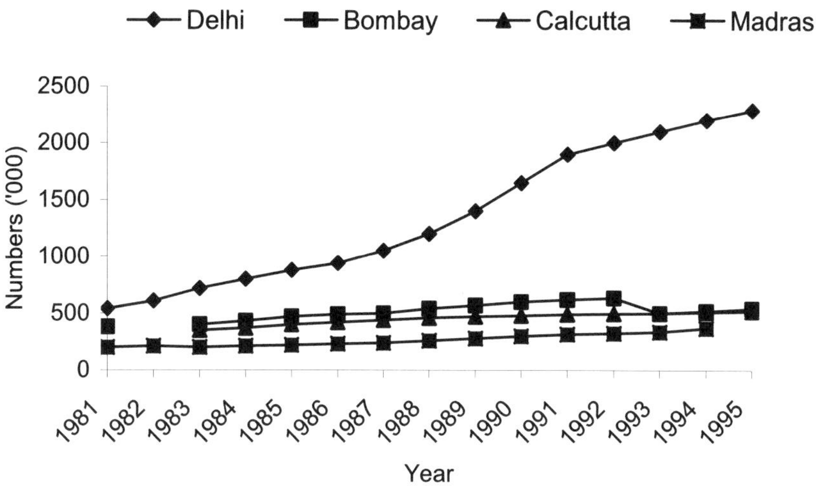

Figure 14: Growth of motor vehicles in metropolitan cities in India (MoST [24]).

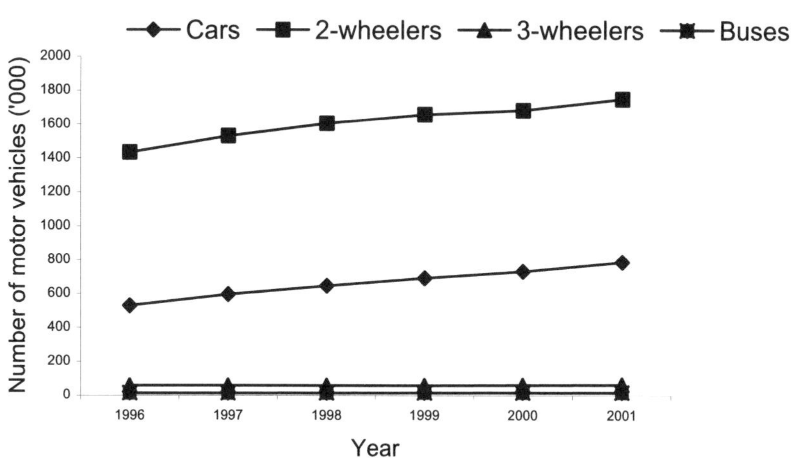

Figure 15: Number of vehicles on Delhi roads in 2000 (Mahajan [25]).

Despite the rise in total road length in Delhi by about 3500 km during 1991–1996, the average vehicle speed has decreased from 18 km/h to 15 km/h in the last decade. This is expected to drop further to 5 km/h by the end of next decade (DUEIIP [10]). The per capita trip rate (PCTR) has also been increasing steadily from 0.49 in 1969 to 0.79 PCTR in 1993 and is likely to exceed 1 by 2011 and 1.2 by 2021 (DUEIIP [10]). Thus, apart from the number of vehicles, other reasons for increase in vehicular pollution include: 2-stroke engines, old vehicles, congested traffic, poor roads, outdated automotive technology and traffic-management system. The cumulative impact of all these factors leads to a much higher vehicular pollution load in Delhi. It was estimated that vehicular traffic emitted about 923 tonnes of pollutant per day during the year 2001 (Mahajan [25]).

7.1.1 Current and future strategies

The impact of vehicular pollutants is greater as these are ground-level emissions. The steps taken by the government to curb Delhi's vehicular pollution in recent years are given in Table 6 (Sengupta [26]). In addition to these, in December 2002, following the Supreme Court (apex court in India) order, Delhi has achieved the distinction of having the largest fleet of CNG (compressed natural gas) buses in the world, numbering over 7400 buses and over 4000 mini-buses. The entire fleet of Taxis (15000) and 3-wheelers (about 45,000) have already been converted into CNG powered engines. Following Delhi, are the cities of Beijing and Seoul where the number of CNG buses are around 1600 and 1000, respectively. Furthermore, the Supreme Court directives reads: 'The Union of India will give priority to the transport sector, including private vehicles all over India with regard to the allocation of CNG'. This means that in Delhi and other cities of India, CNG will be allocated on priority basis and made available for the transport sector.

As a result of these measures, the ambient lead concentration showed a decline in 1998 in the range of 11–60% (Fig. 5). Concentrations of SO_2 and NO_2 are in general within the permissible limits set by CPCB (Section 5.2). Furthermore, the SPM concentration also showed a general declining trend between 1994–2000 (Fig. 16).

Furthermore, one of the important actions taken by the Delhi government is the implementation of DMR (Delhi Metro Rail) project. It comprises a network of underground, elevated, and surface corridors, aggregating 198.5 km, and the entire project will be completed in three phases by 2021. Phase I of the project (62.5 km) has an estimated cost of Rs 105.7 billion and is expected to be completed by 2005 (Table 7). The project is expected to shift the travel demand from personalized vehicles to public transport, reducing waiting and travel time and thus making public transport, services more reliable. In phase I of DMR project, 21.82 million commuter trips and removal of 2600 buses are expected.

Table 6: Measures taken to control vehicular pollution.

Year	Measures taken		
	Fuel quality	Emission norms	Other measures
1991	--	Mass emission norms for CO & hydrocarbons (hot start)	--
1994	Low-leaded gasoline (0.15 g/l)	--	--
1995	Unleaded gasoline (0.013 g/l)	--	Catalytic converter for passenger cars
1996	Fuel quality norms	Evaporative emissions norms introduced.	--
1998	Low sulfur diesel (0.25%)	Cold-start emission norms.	15-year-old commercial vehicles, phased out.
	Leaded gasoline, phased out	Emission norms for vehicles with catalytic converter	Construction for first phase of Delhi Metro Rail Project (DMRP) started.
1999	Low smoke 2T oil for 2-wheeler	Euro I norms	Introduction of CNG powered engines in buses, taxis and other commercial vehicles.
	--	Norms for CNG powered vehicles	Independent fuel-testing laboratory
2000	0.05% sulfur diesel	Euro II norms	Comprehensive inspection and maintenance facilities and Pollution under control (PUC) certificate made mandatory
	No leaded gasoline	--	--
	Gasoline with 1% benzene content in Delhi as against 3% for rest of the country	--	--

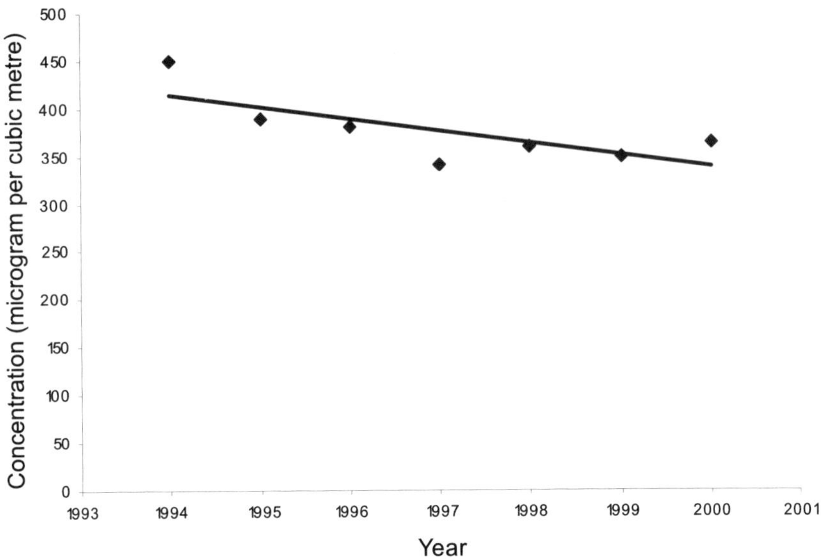

Figure 16: Annual average concentration of suspended particulate matter in Delhi during 1994–2000 (CPCB [5]).

Table 7: Targets for commissioning sections of DMR project Phase I.

Route	Distance	Completion schedule
Shahdara–Tis Hazari	8.3 km	December, 2002
Tis Hazari–Tri Nagar	4.8 km	September, 2003
Tri Nagar–Rithala	8.8 km	March, 2004
Rithala–Barwala	6.1 km	March, 2005
Vishwa Vidhalaya–ISBT	4.0 km	December, 2004
ISBT–Central Secretariat	7.0 km	September, 2005
Connaought Place–West Patel Nagar	7.0 km	September, 2005
West Patel Nagar–Dwarka	16.5 km	September, 2005

In order to realize the benefits of this project fully, due recognition is to be given to the fact that *'an adequate response to the building up of the system can take place only if it really provides a viable alternative'*. For instance, importance will have to be paid to the interchange points, i.e. change points from rail to road. The inter-linkages have to be well developed. Fiscal interventions such as increased parking charges, fuel rationing, congestion charges and area-licensing schemes could also promote effective use of such public transport systems.

In order to control emissions from each new vehicle, India has planned to implement EURO-IV norms by 2005. However, controlling emissions from old vehicles on the road would still remain a challenging task. A policy of quick replacement of on-road vehicles, based on outdated and polluting technologies, may warrant backing of fiscal incentives or disincentives in developing countries, like India.

To conform to the stricter emission norms, the quality of fuel has to be improved and brought in line with European specifications. These may include, near complete elimination of sulfur, aromatic and heavy oils in diesel and lead in petrol by using oxygenated supplement like methyl tertiary butyl ether. Oxygenates boost octane value of the fuel and provide an effective means of reducing harmful emissions such as benzene, toluene, and xylene. Vehicular technology, emission norms and fuel quality should be addressed as a system, and optimization activity needs to be jointly initiated by auto and oil industries.

However, the missing ingredient in current air pollution strategies is the travel-demand management. It is desirable to complement the 'supply-side interventions' with 'demand-management measures', if the ultimate objective is to secure improved levels of air quality. They range from simple traffic-engineering interventions (coordinated signals, reversible lanes, one-way street pairs, and other traffic control devices) to traffic restraints (area-licensing schemes, parking controls, exclusive pedestrian zones, vehicle bans, special bus and high-occupancy vehicle lanes and so on), advance traffic control techniques, and provision of facilities and services to encourage modal shifts (such as sidewalks, bicycle lanes, light and rapid rail transit, and commuter rail).

All these measures would require a policy framework encompassing regulatory, pricing, and taxation mechanisms, and are to be reinforced with effective enforcement so as to encourage the use of clean vehicles and fuels and to modify travel behavior and transport demand.

7.2 Industrial and power sector

Delhi has changed from a city of governance to a multi-functional city, having large-scale industrial as well as a wide range of commercial service-oriented activities. However, the growth of all such activities has been unregulated. Uncontrolled industrial development began mainly in non-confirming areas in form of small-scale units. According to the government of NCT (National Capital Territory) Delhi, there are about 126000 such industrial units in Delhi (GNCTD [27]). According to another survey conducted by the Delhi Government, out of a total of 125000 units, there are 98000 units in the areas categorized as non-conforming by the master plan of Delhi (MoEF [23]).

The Supreme Court of India has directed a number of polluting industries in Delhi to close down. It has also ordered shifting of certain categories of industries from non-conforming to conforming areas. However, the Municipal Corporation of Delhi survey in December 2000 still shows the presence of more than 71000 small-scale manufacturing units in non-confirming zones (DUEIP

[28]). These are namely food, metal, electroplating, textile, paper, paper products and repair services industries.

Authentic information such as number of industries in each sector, scale of production and types of technologies used, pollution control status etc. is not available. Therefore, the air pollution emission scenario from these industries is not clear to-date. However, low combustion energy efficiency and use of inferior fuel (such as combustion of high sulfur-content furnace oil) are the typical characteristics of small-scale industries in India. Therefore, it may be inferred that this sector may have a significant contribution towards Delhi's pollution load, especially for SO_x emissions.

Major air pollution load from the industrial sector in the city comes from various coal-based and gas-based thermal power plants. Three coal based thermal power plants, namely, Rajghat, Indraprastha and Badarpur and one gas-based, namely, DVB GT, are within the city limits. The other two coal-based power plants namely, Faridabad and Dadri and one gas-based (Dadri GT), are adjacent to the Delhi boundary. Table 8 shows the power generation from these plants that contribute to air pollution load in Delhi. Table 9 gives the specific fuel consumption in these plants and Table 10 provides the emission factors. Estimated fuel consumption in power plants is given in Table 11. Table 12 shows the total emission from power plants.

Table 8: Power generated by different thermal power plants in Delhi and adjoining areas (GWh/year) (CEA [29–31]).

Power stations	Fuel	1992/93	1993/94	1994/95
Badarpur 1-5	Coal	4581	4449	4508
I.P. Station	Coal	1380	1146	1112
Rajghat 1-2	Coal	564	618	831
Faridabad 1-3	Coal	894	741	786
Dadri 1-4	Coal	450	1548	2567
Dadri GT	Natural gas	1485	1350	2296
DVB GT	Natural gas	807	786	576

Table 9: Specific fuel consumption in power plants (CEA [32]).

Power plant	Specific fuel consumption
Steam	
Badarpur	0.84 kg/kWh
I.P Station	0.79 kg/kWh
Rajghat	0.82 kg/kWh
Faridabad	0.80 kg/kWh
Dadri 1-4	0.72 kg/kWh
Gas	
Dadri GT	0.30 m^3/kWh
DVB GT	0.36 m^3/kWh

Table 10: Emission factors for power plants (CPCB [33]).

Pollutant	40% ash coal (g/kg)	Natural gas (g/m^3)
SPM	320 (1-eta)	0.08
SO_2	16.60S	0.0096S
CO	1	0.272
HC	0.5	0.016
NO_x	2.64	2.8

Note: eta = efficiency of electrostatic precipitator; S is sulfur content of coal.

Table 11: Estimated fuel consumption in power plants (Coal in thousand tonnes, natural gas in million m^3).

Power station	Fuel	1992/93	1993/94	1994/95
Badarpur 1-5	Coal	3993	3878	3929
I.P. Station	Coal	1131	939	912
Rajghat 1-2	Coal	478	524	705
Faridabad 1-3	Coal	715	593	629
Dadri 1-4	Coal	324	1115	1848
Dadri GT	Natural gas	446	405	689
DVB GT	Natural gas	291	283	208

Table 12: Estimated emissions of air pollutants from plants (tonnes/year).

Pollutant	1992/93	1993/94	1994/95
SO_2	55131	58510	66596
CO	6841	7236	8266
HC	3333	3534	4025
NO_x	19594	20535	23689
SPM	106322	112832	128431

7.2.1 Current and future strategies

Some assorted measures that have been taken so far include: closure and/or shifting of polluting industries, strict enforcement of pollution-control measures in the remaining industries, compulsory use of beneficiated/blended coal with ash content less than 34% in all coal-based thermal power plants from June 2001 (MoEF [34]), increase in green cover, etc. However, the impacts of such measures are not perceptible due to lack of a credible database.

Development and maintenance of a database of industries operating in Delhi would be the first task that needs to be undertaken. This would provide an indispensable input for exploring future policy options that may include a combination of measures such as appropriate industrial siting, strengthening emission standards (more mass-based standards), promotion of cleaner technologies (waste minimization and waste utilization), energy conservation and promotion of energy-efficiency measures and use of market-based instruments for pollution control.

7.3 Domestic Sector

The domestic sector too has its share, although small, in contribution to overall pollution load in Delhi. Table 13 shows the estimated domestic cooking fuel consumption in Delhi as of 1995/96. Based on the emission factors for Indian cooking *chulhas* (stoves) (Table 14), the contribution of domestic sector to air pollution includes 3338 tonnes/yr of SPM; 6319 tonnes/yr CO; and 859 tonnes/yr NO_x (TERI [19]).

Table 13: Estimated domestic cooking fuel consumption (tonnes/year).

Fuel	Rural	Urban
Dung	172112	224277
Coal/coke	709	15467
Charcoal	914	4985
Cooking gas	4634	165065
Wood	59361	171989
Kerosene	17077	235158
Others	1058	13448

Table 14: Emission factors for cooking stoves (g/kg).

	SPM	CO	NO_x
Dung	4.90	31.00	--
Coal/coke	20.00	17.00	--
Charcoal	1.22	10.88	--
Cooking gas	Negligible	0.02	5.25
Wood	1.90	17.00	--
Kerosene	2.80	41.00	--
Crop residue	3.00	27.00	--

According to health-impact studies done by the World Bank, indoor air pollution has emerged as one of the prime environmental health concerns in India (Carter and Hommann [21]). The report points out that biomass fuels combine with open *chulhas* (stoves) and inadequate ventilation, to create problems with the lives of two major vulnerable groups: children and women. The situation is further aggravated by the fact that the use of inefficient and highly polluting fuels in the poorer households that have low nutritional security and low capacity to pay for health care.

7.3.1 Current and future strategies

A positive development in the domestic energy consumption is that liquefied petroleum gas is fast becoming the most popular fuel in Delhi. Figure 18 shows the consumption of LPG in the domestic sector has almost doubled in 1998/99 when compared to 1990/91. Kerosene consumption has dropped by 20% in the same period, reflecting the typical fuel shift towards LPG along with urbanization. Such a shift has a definite impact on the total emission loads from domestic sources as well as on the public health because of reduced indoor and outdoor exposure.

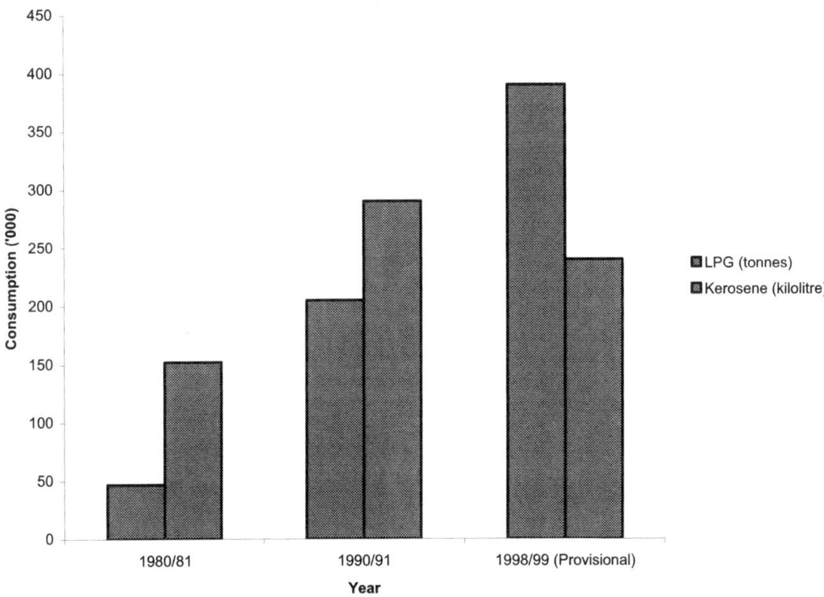

Figure 18: Households using different fuels for cooking (TERI [35]).

In the early 1980s, the Ministry of Non-Conventional Energy Sources (MNES) launched programmes such as the national biogas development programme (NPBD), the national programme for improved cookstoves (NPIC), and the programme for development and promotion of solar cookers (MNES [36]). The intention of these programmes was to promote energy efficiency by providing clean fuels, conserving fuel wood, reducing smoke levels in the cooking space and removing the drudgery of women. A number of NGOs were involved along with the ministry in implementing these programmes and disseminating the technologies. However, these programmes have met with a mixed degree of success and failure due to lack of involvement of local people, especially women. Furthermore, these technologies have a limited application in congested areas of urban slums.

Apart from increased access to clean fuel such as LPG, biogas, and solar, other options to improve the ambient air quality include better ventilation and improved (smokeless) stoves. Although new rural housing units can be designed to ensure better ventilation, this may not be feasible in small and crowded areas, especially in urban slums. In this context, it is important to promote awareness of the dangers related to indoor air pollution so that people may try to adopt ways of minimizing exposure through better kitchen management and infant protection.

Large numbers of urban poor still rely on unprocessed solid fuels. Better targeting of subsidies on kerosene and LPG to the urban poor is necessary in order to ensure reliable, affordable and secure supply of cleaner fuels.

8 Conclusions

Delhi is amongst the most polluted city in the world and the ambient air quality has a tendency of getting worse due to unrestricted growth of the city. This chapter discussed the ongoing strategies and pointed out future options to minimize air emission from transport, manufacturing and power industries and domestic sectors. Besides, continuing and consolidating the ongoing schemes/programmes, new initiatives and definite programmes need to be formulated for the efficient management of urban air pollution.

The institutional framework that supports these concerns has now included all government bodies that are influencing air quality control programmes. Institutions like MNES are further trying its links with the parallel non-government network that operates at grassroots level, thus forming links outside its own structure.

Delhi has integrated environmental issues with the urban planning process. Measures such as adequate allocation of areas under green cover, reduction of energy usage by appropriate design and planning, enforcement of industrial siting guidelines, etc., have been taken. Data on pollution and related parameters are now available on the Internet for public access. However, there is a lack of epidemiological-related data so as to assess the impacts on human health due to the exposure to various pollutants. An integrated study of air quality monitoring, and exposure assessments needs to be conducted to set more reliable air quality standards and suitable emission norms.

Vehicular pollution control is an area that deserves top priority since it contributes significantly to air pollution load in Delhi. Some of the measures taken to curb vehicular pollution are promotion of the public mass transport system (to discourage use of private vehicles), promotion of alternative and cleaner fuels, scrapping of old vehicles and gradual tightening of emission norms, etc. These efforts could further be strengthened by adopting measures such as traffic management (traffic-signal-synchronization), road maintenance, fuel-quality improvement (introduction of ultra-low sulfur diesel) and fiscal measures (such as area licensing, personal vehicle taxation, congestion charges, etc.).

Industrial air pollution control too merits serious attention. The Delhi government has already initiated the process of appropriate location (or shifting) of industries following the Supreme Court orders in 1997. In order to have an effective strategy for existing industries, a credible database of such industries needs to be created. In addition, fiscal measures (for the development and adoption of clean technology and emission reduction) should be promoted along with the strict enforcement of environmental law.

In order to reduce air pollution from domestic fuel combustion, fuel-efficient stoves and cleaner fuels like LPG should be actively promoted. Since access to commercial fuels in rural areas is poor, it is imperative to re-orient and widen the improved cook-stoves and the biogas programs.

The air quality strategies in developed economies are based on a search for technological fix, which is a supply-side option. However, in a developing economy like India, there are limitations to a strategy focused primarily on affecting technology for various reasons that also includes lack of staff for design and enforcement and political unwillingness to implement. It is desirable, therefore, to complement the supply-side interventions with demand-management measures, if the ultimate objective is to secure improved levels of air quality. For this to happen, programmes should be drawn for the effective involvement of non-governmental organizations (NGOs), and citizen's groups.

References

[1] MoEF. *National conservation strategy and policy statement on environment and development,* Ministry of Environment and Forests (MoEF), Government of India: New Delhi, pp. 1–36, 1992.

[2] MoEF. *Policy statement for abatement of pollution*, Ministry of Environment and Forests (MoEF), Government of India: New Delhi, pp. 1–14, 1992.

[3] MoEF. *Environment action programme-India,* Ministry of Environment and Forests (MoEF), Government of India: New Delhi, pp. 1–220, 1993.

[4] MoEF. *Annual report 2000/2001*, Ministry of Environment and Forests (MoEF), Government of India: New Delhi, pp. 1–212, 2001.

[5] CPCB. *Air quality status and trends in India NAAQMS/14/1999-2000,* Central Pollution Control Board (CPCB), Government of India: New Delhi, pp. 1–163, 1999/2000.

[6] CPCB. *Annual report 1995/96,* Central Pollution Control Board (CPCB), Government of India: New Delhi, pp. 1–165, 1996.

[7] IMD. *Climatological tables of observations in India 1951-1980,* Indian Metrological Department (IMD), Government of India: New Delhi, 1999.

[8] Census of India. *Census 2001,* http://www.censuindia.net./results/statedata.html (accessed in January 2002).

[9] DoES. *Delhi statistical handbook,* Directorate of Economics and Statistics (DoES), Government of National Capital Territory of Delhi: New Delhi, pp. 1–135, 1998.

[10] DUEIIP. *Interim status report,* Delhi urban environment and infrastructure improvement project (DUEIIP), Government of Delhi: New Delhi, 2000.

[11] GNCTD. Economic survey of Delhi (1999-2000), Government of National Capital Territory, Delhi (GNCTD): Delhi 2000.

[12] NCRPB. Delhi 1999: a fact sheet, National Capital Region Planning Board (NCRPB): New Delhi 1999.

[13] CPCB. http://envfor.nic.in//cpcb, (accessed in January 2002).

[14] CPCB. *Air quality at major traffic intersections of Delhi (NAAQMS/11/1998-99),* Central Pollution Control Board (CPCB), Government of India: New Delhi, pp. 1–88, 1999.

[15] CPCB. *Parivesh highlights 1999*, Central Pollution Control Board (CPCB), Government of India: New Delhi, pp. 1–42, 1999.

[16] TERI. http://www.teri.res.in, (accessed in January 2002).

[17] Banerjee. S., The enemy within. *Down to Earth,* **5(4)**, pp. 27–32, 1996.

[18] VHAI. *State of India' Health,* Voluntary Health Association of India (VHAI): New Delhi, pp. 1–414, 1992.

[19] TERI. *Natural resource accounting in the Yamuna sub-basin (Report no. 95/EM/61),* Tata Energy Research Institute (TERI): New Delhi, pp. 1–230, 1997.

[20] GHK International Ltd. *Interim status report-Delhi 21,* Delhi Urban Environment and Infrastructure Improvement Project (DUEIIP), May 2000 for MoEF and GNCTD, GHK International Ltd in association with Operations Research Group: New Delhi, 2000.

[21] Carter, B. & Homman, K., *The cost of inaction: valuing the economy-wide cost of environmental degradation in India.* The World Bank: Asian Environment Division, pp. 1–20, 1995.

[22] CPCB. Air pollution and its control, *Parivesh Newsletter,* **2(1)**, Central Pollution Control Board (CPCB), Government of India: New Delhi, pp. 13–15, 1995.

[23] MoEF. *White paper on pollution in Delhi with an action plan*, Ministry of Environment and Forests (MoEF), Government of India: New Delhi, pp. 1–67, 1997.

[24] MoST. *Motor transport statistics of India-1995.* Ministry of Surface Transport (MoST), Government of India: New Delhi, Transport Research Wing, pp. 1–125, 1996.

[25] Mahajan, R., *Integrating health costs and benefits and fuel savings in framing fiscal instruments to control vehicular pollution-a case study of Delhi.* M.Sc. thesis: GGS Indraprstha University, Delhi, pp. 1–95, 2001.

[26] Sengupta, B., Steps taken to control vehicular pollution in India. Paper presented in International Workshop by SIAM 4–5 Dec 2000, Ministry of Environment and Forests (MoEF), Government of India: New Delhi, 2000.

[27] GNCTD. *Delhi statistical Handbook 1999,* Government of National Capital Territory, Delhi (GNCTD): New Delhi, Directorate of Economics & Statistics, pp. 1–325, 1999.
[28] DUEIIP. *Base study for white paper for industries in Delhi,* Delhi urban environment and infrastructure improvement project (DUEIIP), Government of Delhi: New Delhi, 2001.
[29] CEA. Thermal and nuclear generation, *Vidyut Bharti* **15**, Central Electricity Authority (CEA), Government of India: New Delhi, 1992.
[30] CEA. Thermal and nuclear generation, *Vidyut Bharti* **16**, Central Electricity Authority (CEA), Government of India: New Delhi, 1993.
[31] CEA. Thermal and nuclear generation, *Vidyut Bharti* **17**, Central Electricity Authority (CEA), Government of India: New Delhi, 1994.
[32] CEA. *Performance review of thermal power plants,* Central Electricity Authority (CEA), Government of India: New Delhi, pp. 1–203, 1992.
[33] CPCB. *Development of emission factors for thermal power plants,* Final report submitted to Central Pollution Control Board, New Delhi by Environmental Engineers Consortium, Calcutta, pp.1–165, 1994.
[34] MoEF. *Annual report 2000/2001*, Ministry of Environment and Forests (MoEF), Government of India: New Delhi, pp. 1–212, 2001.
[35] TERI. *TERI Energy Data Directory & yearbook 2000/2001,* Tata Energy Research Institute (TERI): New Delhi, pp. 1–452, 2000.
[36] MNES. Annual report 1997-98, Ministry of Non-conventional Energy Sources (MNES), Government of India: New Delhi, pp. 1–110, 1998.

CHAPTER 8

Mexico City's air pollution problem

E. Jauregui
Center of Atmospheric Sciences, National University Mexico. Circuito Exterior Cd. Universitaria C.P. 04510, México D.F.

Abstract

Mexico City is an extreme example of massive urban growth with high population densities in some of its central areas. The associated traffic problems and industrial emissions are exacerbated by the geographical setting of the capital city located in an elevated inland valley leading to poor ventilation and incomplete fuel combustion in vehicles. Anticyclonic weather systems prevailing in winter result in clear skies, thermal inversions and light winds to disperse pollutants. This situation, together with abundant insolation, often leads to elevated levels of air pollutants. The transportation sector contributes with the largest proportion of contaminants, with microbuses, private cars and lorries being the most significant sources. The city has adopted several plans to improve air quality since the late 1970s. Some achievements of these programs have been emissions reduction in vehicles and in the industry. The automobile industry began the introduction of catalytic converters in new vehicles. This allowed a reduction of vehicle emissions despite the increase in the number of vehicles in circulation. Even though substantial progress has been achieved by authorities over the years to clean the air, some programs have lagged such as the conversion of smoke-belching buses and lorries to use cleaner burning fuels. While there is no doubt that Mexico City's atmosphere remains badly polluted, some significant advances have been achieved. Since 1993 not only lead but sulfur dioxide and carbon monoxide have been dramatically reduced to concentrations below the standard for health. This progress in improving air quality suggests that even huge cities in the developing-world cities can begin to reduce pollution.

1 Introduction

Mexico City has been plagued by high levels of pollution for several decades. The capital city is located in an elevated valley (approximately 2250 m above sea level) in central Mexico. The climate is tropical, tempered by altitude. According to the 2000 census the population of the Federal District was 8.6 million from the 16 delegations plus 9.1 million from 18 conurbated municipalities. The total population of the metropolis being 17.7 million extended over an area of 1547.1 km^2 [1]. According to current trends the population of Mexico City will increase by 2% between 1995 and 2020 faster than the nation's expected growth of 1.7%/year. While the central area population has declined, the more distant areas are experiencing the most rapid growth. Anticyclonic weather systems prevailing in winter/spring result in clear skies, thermal surface inversions and light winds to disperse pollutants. The surrounding mountains are another unfavorable factor in reducing ventilation.

The abundant solar irradiation during the cool season together with vehicle and industrial emissions provide the conditions for high levels of ozone that in conjunction with fine particles, constitute the main offenders to the health of the population. The capital city of Mexico is a typical example of massive urban growth and development with extremely high population densities in some of its central districts experiencing high motor vehicle congestion. The situation often leads to elevated levels of air pollution emission. The emissions inventory updated to 1998 for the metropolitan area indicated that total emissions were of the order of 2.5 million tons of contaminants per year [2].

The transportation sector contributes with the largest proportion of contaminants (CO, HC, and No_x) with microbuses, private cars, and lorries being the most significant sources. Transport sources account for nearly all CO, more than 80% of nitrogen oxides, 40% of VOCS (volatile organic compounds), 21% of sulfur dioxide and 36% of fine particles [2]. Table 1 shows emissions inventory in the Mexico City Metropolitan Area (MCMA) for the various sectors.

Table 1: Emissions inventory in the Mexico City metropolitan area in 1998 in percent (RAMA, 2001)

Sector	**PM10**	**SO_2**	**CO**	**NO_X**	**HC**
Industry	15.6	55.4	0.5	13.1	5.0
Services	8.4	23.8	1.5	4.8	52.1
Vegetation/soil	40.1	NA	NA	1.6	3.3
Transport	35.9	20.8	98	80.5	39.5

Industry (including power generation) and services are the main sources of sulfur dioxide and hydrocarbons. Sources of fine particles (those particles measuring less than 10 microns affecting human health) are estimated to be either from natural sources (from soil erosion) as well as from diesel lorries. Public transport is accomplished mainly by minibuses (55.2%), buses (6.7%) and the underground system (13%), while private cars constitute 17.6% of all transportation. Together they consume 44 million litres of fuel per day, INEGI [3]. In terms of the various transport means, minibuses are responsible for the largest proportion of trips with 55% followed by the private car (17.6%) (see Fig. 1).

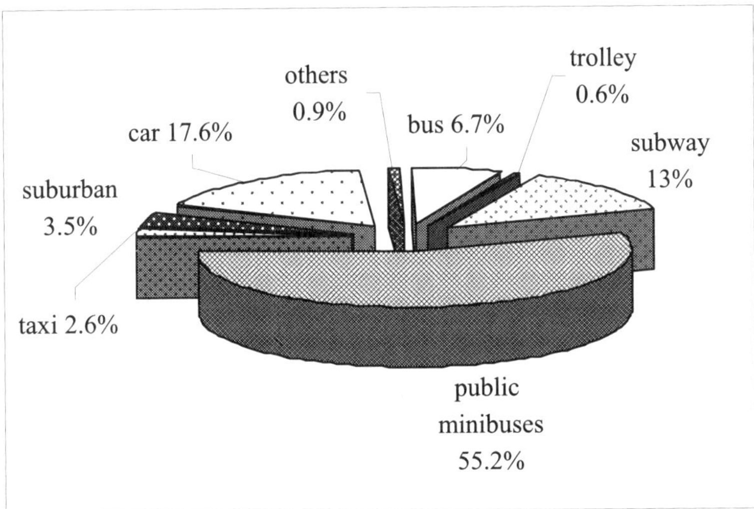

Figure 1: Number of trips made by various means of transportation (in %).
Source: INEGI, DDF–GEM, 1994.

The city has adopted several plans to improve the air quality. The first program addressing the problem of air pollution was installed in 1979. Through this plan local authorities intended to attain standards for the main pollutants. In 1988 a new program was implemented to fight air pollution (PICCA). In 1996 another program to improve the air quality was created (1996–2000) named PROAIRE; the elaboration of this program also included the integrated participation of other government institutions. Some achievements of the first PROAIRE program were emission reduction in vehicles. The automotive industry began the introduction of catalytic converters in new vehicles. At the same time the state oil industry began distribution of unleaded gasoline required by the operation of the catalytic converter. This allowed a reduction of vehicle emissions despite the increase in the number of vehicles in circulation. Finally, in 2002 yet another air quality management plan for the Mexico City metropolitan area was implemented: PROAIRE 2002–2010. The main objective of the new PROAIRE was, as in the previous program the protection of public health. The

89 measures covering a full range of relevant sectors were developed with the aim to achieve reductions in the indices of ozone and fine dust (PM10), to lower the number of days that exceed the norm and limit concentrations of carbon monoxide, sulfur dioxide, lead and nitrogen oxide to established standards. Critics of the program [4] have noted that more than half of the options in the new PROAIRE are old ones that have never been implemented and that many times federal and local government fail to implement policies due to lack of coordination. In addition the current system of industry auditing needs to be improved by creating a surveillance program over a period of years.

2 The air pollution-monitoring network

The problem of air pollution has been well recognized by both the local authorities and citizens since the 1970s and air pollution has been measured routinely since the mid 1980s when it became a social concern. Mexico City maintains a comprehensive air quality monitoring network providing real–time data on all major air pollutants. Air quality measurements in the city are conducted through a monitoring system: the RAMA network recently renamed Sistema de Monitoreo Atmosférico (SIMAT) of automatic measurements is one part of the system. In addition, there is a manual network of (19) stations where measurements of particles (TSP) and lead are taken as well as dry deposition (for acid rain detection). There are also frequent measurements of UV radiation and basic meteorological parameters such as temperature, relative humidity and wind. The monitoring system is designed for surveillance in order to alert the population to dangerous levels of pollutants.

Air quality measurements are presented in terms of the air quality index (IMECA or Air Quality Metropolitan Index) which are reported to the public every hour and are used to determine environmental contingencies. Pollution levels in IMECA units are reported to the public on a scale 0–500. For this purpose the city is divided into 5 zones: downtown, NE, NW, SW and SE districts each one containing at least 2 monitor stations.

Table 2: The air quality index (IMECA) of Mexico City.

IMECA	Condition	Impact on health
0–100	Healthy	None
101–200	Unsatisfactory	Stinging eyes, burning throat in susceptible persons
201–300	Bad	Avoid open air activities. Respiratory symptoms on large numbers of the population. Schools close.
301–500	Very harmful	Respiratory symptoms become critical in children and the elder.

The Air quality index or IMECA varies most of the time within the limits of the class labeled as unsatisfactory (101–200) with ozone being the principal offender. Since the prevailing winds blow from the NE and the main sources of emissions are located in the northern sector, the cloud of precursors (nitrogen oxides, hidrocarbons) drifts toward the residential quarters in the SW where the higher concentrations of ozone are observed.

In addition to the industrial emissions the northeastern sector is plagued by high levels of dust coming from the dried Texcoco bed on the valley plains to the east [5].

Since 1998 a smog alert (or atmospheric contingency program, (PCAA)) is declared in the whole metropolitan area of Mexico City when ozone concentrations reach beyond 0.282 ppm or 240 IMECA units (Phase I). Also, since 1998 a smog alert is declared in one of the 5 zones when PM10 fine particles in that area reach concentrations higher than 300 micrograms/m^3 (24–hour moving average).

3 The public transport problem

Even though substantial progress has been achieved by authorities over the years to clean the air in the capital city some programs have lagged such as the conversion of smoke–belching buses and lorries to use cleaner burning fuels. Currently there are 22,834 microbuses circulating in the Federal District alone of which 36% are 1991 models and older. Furthermore, no new microbuses have been built since 1994. Microbuses were designed as cargo vehicles and were not intended for use in public transportation. Consequently many are in poor shape[[6]]. There is a trust fund to renew the microbus fleet. In 2001 with a budget of 8 million dollars, 800 units were substituted. The program is moving slowly and will take many years to complete at this pace [6].

4 The taxi fleet problem

There are 103,694 taxis in the Federal District plus 7,000 in the conurbated municipalities out of which 49% are older than 10 years that should be off the road according to regulations. Most taxi owners (70%) have only one vehicle. There are a few formal taxi companies. In addition taxi theft is a problem. It is apparently easy to steal a car and turn it into an illegal taxi [7].

The taxi–substitution program aims to substitute old, polluting vehicles with newer, cleaner and safer vehicles and at the same time promotes organized taxi stands to replace the free taxis that cruise the streets looking for passengers thereby generating extra pollution. Since safety is an issue for taxi riders, attempts by the authorities are being made to gain control over taxi operators along with the renewal program so that environmental goals (taxi-stand taxis pollute less than free taxis that cruise the streets) and safety goals (taxi-stand taxis are safer for passengers) could by achieved simultaneously [7].

5 Air pollutant trends

Some achievements of PROAIRE have been significant: while from 1995 to 2000 the number of days when the daily maximum ozone concentrations was exceeded slightly declined from 319 to 273 days. The number of smog alerts or environmental contingencies, which peaked at 113 per year in 1991, was reduced to 2 in 2001 (see Fig. 2). But still the number of days when the norm for ozone was exceeded was very high and constituted 75% of the days (or 273 days). In other words, ozone concentrations declined from 1990 to 2000 but the standard for ozone (0.11 ppm) was still exceeded 75% of the time in 2001 (see Fig. 2). Local experts agree that the trend is strongly in the right direction compared with the worst days from 1990 to 1992 when ozone hit emergency levels on as many as 110 days per year. Since 1999 even after a stricter threshold had been introduced the year before, ozone levels have triggered emergencies on less than 5 days per year (Fig. 2).

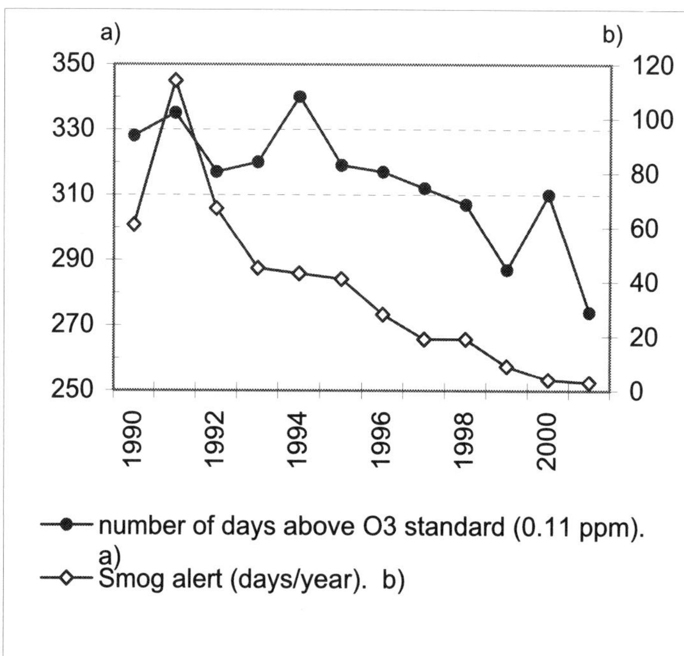

Figure 2: Number of days on which the standard for ozone (0.11 parts per million, ppm) was exceeded and number of days per year when the smog alert (0.257 ppm) was declared. Source: RAMA.

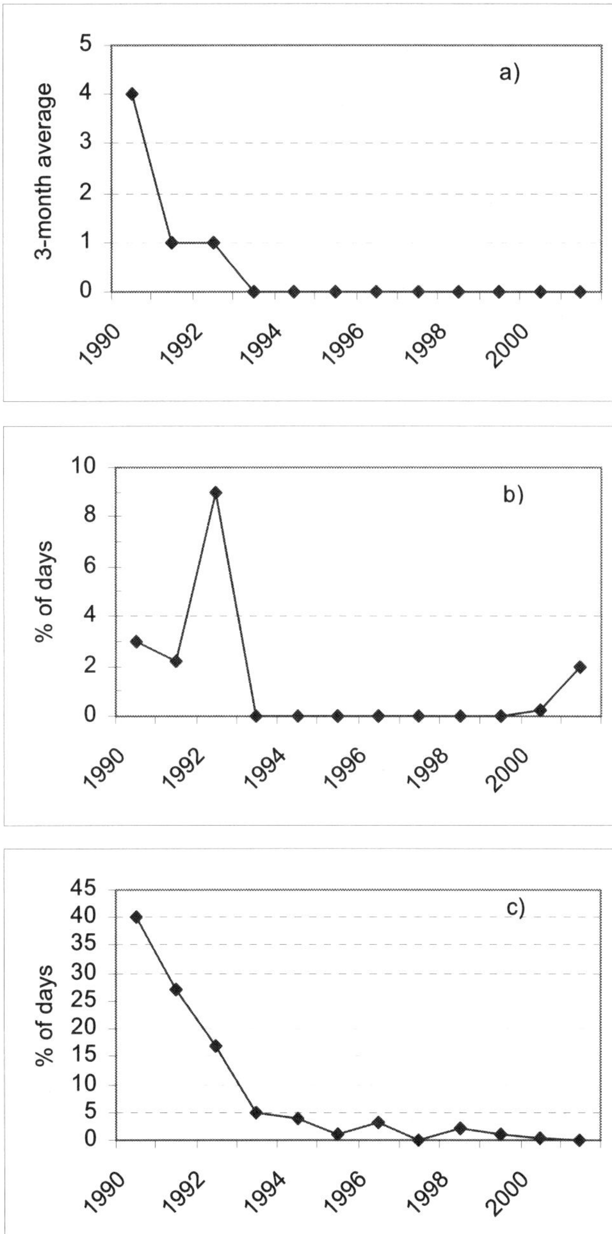

Figure 3: a) Number of 3-month average above the standard for lead b) Percentage of days/year when standard for sulfur dioxide (0.13 ppm) was exceeded and c) Percentage of days/year when standard for carbon monoxide was exceeded.

While there is no doubt that the Mexico City atmosphere remains badly polluted; some significant advances have been achieved and the real test whether pollution has declined (for some of the pollutants) is convincing. One of the most relevant achievements of the air quality program in the MCMA has been the elimination of lead emissions in vehicles. The excessively high concentrations of lead in the air that were common in the 1980s (three times the norm) have been drastically reduced to levels well below the acceptable standard. Since 1993 not only lead but, sulfur dioxide and carbon monoxide have also been dramatically reduced to concentrations below the standard for health (see Fig. 3).

Similar progress although not nearly as striking has occurred during recent years in reducing small particles (PM10) which can penetrate the lungs and cause serious health problems. Jumps in small-particulate pollution have occurred during the 1990s with decreasing frequency and the number of days when the norm was exceed has decreased from 262 days in 1990 to 30 days in 2001 (Fig. 4).

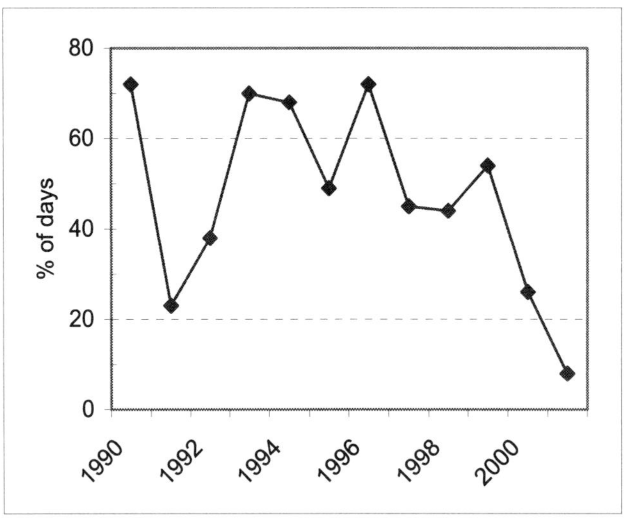

Figure 4: Percentage of days above the PM10 norm (150 micrograms/m3) in 24 hrs. In the metropolitan zone of Mexico city. Period 1990–2001. Source: RAMA.

However, as recently as in 1999 fine dust (from natural sources and vehicle emissions) exceeded the norm in 197 days or 54% of the time. The high variability in concentrations of fine dust (PM10) from one year to the next is likely to be due to natural factors such as forest fires and weather conditions leading to blowing dust. Efforts are being undertaken by the local authorities to reduce the sources of dust (including the fine fraction) by recuperation, conservation and extension of green urban areas as well as by recuperation of eroded areas in the valley plains to the east of the city. One accomplishment of the Water Commission for the Valley has been the reduction of dust sources by reclamation of 10 000 hectares of salty semiarid waste land of ex–Lake Texcoco. The regeneration of this area of the valley by revegetation and construction of lakes has helped favourably modify the urban climate and air quality in the whole metropolitan area [8].

Concentrations of nitrogen oxides from vehicular emissions, the so-called precursors for the formation of ozone have shown a slight decreasing trend since 1996. However, this trend is not significant and the local authorities are aware of the need to reduce both precursors: VOC and nitrogen oxides. However, experts caution that under some circunstances reduction in emission of nitrogen oxides migth actually lead to higher ozone levels in some parts of the MCMA [9].

6 Impact of air pollution on health

The major concern over air pollution in the MCMA and the principal driver for policy actions, is human health. Perhaps the most important concern is the possibility that long–term exposure to fine particles (PM10) could increase the chronic mortality caused by respiratory or cardiovascular disease. Ozone has strong and clear effects on respiratory functions such as eye irritation and coughs and on hospital admissions for asthma.

Preliminary results from a study [10] on long-term health effects of air pollution in the MCMA, reveal that in a recent survey of 4500 students over a period of five years show a higher incidence of coughing with phlegm in the northeast part of the city (where dust particles prevail) and asthma in the southwestern sector (where ozone concentrations are usually high). It is concluded that the average pulmonary function and growth of children appear to diminish with ozone and PM10 exposure. Many ozone health effects come from its impact on lung membranes. Ozone also affects lung secretions and reduces tracheobronchial diameter. Exposure results in a significant reduction in breathing capacity that can cause severe problems for persons with asthma. According to a report by a group of specialists from the Ministry of Environmental Health [11] at base level (or 100 IMECA points) 8% of the population experience symptoms. At 250 IMECAs this increases to about 15% of the population. As ozone levels increase symptoms increase from eye irritation to tear secretion, headaches, sore throats and respiratory difficulty. For each smog contingency episode direct health costs amount to 5 million dollars. Similar effects from ozone were found in the case of PM10 only with a lag of 7–10 days (3–5 days for children) between exposure and health effects [7].

7 Conclusions

The air pollution problems facing Mexico City are similar to those of large cities in the developing world. Their pollution problems are the result of rapid growth of the population and limited resources for adequate implementation of pollution control policies. Against these obstacles many of the measures taken by the local authorities in Mexico City have been successful in significantly reducing its infamous air pollution thanks to one of the world's most aggressive anti-smog initiatives [12]. Some aspects of air pollution have been addressed successfully through the policies and actions enacted in the 1990s. However, some pollutants persist with concentrations exceeding the norm: ozone, particulates and nitrogen oxides.

Since 2000 an initiative of MIT has brought together participation of an interdisciplinary group of Mexican and U.S. researchers (called the Mexico City Program; MCP) with the purpose "to improve understanding of environmental problems in the Mexico City Metropolitan Area". The group (coordinated by Mario and Luisa Molina) has convened a series of workshops in close collaboration with local environmental authorities (the Comision Ambiental Metropolitana, or CAM). The main goal of the program is "to provide objective and balanced assessments of the causes and possible solutions to local, regional and global atmospheric pollution problems". Recommendations emanated from the MIT workshops have been of great help in the formulation of the new PROAIRE pollution control policies. In addition local authorities (SIMAT) organize periodic meetings to promote collaboration on air pollution issues and maintain a page to enable the exchange of data for the academic and public communities. Although much remains to be done, it is clear that Mexico City has made significant advances in reducing its notorious air pollution problem during the 1990s. This progress in improving air quality suggests that even huge developing world cities can begin to reduce pollution.

References

[1] Garza, G., 2000. *Mexico City's territorial extension.* In: La ciudad de México al final del segundo milenio, G. Garza (coordinator), pp. 237–246 in 1998.

[2] Ramos, R., Status of Air Quality Monitoring in Mexico City. 5th Workshop on Mexico City's Air Quality. MIT report No. 28 Cambridge, Mass, p. 32., 2001.

[3] INEGI, Instituto Nacional de Geografía e Informática.

[4] Fernández. SEMARNAT. Financing the new air quality management program. In: Fifth Workshop on Mexico City air quality. Ixtapan, 21–24 Jan. 2002. MIT. Cambridge.

[5] Jauregui, E., The dust storms of Mexico City. *Int. J. of Climatology*, **9**, pp. 169–180, London, 1989.

[6] Mostache, A., *Substitution of microbuses by new ones. G DF.* In: 5th MIT Workshop on Mexico City Air Quality, p. 8, 2002.

[7] Gonzalez, M.E., *Design of the taxi fleet renewal program. Ministy of the Environment.* In: 5th MIT Workshop on Mexico City's Air Quality. MIT Report. Cambridge, p. 8, 2001.

[8] Jauregui, E., Effects of revegetation and new artificial water bodies on the climate of northeast Mexico City. *Energy and Buildings*, **15–16**, pp. 497–455, 1990/91.

[9] Molina, L. & Molina, M. (coordinators), Integrated strategy for air quality management in the MCMA. 5th MIT Workshop, Ixtapan.

[10] Hernández, M., *Long-term health effects of air pollution in the MCMA: the EMPECE study*. National Institute of Public Health. In: 5th MIT Workshop. Ixtapan. p. 18, 2002.

[11] Olaiz, G., *Health effects*. In: Integrated Program on Urban, Regional and Global Air Pollution: Mexico Ctiy Case Study. Planning Workshp. Feb. 16, 1999.

[12] Smith, J., *Little by little breathing easier in Mexico City*. Los Angeles Times. Jan. 15, 2002.

CHAPTER 9

Air quality management in Greater Tehran Metropolitan Area

A. Salahi[1] & Sh. Geranfar[2]
[1] Department of Natural Resources Protection, Research Institute of Forests and Rangelands, Iran.
[2] Polytechnic University, Iran

Abstract

Investigation of air quality in the Tehran Metropolitan Area between 1992–2002 indicated that there are significant amounts of nitrate ion (NO_3^-) over 30 kg/ha/year, and sulfate ion (SO_4^{2-}) deposited in some southern areas of Tehran is approximately 300 kg/ha/year. Such huge amounts of air pollutants come down with snow and rain. Tehran's weather is still polluted and getting worse everyday. There are approximately 12.5 million liters of petrol fuels consumed on a daily basis in this city. Out of that total 9400 tons are carbon monoxide (CO), 145 tons nitrogen oxides (NO_x) and 1100 tons unburned hydrocarbons. The primary source of poor quality of air and pollution in Tehran is passenger automobiles. Air-borne particles account for 75% of Tehran's awful air quality. A total of 10200 tons of dirty particles is produced by passenger cars on a daily basis in Tehran. A further 6100 tons is produced by public transportation vehicles reaching a total of 16300 tons of pollutants per day in Tehran. In Tehran, 70% of trucks and buses and 21% of minibuses emit sulfur dioxide (SO_2) gas. Every 280 days or so, the air inversion phenomenon grips the city of Tehran. This occurrence becomes more critical during the winter months. In addition to dry deposition monitoring systems, acid rain and wet deposition measurement were carried out in Tehran.

1 Introduction

The goal of air quality management is to maintain a quality of air that protects human health and welfare. This goal recognizes that air quality must be maintained at levels that protect human health, but must also provide protection for animals, plants, (crops, forests and natural vegetation), ecosystems, materials and aesthetics, such as natural levels of visibility. And to achieve this air quality goal, it is necessary to develop policies and strategies. Government policy is the foundation for air quality management. A policy framework refers to policies in several areas, including transport, energy, planning, development and the environment. Achievements in air quality management underlie increased economic and social welfare in many developing countries. Sound air quality management is also a proven way of enhancing public health, since air pollution is associated with increase in outpatient visits due to respiratory and cardiovascular diseases in hospital admissions and in daily mortality. It is suggested that without clean air, a sound economic development becomes virtually impossible and social conflicts inevitable. Enormous progress is needed in developing a clean air implementation plan for urban areas, especially in the Tehran metropolitan area. Many people are still exposed to enhanced concentrations of gaseous and particle compounds in Tehran's air. And the use of open fires for indoor cooking and heating exposes people to quite substantial concentrations of suspended particulate matter higher than ambient concentrations according to the limited measurements available. Other sources of air pollution include industrial and vehicular emissions as well as vegetation and waste fires. In recent years many projects for emissions reduction in Tehran have included electrical buses, subway and metro, unleaded petrol, gas consumption in vehicles and residential sites, traffic restriction plan, transferring the pollution emitter industries to out of Tehran, and solar energy. These experiments also used method of wet deposition by precipitation. For measurement and sampling, specific instrument has been designed and were positioned in many stations in the south and west of Tehran.

2 More about air pollution chart in Tehran

2.1 Carbon monoxide (CO)

Carbon monoxide is a colorless, odorless, tasteless gas that is by far the most abundant of the criteria pollutants. It is produced when carbonaceous fuels are burned under incomplete combustion. Carbon monoxide interferes with the blood's ability to carry oxygen to the cells of the body when inhaled; it readily binds to hemoglobin in the blood stream to form carboxyhemoglobin (COHb). Carbon monoxide, in fact, has a much greater affinity for hemoglobin than does oxygen, so that even small amounts of CO can seriously reduce the amount of oxygen conveyed throughout the body. With the bloodstream carrying less oxygen, brain function is affected and heart rate increases in an attempt to offset the oxygen deficit.

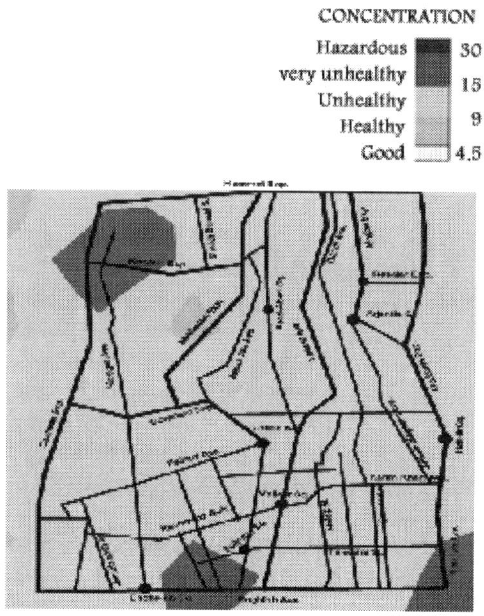

Figure 1: Carbon monoxide (CO) pollution in 6th District of Tehran Municipality.

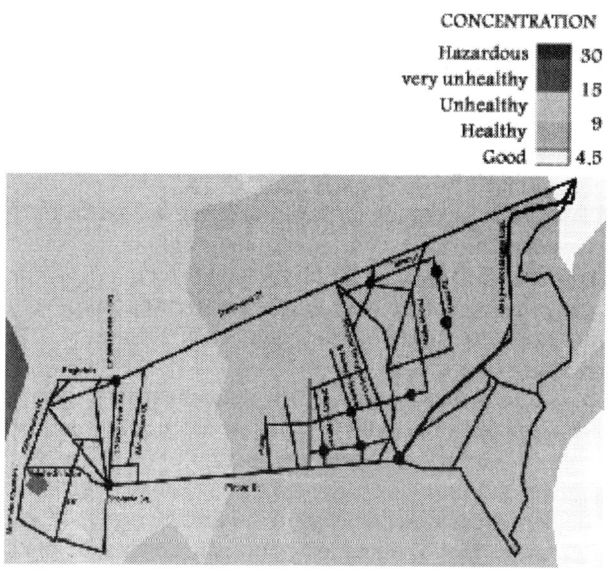

Figure 2: Carbon monoxide (CO) pollution in 13th District of Tehran Municipality.

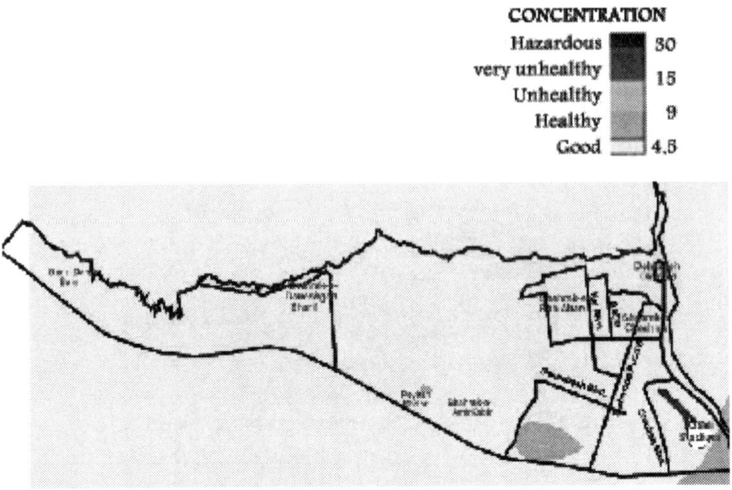

Figure 3: Carbon Monoxide (CO) pollution in 22th District of Tehran Municipality.

2.2 Oxides of nitrogen (NO_x)

Although 7 oxides of nitrogen are known, nitrogen dioxide is the only one with an ambient air quality standard. The main source of nitrogen oxides (NO_x) is combustion of fossil fuels. The direct health effects of NO_2 itself are still uncertain. At higher concentrations than are normally found in the atmosphere, it is an acute irritant. Prolonged exposure to relatively low concentrations, typical of some polluted environments, has been linked to increased bronchitis in children. When air is humid, NO_2 reacts with water vapor to form nitric acid (HNO_3). This corrosive substance also contributes to the acid rain problem and can cause damage to plants and corrosion of metal surfaces. Moreover, because nitrogen oxides are highly reactive gases, they play an important role in the formation of smog. Photochemical smog and ozone (O_3) When oxides of nitrogen, various hydrocarbons, and sunlight come together, they can initiate a complex set of reactions that produce a number of secondary pollutants known as photochemical oxidants.

Ozone (O_3) is the most abundant of the photochemical oxidants, and it is the one for which an ambient air quality standard has been written. Although it is responsible for many of the undesirable properties of photochemical smog, from chest constriction and irritation of the mucous membrane in people, to the cracking of rubber products and damage to vegetation. Ozone is not itself a cause of the eye irritation that is the most common complaint about smog. Eye irritation is caused by other components of photochemical smog. But usually ozone is considered a good indication of photochemical smog intensity.

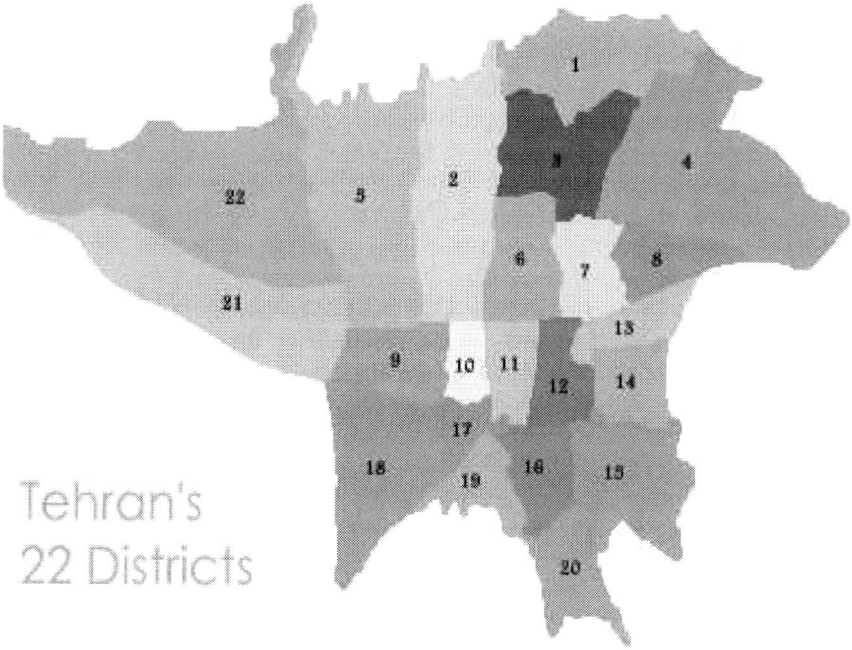

Figure 4: Tehran's 22 Districts.

Sulfur dioxide (SO_2) is a colorless and corrosive gas that originates largely from the combustion of sulfur-containing fuels, primarily coal and oil. Important sources include power plants, smelters, petroleum refineries, and pulp and paper mills together with motor vehicles. Once SO_2 is in the air, it is frequently transformed into sulfur trioxide (SO_3), which reacts with water vapor or water droplets to form sulfuric acid (H_2SO_4). Very tiny particles act as a medium on which the acidic sulfate ion (SO_4^{2-}) is carried over long distances in the atmosphere. When it is "washed" out of the air or deposited on surfaces it contributes to a serious environmental problem known as acid precipitation. The tiny acid-bearing particles can also be drawn deep into the lungs where they can be very damaging. Next to smoking, sulfur dioxide and sulfate ions are probably the most serious contributors to air-pollution-related health problems.

2.3 Particulate matter (PM -10)

A number of terms are used to categorize particulates depending on their size and phase (liquid or solid). The most general term is aerosol, which applies to any tiny particles, liquid or solid, dispersed in the atmosphere. By sources such as factories, power plants, motor vehicles, construction activity, fires and natural windblown dust. Smoke, soot, dust, ash, pollen and many other materials are included. Particulates are frequently the most obvious form of air pollution because they reduce visibility and leave deposits of dirt on the surfaces with which they come in contact. Originally, total suspended particulate (TSP) was the indicator used to represent this category. It includes all particles up to 45 micrometres in diameter. In 1977 the US EPA set new standards that related to particulate smaller than 10 micrometres (PM-10). Emphasis has been placed on these smaller particles because they are responsible for most adverse health effects caused by particulates. These particles are so tiny that they can be drawn into the lungs and damage respiratory tissues.

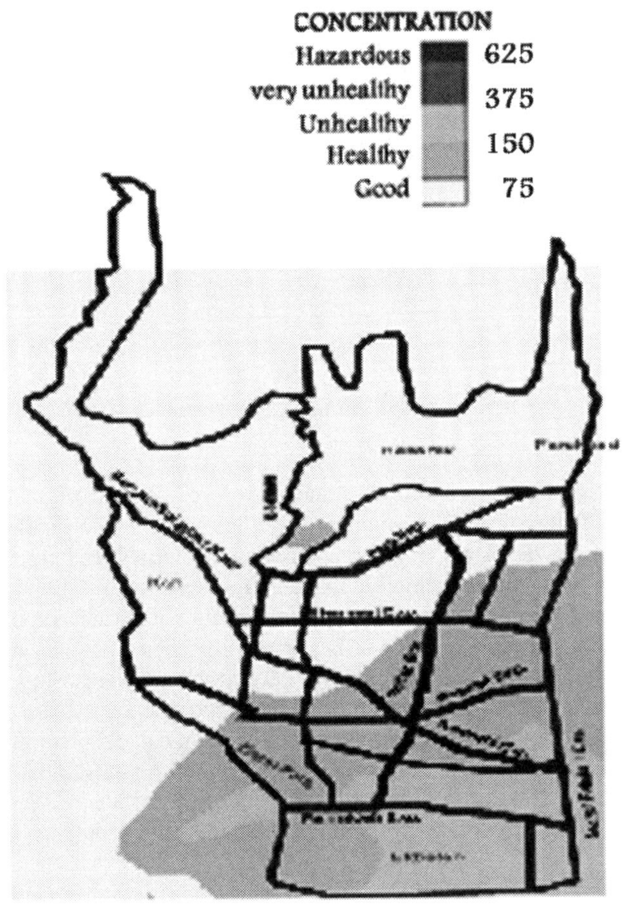

Figure 5: Particulate matter (PM -10) in 5th District of Tehran Municipality.

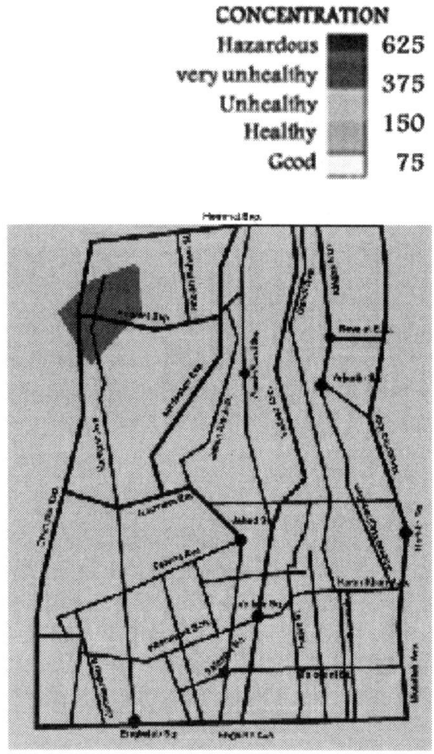

Figure 6: Particulate matter (PM -10) in 6th District of Tehran Municipality.

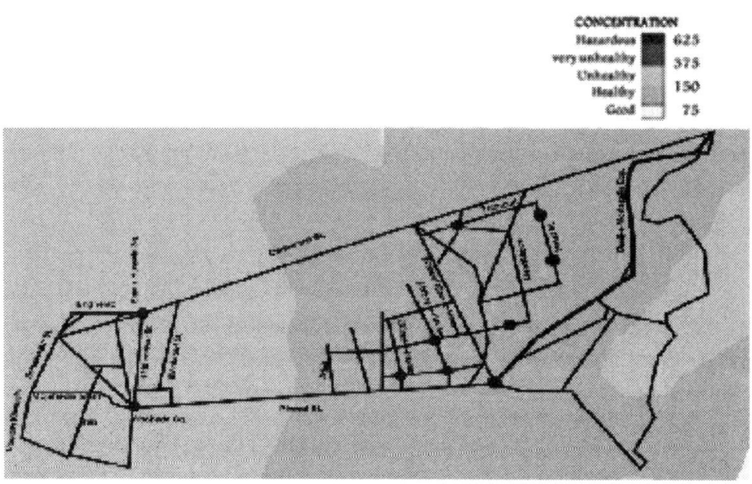

Figure 7: Particulate matter (PM-10) in 13th District of Tehran Municipality.

2.4 Air Quality Data

"Air Quality", by definition, refers to ambient concentrations of air pollutants, when these figures are compared to the Air Quality Standard for each pollutant. There are hundreds of air pollutants in the atmosphere, but six ubiquitous pollutants with potential health hazard have been identified as criteria pollutants. The Air Quality Data for 5 pollutants out of the 6 criteria pollutants (with the exception of lead for which a continuous monitoring method does not exist), i.e. carbon monoxide, sulfur dioxide, nitrogen dioxide, particulates and ozone are measured. The data is gathered continuously in 4 ambient air-monitoring stations in Tehran. The stations are located in various locations in Tehran. The monitoring (sampling and analysis) is carrying out continuously but the data for each day covers a 24-hour period ending at 10 a.m. of the same day.

2.5 Air Quality Standards

Air Quality Standards have been established at two levels: primary and secondary. Primary standards are required to be set at levels that will protect public health and include on "adequate margin of safety" regardless of whether the standards are economically or technologically achievable. Primary standards must protect even the most sensitive individuals, including the elderly and those with respiratory ailments. Secondary Air Quality Standards are meant to be even more stringent than primary standards. Secondary standards are established to protect public welfare (e.g. structures, crops, animals, fabrics). Given the difficulty in achieving primary standards, secondary standards have played almost no role in air pollution control policy and, in fact, they have usually been set at the same levels as primary standards. Urban atmosphere may be contaminated by hundreds of different pollutants. Only a relatively few of these have been identified as having the potential for causing adverse health effects from exposures normally encountered in the ambient environment.

Air Quality Standards now exist for six criteria pollutants: carbon monoxide, nitrogen dioxide, ozone, sulfur dioxide, lead and particulates with aerodynamic diameter less than or equal 10mm (PM-10). Substances that are emitted directly from sources are called primary pollutants. They are certainly not innocuous, however, these pollutants do not, of themselves, produce all of the adverse effects of air pollution. The other pollutants manufactured in the air are called secondary pollutants; they are responsible for most of the smog, haze and eye irritation and for many of the forms of plant and material damage attributed to air pollution.

2.6 PSI

Air Quality indices have been devised for categorizing the air quality measurement of several individual pollutants by one composite number. The index has been adopted by US EPA and is used to report to the public an overall assessment of a given day's air quality. It is called the Pollutant Standard Index

(PSI). The PSI converts air pollution concentrations to a simple number between zero and 500 and assigns a descriptive term such as "good" or "moderate" to that value. A PSI value of 100 indicates that at least one pollutant reached its ambient air quality standard on that day. A PSI of 200 corresponds to a first-stage alert during which time elderly persons with existing heart or lung disease are advised to stay indoors and reduce their physical activity. A second stage alert is called when the PSI reaches 300, at which point the general public is advised to avoid outdoor activity. The descriptive terms corresponding to PSI numbers are given in the following table.

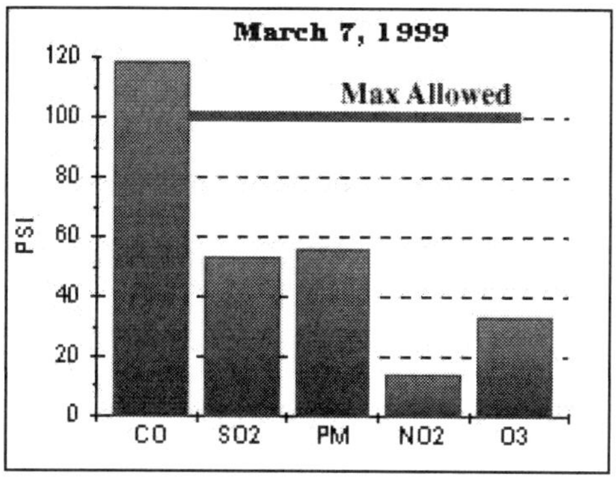

Figure 8: Pollutant Standard Index (PM=Particulate Matter).

Table 1: Air pollutant value

PSI	Descriptor	CO(8 hr) m/m^3	SO_2 (24 hr) mg/m^3	PM-10 (24 hr) mg/m^3	O_3 (1 hr) mg/m^3	NO_2(1 hr) mg/m^3
0–49	Good	0–4.9	0–79	0–50	0–119	-
50–99	Moderate	5–9.9	80–364	50–150	120–235	-
100–199	Unhealthful	10–16.9	365–799	150–380	235–400	-
200–299	Very unhealthful	17–33.9	800–1599	380–420	400–800	1130–2259
>=300	Hazardous	>=34	>=1600	>=420	>=800	>=2260

3 Nitrogen deposition in the greater Tehran metropolitan area

Investigation on air pollution in Tehran metropolitan area between 1992–2000, indicated that there are significant amounts of nitrate ion (NO_3^-) over 30 kg/ha/year, deposited as wet deposition compared to 13 kg/ha/year in the Chitgar Parkland near the Tehran metropolitan area. The amount of NO_3^- in warm seasons is twice that of cold seasons (Fig. 9), and there was a significant difference between cold and warm seasons (Table 2). Annual wet deposition of ammonia (NH_3) was 10 kg/ha/year in the Chitgar Parkland [1].

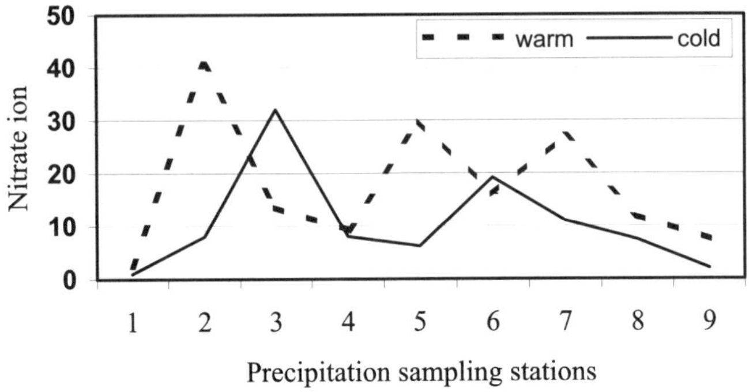

Figure 9: Annual wet deposition of nitrate ion (NO_3^-) in kg/ha/year inside Chitgar Parkland during warm and cold seasons.

Table 2: Dependent variable: pH; independent variable: NO_3 regression analysis

Parameter	Estimate	St. Error	T- Value	Prob.	
Intercept	6.048	0.199	30.457	0.0000	
Slope	0.071	0.035	2.017	0.0522	
	Analysis	of	variance		
Source	S.S	D.F	M.S	F	Prob.
Model	2.240	1	2.240	4.07	0.0966
Residual	17.620	32	0.551		

R = 0.34*

3.1 Introduction

Human activities have recently caused the fall of acid rain and polluted precipitation, containing nitrogen dioxide (NO_x) and sulfur dioxide (SO_2) which result from the consumption of fossil fuels, condensation of mineral materials, and oil refineries combined with the atmospheric acids. Acid rain has destructive effects on forests, causing a decrease in organic matters of soils and increasing soil erosion as well as loss of underground water reservoirs. Acid deposition on vegetation causes their higher susceptibility to pests and pathogen factors [2].

Investigation on acid rain in the area south of Tehran indicates that annual deposition of nitrate ion (NO_3^-) in some Southern areas of Tehran is more than 30 kg/ha/yr [3], and sulfate ion (SO_4^{2-}) in some Southern areas of Tehran is approximately 300 kg/ha/yr [4], compared to 80 kg/ha/yr in some sampling stations of Chitgar Parkland (Fig. 10). Such huge amounts of air pollutants come down with snow and rain [5].

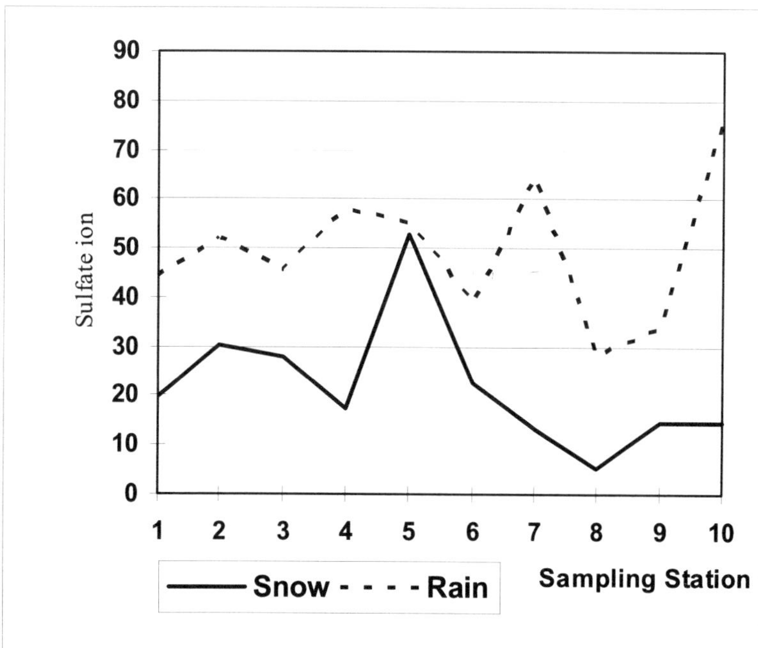

Figure 10: Average amount of sulfate ion (SO_4^{2-}) wet deposition as kg/ha/year inside precipitation of Chitgar Parkland.

3.2 Experimental method/procedures

This experiment and study used the method of wet deposition by precipitation. For measurement and sampling, specific instruments were designed and established in different stations in south Tehran, and also on the east and west sides of Tehran. Samples were analyzed in the Analytical Chemistry Laboratory using a flame photometer for potassium and sodium, a spectrophotometer for SO_4^{2-} and titration for NO_3^-.

3.3 Results and discussion

Monthly rainfall indicates that, in spite of the production of (NO_x) in the cooling season, total nitrate is decreased during winter (Fig. 9). The reason is that production of NO_3^- originates from photochemical processes, and the conditions for photochemical reactions are not provided in winter. Therefore, production of NO_3^- is decreased (Tables 3 and 4). Amounts of NO_3^- as wet deposition in a number of sampling stations in Tehran was over 30 kg/ha/yr, and 13 kg/ha/yr for the Chitgar Parkland, 15 kilometre west of the Tehran metropolitan area.

Table 3: Chitgar Parkland. Analysis of Variance for acidity (pH) related to precipitation.

K value	Source	Degrees of freedom	Sum of squares	Mean square	F value	Prob.
1	Replication	5	213.816	42.763	1.6973	0.1721
2	Factor A	2	50.378	25.189	0.9998	
4	Factor B	1	130.721	130.721	5.1884	0.0316
6	AB	2	99.369	49.684	1.9720	0.1602
-7	Error	25	629.868	25.195		
	Total	35	1124.152			

Table 4: Different seasons (time) of precipitation sampling. Duncan's multiple range test related to NO_3^-.

Warm seasons	7.551	**a**
Cold seasons	3.739	**b**

The amount of NO_3^- in warm seasons was more than twice that of cold seasons (Fig. 9). Seasonal sampling of NO_3^- showed that there was significant difference (P= 0.01) between cold and warm seasons (Table 4) [1]. In the cold period, increased production of NO_3^- does not parallel the increased consumption of fossil fuels. This means that conditions for photochemical reactions do not occur during winter and cold seasons. Ammonia (NH_3) is formed as a result of the decomposition of most nitrogenous organic materials, used as fertilizer and as a chemical intermediate. Evaporation from animal manure accounts for most of the emissions of NH_3 to the atmosphere. It is however, the surplus of N in the farming cycle that lies at the bottom of the problem, produced as a result of industrialized farming methods based on the use of artificial fertilizer [6].

Annual wet deposition of NH_3 was 9 kg/ha/yr (Fig. 11). Acidity (pH) of precipitation is neutralized by suspended particulate matter (S.P.M). However, many samples of precipitation showed acid rain (pH = 4) in Tehran. This minimum measured acidity in Tehran is 50 times as acidic as the atmospheric background acidity. Analysis of variance for acidity (pH) related to precipitation in Chitgar Parkland showed that there was significant difference (*p* =0.01) for different blocks of precipitation sampling (Tables 5 and 6). The regression equation for the independent variable of NO_3^- and dependent variable of acidity (pH) was computed as Y= 6.048 + 0.071 X (Fig. 12). The coefficient of determination was 0.34 (r=0.34). The coefficient of determination for correlation between NO_3^- and SO_4^{2-} was 0.64, and between NO_3^- and the electro-conductivity (EC) was 0.74, also between NO_3^- and Na^+ it was 0.57 [7].

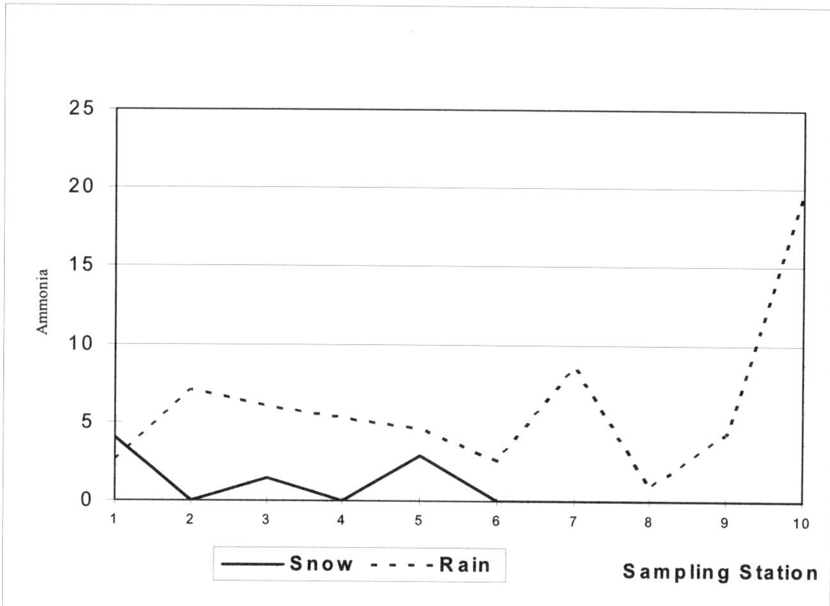

Figure 11: Average amount of ammonia (NH_3) inside precipitation of Chitgar Parkland (Kg/ha/year).

Table 5: Analysis of variance for acidity (pH) related to precipitation in Chitgar Parkland.

k value	Source	Degrees of freedom	Sum of squares	Mean square	F value	Prob.
1	Replication	5	4.836	0.967	2.0430	0.1070
2	Factor A	2	6.286	3.143	6.6390	0.0049
4	Factor B	1	0.912	0.912	1.9265	0.1774
6	AB	2	5.097	2.548	5.3830	0.0114
-7	Error	25	11.835	0.473		
	Total	35	28.966			

Table 6: Duncan's multiple range test related to pH for different treatment of Precipitation sampling.

Near freeway	6.788	**a**
Centre of Parkland	6.488	**ab**
Inner part of Parkland	5.791	**b**

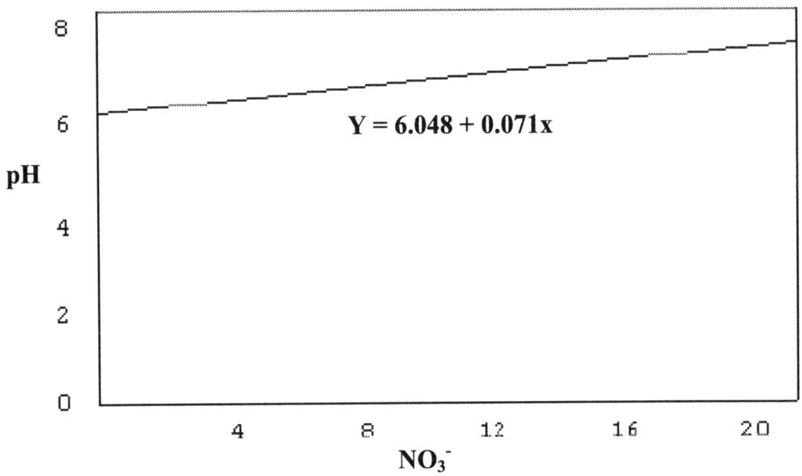

Figure 12: Regression line for pH (dependent variable) and NO_3^- (independent variable) studied in South of Tehran.

3.4 Conclusion

The main contributor to the concentration of NO_x in urban surroundings is usually road traffic, although in some cities combustion plants make a significant contribution. Animal manure consequently becomes a waste product on farms with intensive stock raising, while those concentrating on grain growing have to make up for a shortage of nutrients by using artificial fertilizer. Since so much of the feed-stuff production is now dependent on the use of artificial fertilizer, it can be said that the NH_3 that emanates from animal manure has its origin in the artificial activities. Where farming operations are thus unbalanced, N is continually being added in the form of artificial fertilizer, and leaks out into air and water in great quantities. Hence, the joint strategy puts more emphasis on cutting down NO_x emissions, and more research on health effects is needed.

4 Comparison of air quality in 1999–2002 with regarding to carbon monoxide (CO) and particulate matter (PM-10)

4.1 Carbon monoxide (CO)

As it appears; clean air covers 1% of the days of 1999 and non-clean days and good condition in 2000 view point of carbon monoxide (CO) and 26% of days in 1999 were moderate or healthy days, but in 2000 this decreased to 17% of healthy days (Fig. 13). The number of unhealthy days increased from 69% in 1999 to 79% in 2000. Both years had 4% of days in the very unhealthy level. Air Quality decreased in 2000 compared with 1999 from the view point of CO contaminant. The reason is the severe stability of atmosphere in the first half of the year in 2000 and atmosphere instability was not able to decrease pollution. Also, increased sources of pollution were effective to increase level of CO pollution.

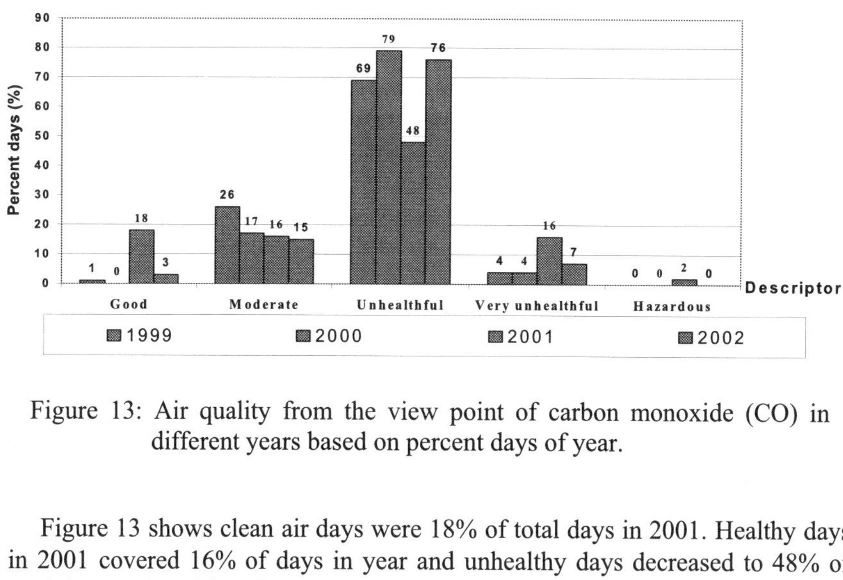

Figure 13: Air quality from the view point of carbon monoxide (CO) in different years based on percent days of year.

Figure 13 shows clean air days were 18% of total days in 2001. Healthy days in 2001 covered 16% of days in year and unhealthy days decreased to 48% of total days. 16% of days in 2001 were positioned in very unhealthy condition. 2% of days in 2001 were at the hazardous level. As it appears, the air quality index of CO in 2001 shows better condition compared with 2000. Increased atmospheric instability in 2001 was the main reason of air quality improvement. A major social factor affecting on distribution pattern of pollution in 2001 was official holidays in spring compared with similar time in 2000. Clean days in 2002 were 3% of total days, and healthy days or moderate condition in 2002 were 15% of the year. Unhealthy days in 2002 were 7% of the year. Air quality in 2002 shows a relative improvement due to instability of atmosphere in spring and early 2002.

4.2 Particulate matter (PM-10)

The level of particulate matter pollution from the view point of PM-10 showed that during 1999 clean days and good condition were 24% of year compared with 22% of clean days in 2000. Healthy days and moderate condition in 1999 were 54% of year compared with 57% healthy days in 2000. Unhealthy days in 1999 were 20% of days compared with 19% unhealthy days in 2000. Hazardous level related to particulate matter in 1999 and 2000 were 2% of days.

With regarding to particulate matter (PM-10) the air quality in 1999 and 2000 were similar. In spite of increased sources of pollution and stability of atmosphere in the first half of 2000, due to more instability of air in the second half of 2000 as precipitation compared with 1999, the index of air quality in 1999 and 2000 were similar.

15% of days in 2001 were positioned in good condition and clean days. Healthy days and moderate condition in 2001 increased to 68% of years compared with 57% healthy days in 2000. Unhealthy days in 2001 were 17% of year. Fortunately there were no very unhealthy days in 2001. Relative improvements of air quality in 2001 compared with 2000 were observed after increased instability of atmosphere as precipitation in 2001. Also, more official and formal holidays in 2001 spring compared with 2000 as a social factor caused improvement of air quality in the Greater Tehran Metropolitan Area related to particulate matter contaminant.

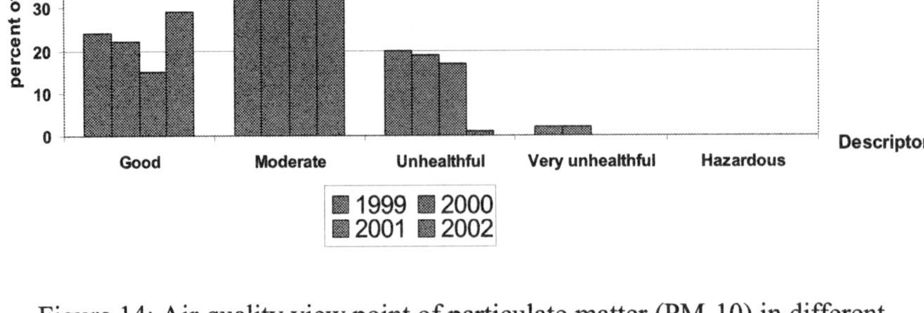

Figure 14: Air quality view point of particulate matter (PM-10) in different years based on per cent days.

The air quality index of PM-10 in 2002 shows that 29% of days were positioned in clean days and good condition. Healthy days and moderate condition includes 70% of days in 2002 and 2001. Unhealthy condition and days were included 20% of days in 2001 and 1% in 2002. There were no very unhealthy days in 2001 and 2002. Air quality from the view point of Particulate Matter (PM-10) show relatively improvement in 2002 compared with 2001. The main reason was increased atmosphere instability in early 2002 and spring.

5 Emission reduction projects in Tehran

5.1 Indoor environments

1) Measuring project of the air within the central building of Bank Mellat.
2) Measuring the environmental pollutants within the central building of Bank Maskan.
3) Measuring the air pollution level within the building of security Network of Central Bank.

5.2 Outdoor environments projects

5.2.1 Tehran transport emissions reduction project

A Project concerning emissions reduction in Tehran, namely the Tehran Transport Emissions Reduction Project has been carried out. The implementing Agency was the World Bank in Iran. The GEF allocation was US$2.0 million and Tehran Municipality co-financing was US$2.0 million. The project took place during October 1993–December 1997. The environmental problem was urban transport's large contributions of greenhouse gases and air pollutants. The project goals were to assess measures that would best reduce greenhouse gas emissions and pollutants from vehicular traffic.

Worldwide, transport accounts for 12 to 15 per cent of greenhouse gas emissions. Although developing countries are only responsible for a small portion of these emissions, their share is growing due to increased use of automobiles with obsolete, fuel-inefficient technologies. Low-cost interventions in a number of interrelated areas, pricing of transport services and fuels, public transit, maintenance and renewal of vehicle fleets, and domestic vehicle technologies could minimize these emissions and associated air pollutants. Tehran, Iran, the focus of this project, suffers greatly from automobile exhaust and pollution. Iran has identified air pollution in Tehran as a top environmental issue. Some 1.4 million motor vehicles including half a million motorcycles operate on extremely congested roads. Consuming an estimated 2 million tons of gasoline and diesel fuel each year, these vehicles severely pollute the air and release 6 million tons of carbon dioxide to the atmosphere. Iran has had low fuel prices, reducing incentives for fuel-efficient technologies. The city's existing fleet of cars has older engine designs, consuming 50 per cent more fuel than more up-to-date technologies, and are not kept well tuned. Public transportation is poor, increasing the reliance on cars. Tehran was selected as the first city in which GEF is addressing urban-transport–related greenhouse gas emissions and air pollutants.

Low-cost interventions could drastically improve environmental quality. In addition, the assortment of transport-related projects Tehran authorities had been planning or developing provided the opportunity to develop a methodological base for introducing emission-reduction concerns into standard urban transport planning practices. The results will be applicable in other developing nations contributions to the project with GEF funding, Tehran authorities have assessed such measures as efficient pricing of inputs and urban transport services that reduce greenhouse gas emissions from vehicular traffic, while improving local air quality. The project identified a variety of measures to achieve target air quality improvement that reduce both greenhouse gases and air pollution at the lowest cost. The information is assisting decision makers in designing a program of local air pollution abatement that also cost-effectively addresses climate change.

Iran and GEF jointly funded this project. The GEF contribution enabled Iran to address global concerns that might not otherwise have received priority from a national perspective. Project components included *(a)* emissions inventory and installation of air–quality–monitoring equipment, *(b)* traffic management and restraint, *(c)* improvement of vehicle fleet and fuels, *(c)* strategic planning for urban transport emissions reduction, *(d)* transport and air quality seminar, and *(e)* review of study's progress by international panel of experts.

The GEF project installed properly functioning emission-monitoring equipment around Tehran. Data collection and analysis was then completed. In May 1997 an international conference on air pollution in Tehran attracted 1,400 people mostly from Iranian ministries, governmental organizations, and universities. Discussions covered many topics relevant to this project, including emission inventories; urban planning; vehicle fleet, fuels, traffic, and transportation management; air quality modelling; and action plans for reducing air pollution.

The project was successfully completed, and the Municipality of Tehran has committed itself to implementing the consultants' major recommendations. It has also instructed the Air Quality Control Company, a municipal subsidiary, to take measures to execute remaining and continuing operations of the implementation plan.

Benefits include substantially improved air quality, reduced greenhouse gas emissions, improved health of residents, energy saving, improved traffic safety, and increased reliability of the traffic system, assisted the Tehran municipality and Iranian government in embarking on urban transport policies that are environmentally sustainable, both locally and globally, guided policy decisions on urban transport development in developing countries and the work of international bodies dealing with climate change issues.

5.2.2 Air pollution affecting Chitgar Parkland

Investigation on the relation between traffic and lead uptake in pine tree (Pinus eldarica) with measurement of lead (Pb) through Pinus eldarica in 5 stripes parallel with the Tehran-Karaj Freeway has been done in spring and summer. The amount of Pb in plants close to the freeway and in summer was 3 times the measured Pb in plants far from freeway. Also, spring Pb uptake in pine trees close to the freeway was 4 times compared with Pb uptake by pine in the inner part of Chitgar Parkland and the average Pb uptake in spring samples was 1.5 times of that summer samples. The amount of Pb in summer samples close to the freeway was more than 20 p.p.m. in semi-faded samples, which is as much as Pb uptake in spring samples, the concentration of Pb in semi-faded summer samples was 2.5 times of the Pb uptake in fresh pine samples in summer. The measurement of Pb in pine needles was carried out by atomic absorption.

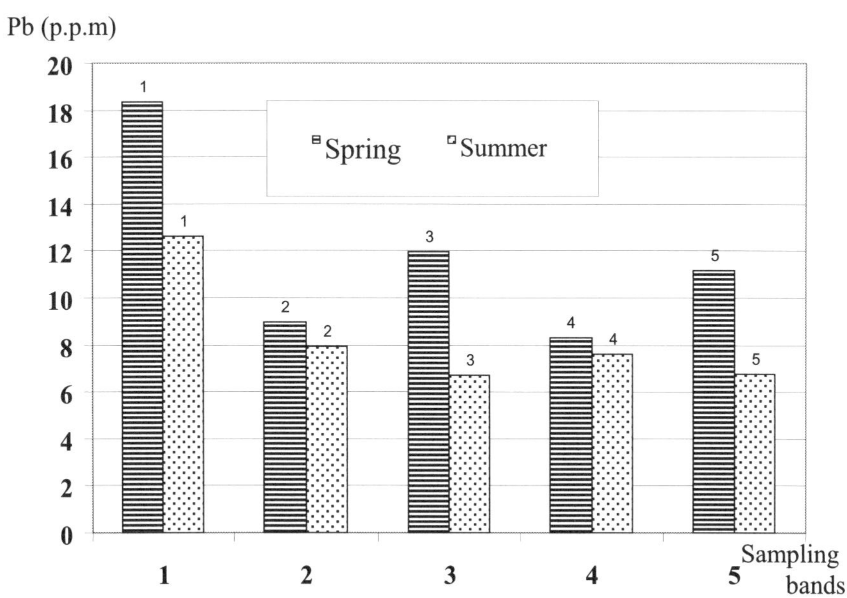

Figure 15: Concentration of lead (Pb) in pine tree samples during spring and summer in Chitgar Parkland west of Tehran.

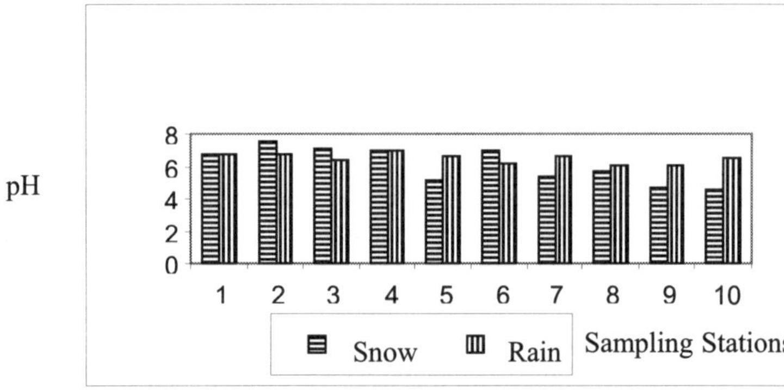

Figure 16: Different pH of rain and snow samples in Chitgar Parkland 15km west of Tehran.

Table 7: Duncan's multiple range test related to pH for different treatment of precipitation sampling in Chitgar Parkland 15 km west of Tehran.

Near freeway	6.788	**a**
Centre of parkland	6.488	**ab**
Inner part of parkland	5.791	**b**

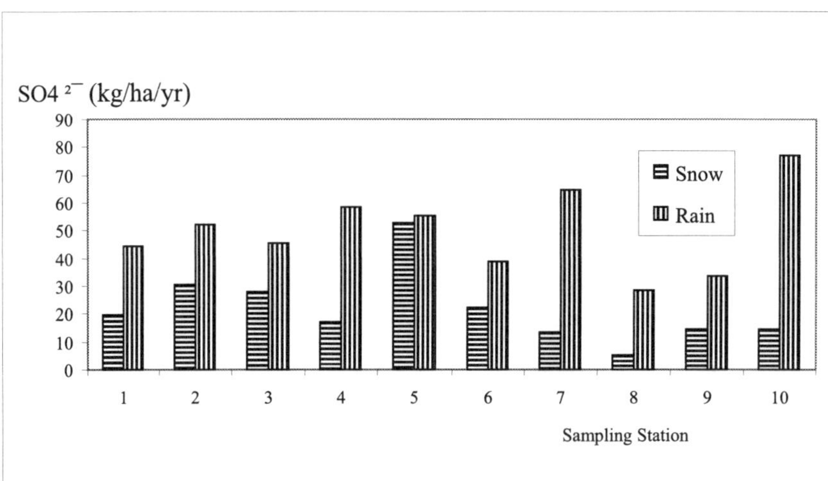

Figure 17: Diagram of the average amount of sulfate ion (SO_4^{2-}) concentration inside precipitation of Chitgar Parkland.

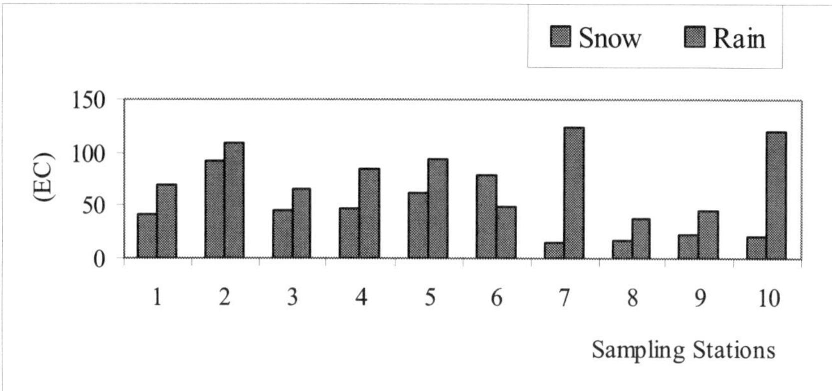

Figure 18: Diagram of the average electro conductivity (EC) in μS/cm inside precipitation of Chitgar Parkland.

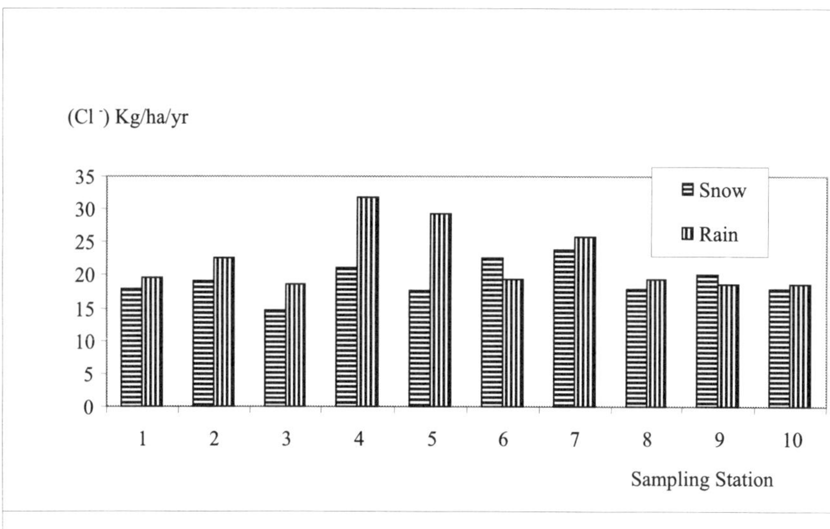

Figure 19: Diagram of the average amount of chloride ion (Cl^-) concentration inside Precipitation of Chitgar Parkland.

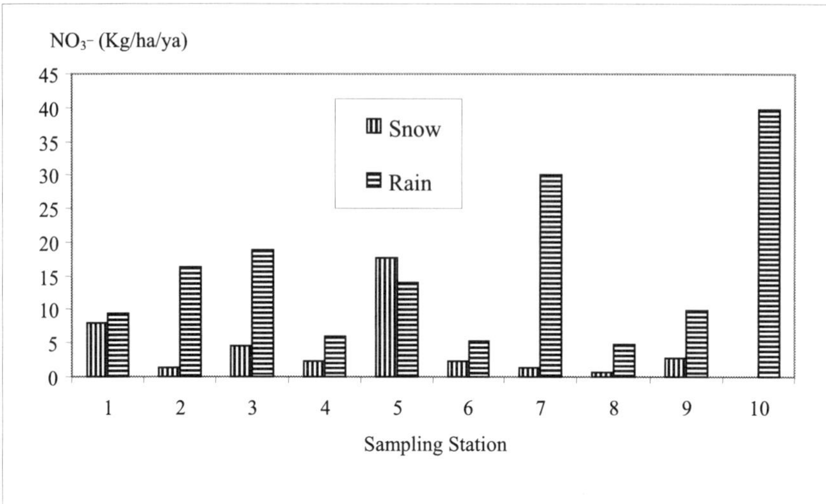

Figure 20: Diagram of the average amount of nitrates ion (NO_3^-) inside precipitation of Chitgar parkland.

Chloride ions (Cl^-) do not show meaningful correlation with pH, but there is a relation between Cl^-, SO_4^{2-} and NO_3^-. It indicates that the origin of Cl^- production is different from SO_4^{2-} and NO_3^-. SO_4^{2-} and NO_3^- are mainly produced from fossil-fuel burning, but the entrance of Cl^- to the atmosphere is from waste burning and PVC, also the use of salt on frozen surfaces of roads.

Annual wet deposition of NO_3^- in kg/ha/year inside Chitgar Parkland in West of greater Tehran metropolitan area during warm and cold seasons indicated that, in spite of production of nitrogen oxides (NO_x) in the cooling season, total nitrates is decreased over winter. The reason is that production of NO_3^- originates from photochemical processes, and conditions are not provided for photochemical reactions in winter, therefore production of NO_3^- is decreased.

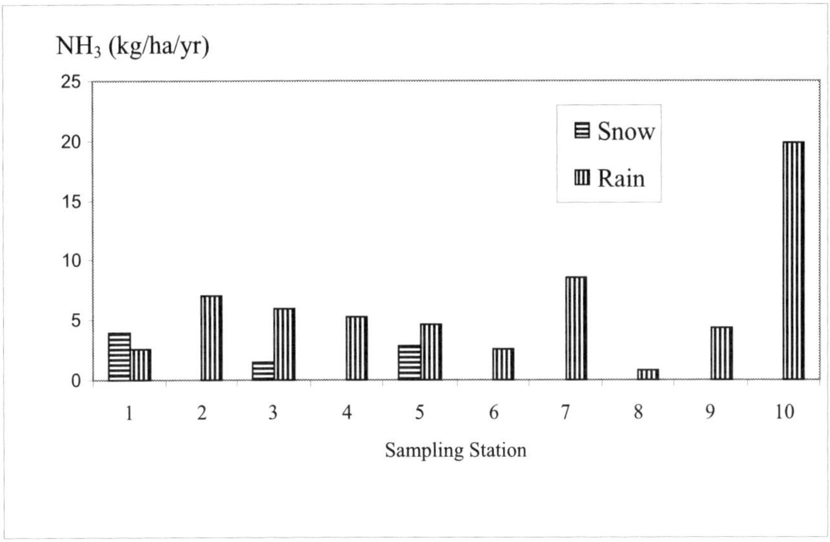

Figure 21: Diagram of the average amount of ammonia (NH3) inside precipitation of Chitgar parkland.

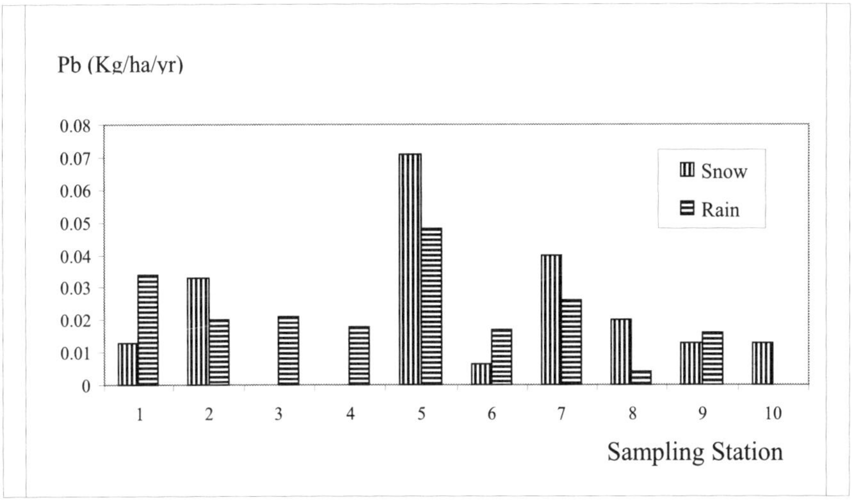

Figure 22: Diagram of the average amount of lead (Pb) inside precipitation of Chitgar Parkland.

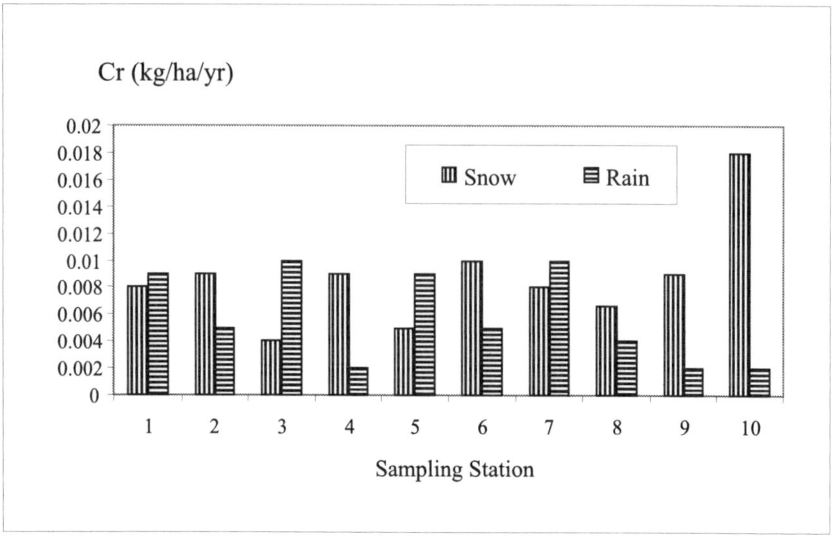

Figure 23: Diagram of the average amount of chromium (Cr) inside precipitation of Chitgar parkland.

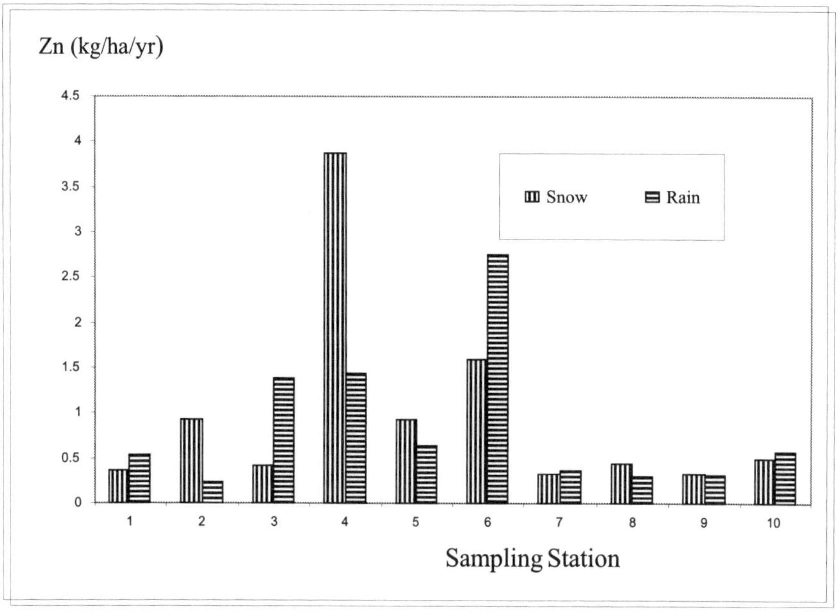

Figure 24: Diagram of the average amount of zinc (Zn) inside precipitation of Chitgar parkland.

Such amounts of zinc originate from vehicle traffic and releasing Zn from car tires and wheel friction.

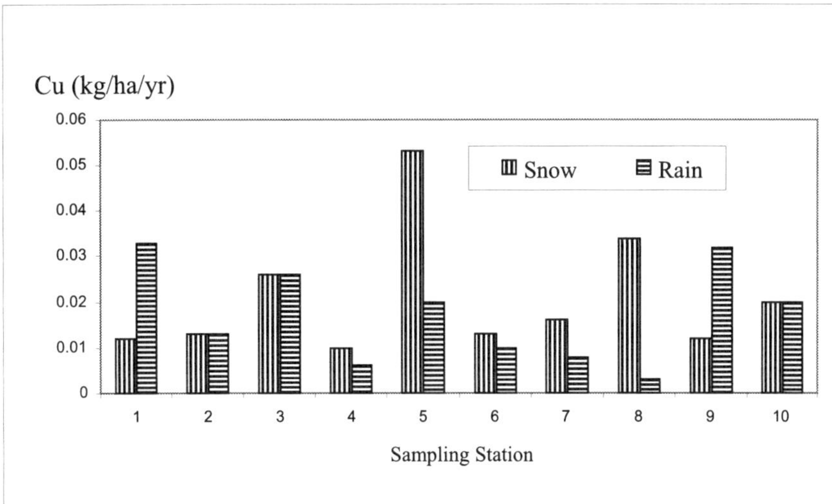

Figure 25: Diagram of the average amount of copper (Cu) inside precipitation of Chitgar Parkland.

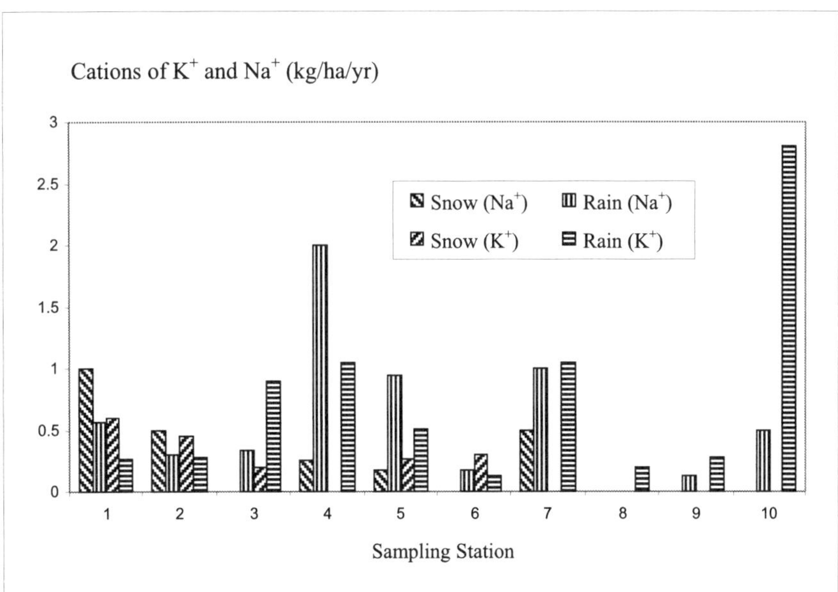

Figure 26: Diagram of the average amount of cations of Na^+ and K^+ inside precipitation of Chitgar parkland.

These cations are derived from suspended particulate matter, dust raised from mining, construction operations, cement factories, etc. The soil and geology of bedrock are the main factors of susceptibility of regional acidification. Calcium and magnesium have strong correlations with alkalinity[4].

Acknowledgements

The authors would like to thank Mr. Hassan Askary for close cooperation from Tehran Municipality, and Mr. Hooman Shamsollahi, Prof. Mahmoud Shariat Head of Environmental Health Group, Tehran University, Hamid Yarmand former head of "Research Division of Forests and rangelands Health and Protection". Also our colleagues Mr. Vahid Moniri, Mr. Rasoul Omid and Mr. Ahmad Mehrabi helped us during our research.

References

[1] Ali Salahi Kojoure, Acid rain and their neutralizing agents in south of Tehran. ACID REIGN'95? *Proc. of the 5th International Conference on Acidic Deposition*, 26–30 June, Goteborg, Sweden, Kluwer Academic Publishers, p. 343, 1995.

[2] Ali Salahi Kojoure, Environmental aspects of acid precipitation quality. ACID SNOW AND RAIN. *Proc. of the International Congress of ICASR* 6–8 October, Niigata, Japan, pp. 293–298, 1997.

[3] Ali Salahi Kojoure, Investigation on air pollution affecting Chitgar Parkland vegetation cover. Final report. Research Institute of Forests and Rangelands Iran. Pp. 1–100, 2000.

[4] Ali Salahi Kojoure, Acid rain in south of Tehran and their intensity and alternation. Master's Degree Thesis. Higher degree Azad University, Tehran, pp. 1–140, 1993.

[5] Johnson, A.H., Acid Deposition and Forest Decline. *Environmental Sciences & Technology, American Chemical Society*, **17(7)**, pp. 294–305, 1983.

[6] Fraser, G.A., The potential impacts of the Long Range Transboundary Air Pollutants on Canadian Forests. Information Report E-X-36, Economic Branch Canadian Forestry Service, 1985.

[7] Kamari, J., *The Rains Model of Acidification*, Kluwer Academic Publishers, pp. 225–231, 1990.

[8] Johannesson, M., Ammonia also enters the pictures. ACID NEWS, No.2, June. The Swedish Society for Conservation of Nature. pp: 20–21, 1998.

[9] Smith, W.H., Air Pollution and Forests. Second Edition, ISBN 0-387-97084-3, 1990.

[10] Salahi, A., Geranfar, S. & Korori, S.A.A., *Nitrogen deposition in the greater Tehran Metropolitan area.* Proc. Of the 2nd International Nitrogen Conference on Science and policy. TheScientificWorld **1(S2)**, pp. 261–265, 2001.

CHAPTER 10

Regional air quality management in Italy

C. Trozzi
Techne Consulting s.r.l., Italy

Abstract

In this chapter the activities taken in the frame of preparation of Regional Air Quality Management Plans in Italy are reviewed and described. The experiences of several regional plan realizations are reported and the main related topics are discussed. Finally, the chapter briefly discusses the complex nature of air quality management planning that involves a complex architecture of information systems to support its realization.

1 Introduction

The chapter describes the methodology used and the problems involved in the realization of Regional Air Quality Management Plans in Italy 0. The methodology is based on several experiences of air quality management plans realized by the author (Trento region, Toscana region, Bolzano region, Liguria region, Friuli Venezia–Giulia region, and Umbria region).

The Italian Ministry of Environment in 1991 (Decree of May 20, 1991) introduced prescriptions and rules for the elaboration of Regional Air Quality Management Plans. In December 1993 the Ministry of the Environment financed the realization of such plans. In 1997 the first regional plan was realized 0 and in 1998–2000 several regional administrations have completed the preparatory studies. Concerning emissions, Italian legislation, through 203/88 Republic President legislative decree dated May 24, 1988, has delegated provincial administrations to realize and to maintain the inventories. Recently, a new decree dated October 1, 2002, was emanated to regulate regional air quality planning.

According to the law, the air quality management plan is the instrument of programming, coordination and control of human activities with atmospheric emissions, having as the primary aim the protection of human health and of the environment.

The following activities are performed in the frame of the plan preparation:

- Emissions inventory (with use of models to estimate emissions from particular sources: road transport, airports, forests, ports and navigation lines, forest fires);
- Air quality and meteorological data analysis;
- Classification of territory and analysis of priority in intervention;
- Mortality data analysis;
- Projection of emissions without emissions reduction measures;
- Planning of measures to reduce emissions, definition of scenario of reduction and projection of emissions in the plan scenario,
- Use of air quality dispersion and photochemical models in the actual situation and in future projection with and without the plan application.

2 Emission inventory

An emissions inventory is a consistent collection of emissions data, grouped according to: economic activity, time interval (year, month, day, etc.), territory unit (province, municipality, grids of 1 km^2, etc.) and fuel (if applicable).

The emission inventory activity in Italy started in 1980 at a national level 0 and has been applied at the local level since 1990 0. The preparation of air pollutants emissions inventories allows characterization of the different roles played by the various emission sources and consequently represents a basic tool to define criteria for air quality management plans 0.

2.1 Nomenclature of activities

The nomenclature used at the local level follows the guidelines of the European Commission CORINAIR working group 0. CORINAIR nomenclature includes about 200 activities grouped in 11 groups:

- Combustion in energy and transformation industries (stationary source),
- Non–industrial combustion plants (stationary sources),
- Combustion in manufacturing industry (stationary sources),
- Production processes (stationary sources),
- Extraction and distribution of fossil fuels and geothermal energy,
- Solvent and other product use,
- Road transport,
- Other mobile sources and machinery,
- Waste treatment and disposal,
- Agriculture,
- Other sources and sinks.

This nomenclature was enlarged to take into account locally relevant activities such as the textile industry, shoes industry, shipyards activities, hot zinc lamination, resin processing and resin application.

2.2 Pollutants

The pollutants included in the inventory at local level are:

- Five main air pollutants, namely nitrogen oxides (NO, NO_2, N_2O), sulfur oxides (SO_2, SO_3), non–methane organic compounds (VOC), carbon monoxide (CO) and suspended particles with diameter less than 10 μm (PM_{10}); regarding VOC speciation has been considered, in order to permit the aggregation into species for using photochemical models;
- Ammonia (NH_3);
- Heavy metals (As, Cd, Cu, Cr, Hg, Pb, Zn);
- Greenhouses gases: carbon dioxide (CO_2), methane (CH_4) and nitrogen protoxide (N_2O).

Some experiences of air toxics emissions inventories were realized at the national 0 and local level 0 while, in current inventory activities, benzene and polycyclic aromatic hydrocarbons (PAHs) are introduced.

2.3 Sources classification

The sources are generally split in four categories: main point sources, minor point sources, area sources and linear sources.

The fixed sources for which the total annual emissions of one pollutant is larger than a fixed threshold value (Table 1), are considered main or minor point sources. Linear sources correspond to the main communication ways (roads, rivers, railways, and seaways) and generally all the highways and all the main extra-urban roads are included. All the other sources are defined as area sources.

Table 1: Thresholds for main and minor point sources.

Pollutant	Minor point sources	Main point sources
Carbon monoxide	50 t/year	250 t/year
Other main pollutants	5 t/year	25 t/year
Heavy metals	50 kg/year	250 kg/year

2.4 Emissions estimate: point sources

For main point sources, information is gathered through a questionnaire that allows collection of general data (identification, location, etc.), structural data (stacks and units characteristics) and quantitative data (pollutant concentrations at the stacks, pollutant emissions, actual production, fuel consumptions). For minor point sources information is gathered through a simplified questionnaire with general data, pollutant emissions and actual production.

Emissions from point sources can be convoyed in stacks or fugitive. When emissions are emitted by stacks it is possible to evaluate, in a preliminary way, the emissions for single point of emission, through:

$$E_i = 10^{-6} P\ g\ h\ C_i$$

where: P, hourly gas flow; i, pollutant; g, working days for year; h, working hours for day; C_i, average concentration of pollutant i. Then the emissions are calculated for the whole plant as the sum of emissions of all stacks.

Data are validated for: completeness (some information such as gas flow, concentrations, hours can be omitted); accuracy (declared concentrations can be higher than the effective ones, the number of hours can be referred to all plant and not to single unit or specific process). Finally, declared data are compared with estimated data through standard emission factors. This procedure is used to validate information and to complete the data through the insertion of pollutant emissions not declared by the industries, but existing.

2.5 Emissions estimate: area sources

Where available, plant data are used, for industrial area sources, to evaluate the emissions and to allocate them on a square grid mesh.

Other area sources (for instance, domestic solvent use and natural sources) are evaluated on a geographical basis, inside each municipality and subsequently spatially disaggregate on a square grid mesh. These sources are characterized using statistical or survey data on suitable activity indicators (for example: paint consumptions, fuel consumptions) and emission factors.

The basic formula to calculate emissions is:

$$E = A\ F$$

where: A, activity indicators; F, emission factor expressed in grams per unit of activity.

In the absence of specific municipal indicators, it is possible to use surrogate variables that, because of their high correlation with the activity, allow almost reliable results to be obtained. The emissions factors can be found in the literature (principally by the EMEP/CORINAIR Task Force on Emission Inventories 0 or United States Environmental Protection Agency 0) and integrated with experimental data, eventually available at local level. Special attention must be devoted to some sectors as discussed in the following sections.

2.6 Emissions estimate: road transport

Emissions from road traffic are estimated using the SETS model 0. The model was realized by extending and adapting at the local level the methodology developed in the frame of the EC CORINAIR working group 0 as updated in the framework of the European Commission MEET Project 0 (Methodologies for estimating air pollutant emissions from transport).

The model estimates the emissions of the main air pollutants, heavy metals and greenhouse gases on a single road or over an area (region, province, district, etc.) subdivided into:

- Hot emissions E^{hot}: the emissions from vehicles whose engines reached their working temperature;
- Cold emissions E^{cold}: the emissions during the warming up of the vehicle; conventionally, they are the emissions that happen when the cooling water's temperature is lower than 70°C;
- Evaporative emissions E^{evap} (for volatile organic compounds only).

Hot emissions are estimated as:

$$E^{hot}_{i} = \Sigma_{jkmn} F^{hot}_{ijk} h_j v_j d_{jk} G_{ijkm} g_{jm} L_{ijkn} l_{jn}$$

where: i, pollutant; j, typology of vehicles (fuel, classes of capacity and normative of the European Communities); k, class of speed; F^{hot}_{ijk}, emission factor; h_j, number of vehicles; v_j, traffic flow (average mileage as km/vehicle); d_{jk}, fraction of mileage at given speed class; m, road gradient class; G_{ijkm}, road gradient correction factor; g_{jm}, fraction of mileage at given road gradient; n, load factor class; L_{ijkn}, load correction factor (reported emission factors are evaluated at 50% of maximum load; correction is computed for heavy vehicles, where load effect is more relevant, and for main pollutants); l_{jn}, fraction of mileage at given load.

Extra cold emissions are evaluated as:

$$E^{hot}_{i} = \Sigma_{jkhment} v_{jl} q_{jm} p_{jmk} \omega_{ij} q^{a}_{hkm} q^{b}_{ne} q^{c}_{htm} (f_{ih} + g_{ijn}-1) h_{ijth} / (10^6 d_t)$$

where: i, pollutant; j, typology of vehicles (fuel, classes of capacity and normative of the European Communities); k, speed class; h, cold speed class; m, month; e, ambient temperature class; n, start–up temperature class; t, trip length class; v_{jl}, traffic flow; q_{jm}, monthly fraction of annual mileage; p_{jmk}, monthly fraction of mileage under cold start or intermediate temperature; ω_{ij}, excess emission at 20 °C and 20 km/h; q^{a}_{hkm}, monthly fraction of the trips traveled at given cold speed class for a given overall average speed class; q^{b}_{ne}, fraction of the trips traveled with a given start–up engine temperature class and a given ambient temperature class; q^{c}_{htm}, monthly fraction of trips started with a cold engine for given trip length class and cold speed class; d_t, average distance of the trips under cold start conditions; f_{ih}, correction coefficient for a given class of cold speed; g_{ijn}, correction coefficient for a given class of start–up temperature; h_{ijth}, dimensionless excess emission.

Evaporative losses are of four types:

- Filling losses (occur when the vehicle's fuel tank is filled and the contents of saturated vapors are displaced and usually vented to the atmosphere); this type of evaporative losses are usually attributed to the fuel–handling chain and is not covered by the SETS model;
- Diurnal breathing losses (are the result of the night–day temperature cycle, causing the contents of the fuel tank to contract and expand, pushing saturated vapor out on expansion); particularly during a short period plastic fuel tanks have been installed in cars; these tanks suffered from diffusion of fuel through the plastic;
- Hot–soak losses (occur when a vehicle is switched off after operation and the equalization of the temperatures leads to the evaporation of the fuel in certain parts of the engine); depending on the temperature of the engine at switch off, one can differentiate between warm–soak and hot–soak losses;
- Running losses (occur during the operation of the vehicle); are the least documented source of evaporative emissions; on modern cars equipped with carbon canisters the canister should capture any running losses but there are reports that show that running losses would occur nevertheless. Evaporative losses from vehicles are known to depend on four major factors:
- Vehicle technology (equipped with or not with carbon canisters and with "sealed" plastic tanks),
- Ambient temperature and its daily variation,
- Gasoline volatility (depending on the temperature variation),
- Driving conditions (average trip length, parking time, etc.).

Evaporative emissions are evaluated as:

$$E_{ij}^{eva} = \Sigma_m (S_{ijm} + D_{ijm} + R_{ijm}) = \Sigma_m \{365 \cdot h_j [x_j [(1-q_j) [(1-w_{jm})e^{s,hot}_{ijm} + w_{jm}e^{s,warm}_{ijm}] + q_j e^{fi}_{ij}] + e^{d}_{ijm}] + m_{jm}[(1-w_{jm})e^{r,hot}_{ijm} + w_{jm}e^{r,warm}_{ijm})\}$$

where: i, pollutant (COV as total or single compound such as benzene); j, typology of vehicles (classes of capacity and normative of the European Communities, only for gasoline); m, month; S_{ijm}, hot-soak emissions; D_{ijm}, diurnal losses; R_{ijm}, running losses; h_j, number of vehicles; x_j, average number of trips of a vehicle per day; q_j, fraction of gasoline-powered vehicles equipped with fuel injection; w_{jm}, fraction of trips finished with cold or warm engine (shorter trips) or with catalyst below its light-off temperature; $e^{s,hot}_{ijm}$, mean emission factor for hot-soak emissions (which is dependent on fuel volatility); $e^{s,warm}_{ijm}$, mean emission factor for cold-soak and warm-soak emissions (which is dependent on fuel volatility and average monthly ambient temperature): e^{fi}_{ij}, mean emission factor for hot-soak and warm-soak emissions of gasoline-powered vehicles equipped with fuel injection; m_{jm}, total month mileage, e^{d}_{ijm}, mean emission factor for diurnal losses of gasoline-powered vehicles equipped with metal tanks, depending on average monthly ambient temperature, temperature variation, and fuel volatility; $e^{r,hot}_{ijm}$, average emission factor for hot-running losses of gasoline-powered vehicles (which is dependent on fuel volatility and average monthly ambient temperature); $e^{r,warm}_{ijm}$, average emission factor for warm-running losses of gasoline powered vehicles (which is dependent on fuel volatility RVP and average monthly ambient temperature).

A summary of data needed to emissions estimate is reported in Fig. 1. SETS considers the following typologies of vehicles: gasoline, diesel and GPL cars, gasoline and diesel commercial vehicles, bus, motorcycles, two stroke cars. Subdivisions are performed according to classes of capacity and to the application of the normative of the European Communities.

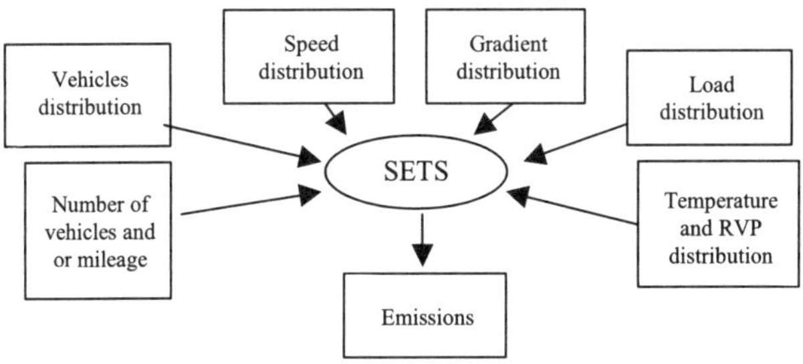

Figure 1: Input/output flow diagram for SETS model.

The model, with the purpose to define in a way more detailed the real ways of employment of the vehicles, uses the subdivision of the total mileage in thirteen classes of speed (10–20, 20–30, …, 120–130).

For each highway and main extra–urban road (linear sources) emissions are estimated collecting (or evaluating through models) data on vehicle flux and speed distribution and using the national and regional vehicle park composition. For traffic other than highway and main extra-urban roads, global area emissions are estimated based on regional vehicle park composition and a global mileage evaluation. A control procedure is applied computing fuel consumption and comparing this figure with data for fuel distributed.

Examples of this approach can be found in literature [14–18].

2.7 Emissions estimate: waterborne transport

Emissions from waterborne transport, in the port environment and in navigation, are estimated using the AIR SHIPS model 0, realized with the methodology developed by the author 0 and updated in the framework of the MEET Project 0 (Methodologies for estimating air pollutant emissions from transport), a project funded by the European Commission under the transport RTD program of the fourth framework program. The methodology was reviewed in two specialty conferences on marine combustion engines 0 and ports 0 and in the conference on EC COST action 0. At the national level the methodology was applied to estimate global emissions from ships traffic 0.

In shipping activity it is customary to distinguish between approaching to and docking in the harbour; hotelling in the harbour; departing from the harbour, and cruising. The approaching phase starts when the ship's deceleration begins and ends at the moment of the docking, while the departing phase starts with departure from the berth and ends when cruising speed has been reached. After its arrival in harbour, a ship continues to emit at the dockside (while in the hotelling phase). Power must be generated in order to supply the ship's lighting, heating, refrigeration, ventilation, etc. A few steam ships use auxiliary diesel engines to supply power but they generally operate one or more main boilers under reduced load. Ships powered by internal combustion engines normally use diesel power generators to furnish auxiliary power. For liquid bulk ships, power requirements of the cargo pumps for tanker off-loading and of the ballast pumps for tanker loading must be also take into account. In smaller tankers, the pumping power requirement will add to the electrical load whereas for larger tanker steam-turbine-driven pumps are generally used (even on motor tankers) with a consequent boiler load. As these power requirements can be relatively high, the emissions must be separately estimated.

The model summarizes shipping activity in one cruising phase, two maneuvering phases (approaching and departing) and one hotelling phase; power requirements of the cargo/ballast pumps are also taken into account for liquid bulk ships.

The total emissions E_i of pollutant i are obtained as:

$$E_i = \Sigma_{jklm}\, C_{jkmt}\, d_{jklm}\, F_{ijlm}$$

where: i, pollutant ; j, fuel (bunker fuel oil, marine diesel oil, marine gas oil, gasoline fuel); k, ship class (solid bulk, liquid bulk, general cargo, container, passenger/ro–ro/cargo, passenger, high-speed ferries, inland cargo, sail ships, tugs, fishing, other); l, engines type class (steam turbines, high-speed motor engines, medium-speed motor engines, slow-speed motor engines, inboard engines–pleasure craft, outboard engines, tanker loading/off-loading engines); m, mode (cruising, manoeuvring, hotelling, tanker offloading); t, gross tonnage class; C_{jkmt}, daily fuel consumption; d_{jklm}, days in navigation; F_{ijlm}, average emission factors.

2.8 Emissions estimate: air transport

Emissions of air pollutants from aircraft can be subdivided into emissions by cruise and emissions by landing and take–off cycles (LTO cycles). The cruise emissions are not relevant for regional air quality management but are to be taken into account for national estimate 0. Emissions from LTO (that can be allocated to airports) are estimated using the AIR AIR model [27–28].

A single LTO cycle includes aircraft operations in flight and on the ground: descent and approach from a height of about 3000 feet above ground level, touchdown, landing run, taxi, idle and shutdown, start–up and idle, checkout, taxi out, takeoff and climbing to 3000 feet above the ground level. In order to estimate aircraft emissions within the airport, the above operations can be divided into four phases: approach, taxi/idle, takeoff and climb-out.

The four phases of the typical LTO cycle have different times of operation depending on aircraft classe (jumbo jet, long-range jet, medium-range jet, commercial turboprop, business turboprop, piston). It is worthwhile outlining that operation times depend on the airport size and traffic.

The emissions are obtained as:

$$E_i = 10^{-3}\, \Sigma_{mk}\, C_k\, \Sigma_l\, F_{ikl}\, t_{lm} / 60$$

where: i, pollutant; m, class of airplanes; k, airplane model; l, operational phases of LTO cycle; C_k, number of LTO cycles; F_{ikl}, emission factor; t_{lm} , typical time in minutes spent in the phase l from the class of airplane m which the model of airplane k belongs.

The model allows introducing specific times for the airport under study and calculating specific emission factors for LTO cycle. Examples of application can be found in the literature 0.

2.9 Emissions estimate: forest

Emissions of VOC from forests are very important in Italy due to vegetation coverage of territory and climatic conditions. The growing interest in this field originates from the relevant impact of the volatile organic compounds emissions in the ozone generation in urban and remote areas. Emissions are estimated using the AIR FOREST model 0 developed on the basis of EMEP/CORINAIR Task Force on Emission Inventories 0 methodology.

Total VOC emissions are:

$$E = \Sigma_{kj}(E^I_{kj} + E^M_{kj} + E^A_{kj})$$

with:

$$E^I_{kj} = 10^{-8}\, B_k S_{kj} F_k^{I,30}\, \Sigma_m(F^{I,Tg_{mj}}\, F^{I,L_{mj}}\, t^g_{m})\delta_{km}$$

$$E^M_{kj} = 10^{-8}\, B_k S_{kj}\, F_k^{M,30} \Sigma_m(F^{M,Tg_{mj}}\, t^g_m + F^{M,Tn_{mj}}\, t^n_{m})\, \delta_{km}$$

$$E^O_{kj} = 10^{-8}\, B_k S_{kj}\, F_k^{O,30} \Sigma_m(F^{O,Tn_{mj}}\, t^g_m + F^{O,Tn_{mj}}\, t^n_{m})\, \delta_{km}]$$

where: j, altimetric class (plain, hill, and mountain); k, vegetal specie; m, month; E^I_{kj}, emission of isoprene; E^M_{kj}, emission of monoterpenes; E^O_{kj}, emissions of other compounds; B_k, default foliar biomass density; S_{kj}, vegetation area; $F_k^{I,30}$ ($F_k^{M,30}$, $F_k^{O,30}$), isoprene (monoterpenes, other compounds) emission factor at 30°C and 1000 μmolm^{-2}s^{-1} PAR (photosintetically active solar radiation); $F^{I,Tg_{mj}}$ ($F^{M,Tg_{mj}}$, $F^{O,Tg_{mj}}$), isoprene (monoterpenes, other compounds) daily temperature correction; $F^{I,L_{mj}}$, correction factor for photosintetically active solar radiation L_j (equal to 45–50% of total solar radiation); $F^{M,Tn_{mj}}$ ($F^{O,Tn_{mj}}$), monoterpenes (other compounds) nightly temperature correction; $t^g{}_m$ ($t^n{}_m$), daily (nightly) number of hours in month m; δ_{km}, equal 1 for conifers, and other vegetation and, from April to September, for deciduous and zero otherwise.

The model estimates emissions for nineteen vegetal classes (grouped in: conifer, deciduous isoprene emitters, deciduous non–isoprene emitters, other vegetation), three altimetric classes and twelve months.

In the model the isoprene emissions are assumed equal to zero in the night and the δ_{km} factor is used to take into account that, for deciduous vegetation, COV are emitted only in the vegetation period (April–September).

Effective temperatures $T^g{}_{mj}$ e $T^n{}_{mj}$ are defined as:

$$T^g{}_{mj} = (\underline{T}^M{}_{mj} + \underline{T}^a{}_{mj})/2 = (3\underline{T}^M{}_{mj} + \underline{T}^m{}_{mj})/4$$

$$T^n{}_{mj} = (\underline{T}^m{}_{mj} + \underline{T}^a{}_{mj})/2 = (3\underline{T}^m{}_{mj} + \underline{T}^M{}_{mj})/4$$

Where $\underline{T}^m{}_{mj}$, $\underline{T}^M{}_{mj}$, $\underline{T}^a{}_{mj}$ are the average of minimum, maximum and average daily temperature for month m.

The species are introduced along the national statistical classes' definition and grouped along the CORINAIR classes' definition.

The emission factors at 30°C and foliar biomass factors are derived from EMEP/CORINAIR 0 and other studies [31, 32].

Several studies have been realized in recent years about the temperature and light dependence of the emissions. The methodology used adopted the algorithms of Guenther 0 for the correction factor (in µmol photons [400–700nm wavelength] $m^{-2}s^{-2}$) for temperature:

$$\frac{e^{\frac{95000(T-30)}{2520.3891*(T+273.15)}}}{1+e^{\frac{230000(T-40.85)}{2520.3891*(T+273.15)}}} \qquad \text{Isoprene}$$

$$e^{0.09(T-30)} \qquad \text{Monoterpenes and other compounds}$$

and for photosintetically active solar radiation (only for isoprene):

$$\frac{0.0028782L}{\sqrt{1+0.00000729L^2}}$$

2.10 Emissions from forest fires

Emissions from forest fires are estimated using the AIR FIRE model 0 developed on the basis of the EMEP/CORINAIR Task Force on Emission Inventories 0 and, for PM_{10}, United States Environmental Protection Agency 0 methodologies. At the national level in Italy forest fires can contribute up to 4.5% of carbon monoxide and up to 1% of nitrogen oxides and carbon dioxide emissions while, at the regional level, forest fires can be up to 5% of carbon monoxide and up to 8% of particulate matter with diameter less than 10 microns emissions.

The model estimates emissions from the surface of different tree species burned, and evaluation of effective burned biomass as shown in Fig. 2.

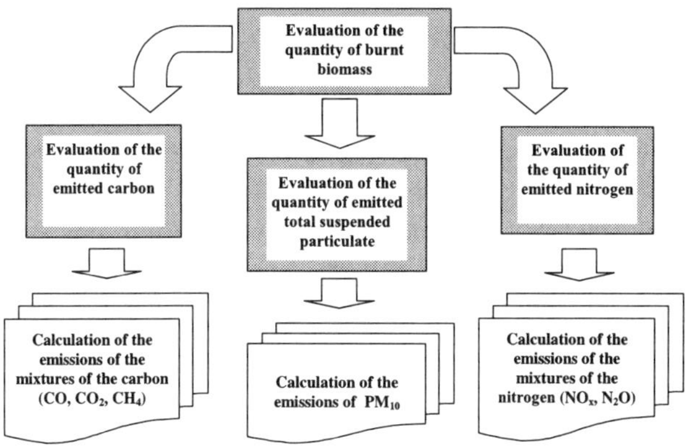

Figure 2: Input/output flow diagram for Air Fire model.

The quantity of dry biomass burnt is derived from:

$$M = \alpha \Sigma_i A_i B_i$$

where: i, vegetation specie; A_i, fired area; B_i , average quantity of dry biomass emerged by the ground; α, efficiency of the fire or fraction of definitely destroyed biomass (in the case of complete fire must be set equal to 1).

In the model are taken into consideration the different crops as defined by the forest statistics: resinous tall stem; deciduous tall stem; mixed tall stem; simplex coppice; mixed coppice; Mediterranean bush. An approximate evaluation of the middle quantity of biomass emerged by the ground for every crop is contained in the Air Fire model.

The quantities of emitted carbon, nitrogen and total particulate are:

$$C = \beta M$$

$$N = \gamma' C = \gamma M$$

$$P = \delta M$$

where β is the carbon's quantity contained in the biomass (can be set equal to 0.45 g/g of burnt biomass); γ' is the proportion among emitted carbon and nitrogen (can be set equal to 0.01) and γ the nitrogen quantity contained in the biomass (can be set equal to 0.0045 g/g of burnt biomass); δ is the quantity of total particulates emitted for unit of biomass (0.0085 g/g of burnt biomass). The quantity remainder of biomass is mainly constituted by hydrogen and oxygen. The emissions of carbon compounds (carbon monoxide, carbon dioxide, and methane), nitrogen compounds (nitrogen oxides, nitrogen protoxide) and particulate matter with diameter less than 10 microns are obtained as:

$$E_C = \varepsilon_j \, \delta_j \, C$$

$$E_{N_j} = \varepsilon_j \, \delta_j \, N$$

$$E_{PM_{10}} = \varepsilon_j \, P$$

where: j, compound; ε_j, fraction of total carbon (nitrogen, total suspended particulate) emitted as compound j; δ_j, factor of passage from the emissions in quantity of carbon (nitrogen, total suspended particulate) to the emissions in quantity of the specific compound.

2.11 Spatial disaggregation

The emissions are at first evaluated at the municipal level or with direct municipal indicators or using a lot of proxy variables known at the municipal level to allocate regional values on a municipality basis. Proxy variables allow obtaining information on a certain spatial resolution assuming that it is known for larger spatial resolutions. The municipal level in Italy corresponds to level 5 of NUTS (Nomenclature of Territorial Units for Statistics) classification scheme established by Eurostat and widely used in Community legislation since 1988.

The municipal level seems to be the most reliably defined to estimate emissions. Inside the municipality a square grid mesh is built up to represent pollutants emissions to allow the application of diffusion models. To disaggregate pollutant emissions from the municipal level to the mesh level, the methodology of proxy variables is used.

When the municipal value is known, it is distributed on the square grid mesh according to the two following cases:

- The data to be distributed is an extensive variable and it is proportional to coverage of a thematism on a single mesh (for example: urban zones); in this case the weight of the mesh coverage on total municipal coverage is utilized;
- The data to be distributed is an intensive variable and depends on the presence of the activity estimated at municipal level on the single mesh (for example the presence of a small industry); in this case the weight of the mesh activity on total municipal activity, is utilized.

The emissions on mesh k are obtained as:

$$E_{ik} = \Sigma_j \left(E_j P_{lkj} / \Sigma_k P_{lkj}\right)$$

where: i, activity; j, municipality; k, mesh; l, proxy variable appropriate to activity i; E_j, total emission; P_{lkj}, value of the proxy variable.

Examples of proxy from CORINE Land Cover are: urban; agricultural; industrial and commercial; modelled artificial; mining, permanent crops, arable land, deciduous forests; coniferous forests; airports; landfills.

2.12 Temporal disaggregation

Annual emissions are at first evaluated. Once annual emissions have been evaluated, it is also important to obtain an estimate of their hourly, monthly and daily distribution especially to allow the application of diffusion models to study short–time pollutant dispersion. For main point sources temporal disaggregation may be evaluated directly at the plant, through the questionnaires. The disaggregation of other sources is estimated through the use of corrective factors that have a similar rule of the proxy variable in the case of spatial distribution (for example: typical working hours, wintertime, temperature, monthly selling of fuels, etc.).

2.13 Sensitivity analysis

Many studies have looked at how uncertainty, or the error related to the evaluation of every single parameter of emission estimate methodologies, affects the emission estimate [30, 35, 36].

The emissions equation can be written as:

$$E = E\,(P_1, P_2, \ldots, P_n)$$

We assume that all the parameters are not correlated with other parameters that appears in emissions equation and that function E is linear or nearly linear (that is, it does not differ very much from a linear function for all possible values of all p_i). In this case, true for small variations of the parameter around the mean value, it is possible to demonstrate that the root–mean–square of E is:

$$s_E = (\partial E/\partial P_1)^2\, s_{P_1} + (\partial E/\partial P_1)^2\, s_{P_n} + \ldots + (\partial E/\partial P_n)^2\, s_{P_n}$$

and going from root–mean–square to relative error:

$$\Delta E/E = c_{P_1}(\Delta P_1/\,P_1) + c_{P_2}(\Delta P_2/\,P_2) + \ldots + c_{P_n}(\Delta P_n/\,P_n)$$

where:

$$c_{P_i} = P_i/E(\partial E/\partial\,P_i)$$

This equation can be used for two complementary aims: to evaluate how an error on different parameters affects emissions estimation and so to evaluate the uncertainty of the estimate itself. c_{P_i} coefficients could be interpreted as the dependence of relative emission variation $\Delta E/E$ on relative variation of parameter $\Delta P_i / P_i$.

For example, in the case of emissions from forests 0, we obtain:

$$E = E(T,L)$$

considering as constant all the other parameters of the equation.

In this case:

$$\Delta E/E = c_T(\Delta T/T) + c_L(\Delta L/L)$$

where:

$c_L = 1/(1+0.00000729L^2)$ for isoprene

$c_L = 0$ for monoterpenes and other compounds

$c_T = 0.09\ T$ for monoterpenes and other compounds

$$c_T = \frac{T}{2520.389(T+273.15)^2}\left[\frac{72220000}{1+e^{\frac{230000(T-40.85)}{2520.3891*(T+273.15)}}} - 43420750\right]$$

for isoprene

As a consequence:

- At 20°C, a variation of 2°C (10%) in temperature gives a variation of about 30% in isoprene (c_T=2.86 at T=22°C); at the same temperature, a variation of 2°C (10%) in temperature gives a variation of about 20% in monoterpenes and other compounds emissions (c_T=1.98 at T=22°C).
- At 1000 $\mu mol m^{-2} s^{-1}$ photosintetically active solar radiation, a variation of 100 $\mu mol m^{-2} s^{-1}$ in photosintetically active solar radiation (10%) gives a variation of only 1% in isoprene emissions (c_L=0.10 at L=1100 $\mu mol m^{-2} s^{-1}$).

The conclusion is that the model is not very sensitive to variation in solar radiation and is much more sensitive to variation in temperature. The model can be used for emissions estimate in situations in which the parameters are average estimates with particular attention to temperature evaluation.

2.14 Results

Figure 3 is an example of an emissions map. In Table 2 the traffic contributions to total emissions are reported.

As a result of case studies and of data from Table 2 it is possible to state that:

- Where no important industry is present in the territory, traffic is the main relevant source of emissions for all pollutants (also sulfur oxides and fine particulates) and is responsible for 9/10 of carbon monoxides, 2/3 of nitrogen oxides, 1/2 of volatile organic compounds and 1/3 of carbon dioxide emissions;

- Where some very big power plant or industry is present in the territory, traffic contribution to carbon dioxide and sulfur oxides is less but the contribution for other pollutants is always relevant;
- The contribution to carbon monoxide emissions is less when big iron and steel industries are present but, as these last sources are point sources, the impact on the territory of traffic emissions is the most important;
- Where important industrial (paint or solvent use installations) or domestic (wood combustion) VOC sources are present, the traffic contribution to volatile organic compounds emissions is less;
- Referring to urban situations, we can also note that the ratio of nitrogen oxides traffic emission to nitrogen oxides domestic heating emissions is about 5 to 1.

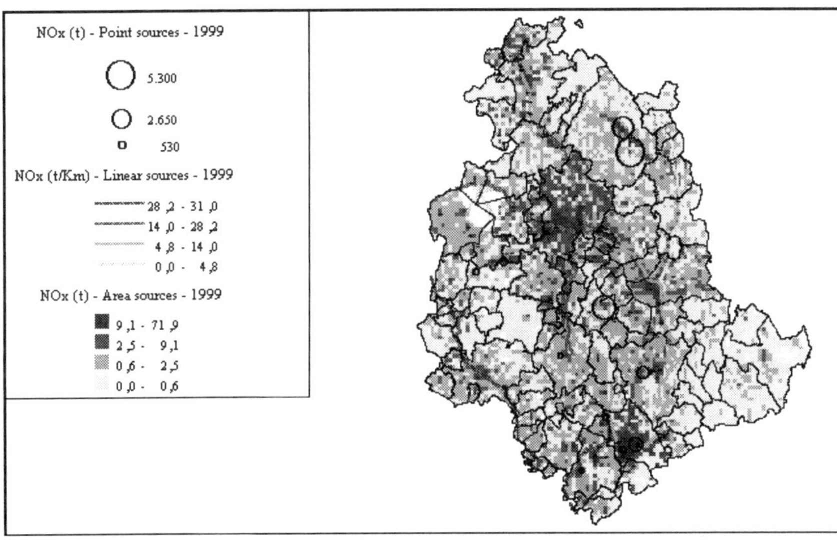

Figure 3: Example of spatial representation of emissions (Umbria Region).

Table 2: Percent of traffic emissions as part of the total emissions for selected case studies in Italy.

	CO	VOC	NO_X	PM_{10}	SO_X	CO_2
Areas with power plants	75–85%	36–60%	17–50%	15–45%	0.5–5%	5–20%
Areas with power plants and iron & steel industry	40–70%	40–60%	17–40%	30–35%	0.3–6%	5–20%
Areas with important COV non traffic emitters	63–92%	27–35%	50–80%	23–82%	12–40%	30–40%
Other Areas	79–95%	49–62%	60–76%	27–63%	17–66%	27–39%
Average value	80%	44%	55%	42%	23%	26%

3 Analysis of air quality data

In the frame of Air Quality Management Plan activities, all data coming from the air quality networks are analyzed.

The analysis is performed at first by comparing the data with actual and future air quality standards. The legislation on air quality is in transition with the adoption of European Directive 96/62/EC 0 and of derived ones [38–40].

The new air quality standards are represented by:

- *Limit value* shall mean a level fixed on the basis of scientific knowledge, with the aim of avoiding, preventing or reducing harmful effects on human health and/or the environment as a whole, to be attained within a given period and not to be exceeded once attained;
- *Target value* shall mean a level fixed with the aim of avoiding more long–term harmful effects on human health and/or the environment as a whole, to be attained where possible over a given period;
- *Alert threshold* shall mean a level beyond which there is a risk to human health from brief exposure and at which immediate steps shall be taken by the Member States;
- *Long–term objective* means an ozone concentration in the atmosphere below which, according to current scientific knowledge, direct adverse effects on human health and/or the environment as a whole are unlikely, to be attained as far as possible in the long term with the aim of providing effective protection of human health and the environment;
- *Information threshold* means an ozone alert threshold for sensitive sections of the population.

The analysis of air quality compared with actual and future standards shows that:

- The carbon monoxide and sulfur oxides standards are generally complied with,
- There is generally a critical situation for nitrogen dioxide compared to the mean value future standard,
- There is generally a critical situation for PM_{10} compared to the mean value future standard,
- There is a critical situation for ozone in urban areas, not directly impacted by high traffic nitrogen monoxide emissions, with reference to the standard for the protection of human health;
- There is a critical situation for ozone in remote areas, with reference to the standard for the protection of vegetation.

In different plans, after the analysis of every pollutant is carried out, an Air Quality Index (AQI), which takes into account different pollutants at the same time, is introduced 0. The index is elaborated from the Pollutant Standard Index (PSI) developed by United States Environment Protection Agency 0.

The AQI is built starting from specific indexes for every pollutant:

$$I_i = 100\, C^*_i / S_i$$

where:

i, pollutant,

C^*_i measured hourly concentration (C_i) for nitrogen dioxide, carbon monoxide and ozone, while it is the 24-hours carried mobile average for sulfur dioxide and particulate matter; particularly in this second case:

$$C^*_i = \Sigma_{j=i,i+24}\, C_i / 24$$

S_i is the concentration value for the attention state (defined in the legislation prior to the new EU directive as a situation of air pollution that, if persistent, determines the risk of alert state: a situation susceptible to determine an environmental and sanitary risk).

The index I_i, so built, is equal to 100 when the concentration measured (or the mobile mean over 24 hours) is equal to the attention state; an index lower than 100 means that the pollutant has a value lower than the attention state.

After the different indexes I_i have been calculated for every pollutant, we select the maximum index I between different indexes:

$$I = \max{}_i I_i$$

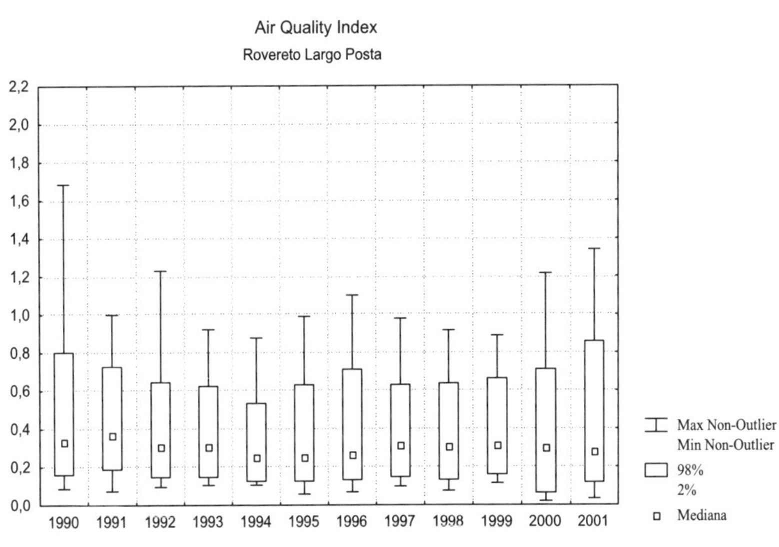

Figure 4: Example of Air Quality Index application (Trento Region).

In this way a characterization of the pollution level a part from the pollutant taken into account is obtained (Fig. 4). For example, in winter the more critical pollutants are CO and NO_x, depending on road traffic and domestic heating. In summer it is possible that for O_3 a higher index is registered depending on temperature.

4 Classification of territory and analysis of priority in intervention

One of the most important problems in air quality management is the location of the areas where the remediation actions are priority. From monitoring data we know if air quality in selected locations meets the national and international air quality standards and from spatially resolved emissions inventories we can have an idea of the contribution of the different sources to air pollution. In the optimal situation we have, with the use of air quality models, an approximate description of air pollution distribution over the territory.

In Italy air quality management plan activities, a special innovative methodology, based on multivariate statistical analysis techniques and particularly cluster analysis, is established to define territorial priorities of remediation actions 0.

The methodology was developed along the following lines:

- Regional territory classification based on environmental sensibility of areas;
- Regional territory classification based on pollutants loads;
- Regional territory classification based on remediation action priorities.

For classification of areas within the regions a methodology based on multivariate statistical analysis techniques was developed. Particularly, automatic classification criteria such as cluster analysis combined with simple statistical procedures were used.

4.1 Environmental sensibility

The classification of territory based on environmental vulnerability (Fig. 5) uses the following indicators:

- Resident population (from census statistics),
- Forests and other vegetation areas (from forest statistics),
- Beds available for tourists (from tourism statistics),
- Biotopes (from European Union funded project Bioitaly: Biotopes Inventory of Italy).

A biotope is defined following the *Council of the European Communities Directive 92/43/EEC of 21 May 1992 on the conservation of natural habitats and of wild fauna and flora* as a site that, in the biogeographical region or regions to which it belongs, contributes significantly to the maintenance or restoration at a favorable conservation status of a natural habitat type of Community interest or of a specie of Community interest.

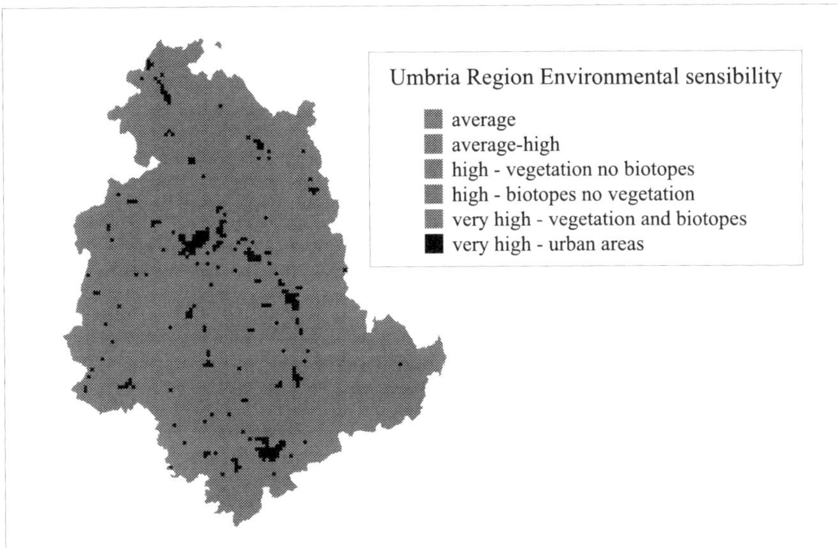

Figure 5: Umbria territory classification based on environmental sensibility.

The following procedure was used for data evaluation on 1 km × 1 km grid:

- Municipal data on resident population and on tourists beds available were assigned to 1 km × 1 km squared grid through topologic queries on CORINE land cover class "Urban fabric";
- Biotopes were evaluated on 1 km × 1 km squared grid through topologic query on Bioitaly sites of regional, national or community interests.
- Forests and other vegetation were evaluated on 1 km × 1 km squared grid through topologic query on CORINE land cover map "Forests".

4.2 Pollutants loads

For the classification of territory based on pollutants loads area, line and point sources from regional emissions inventories were taken into account. Emissions of the following main five pollutants were used: carbon monoxide, volatile organic compounds, nitrogen oxides, fine particulate or suspended particles with diameter less than 10 micron, sulfur oxides.

The methodology uses multivariate statistical analysis for classification of area/line sources, evaluated on 1 km × 1 km, and simple statistical methods for autonomous classification of point sources.

For area and line emissions classification the same methodology was used as in environmental sensibility case; the methodology is based on multivariate statistical analysis techniques, and particularly on automatic classification criteria such as cluster analysis. The cluster analysis was applied using as variables the emissions of 5 pollutants in 11 groups of activities and as cases the grid values.

Point sources are individually treated using simple statistical methods for classification. In particular, an indicator was build for each point source *k*, using the emissions E of the pollutants j, as:

$$P_k = \Sigma_j \left(E_{jk} / (1 + 4 * \delta_{j,CO})\right) \quad \text{where} \quad \delta_{j,CO} = 1 \quad \text{if} \quad j = CO, \; 0 \text{ otherwise}$$

MapInfo "Natural Break" algorithm was used for classification 0; the algorithm distributes the values so that the average of each class is as close as possible to each of the class values in that class.

As an indicator of municipality pollutant load the number of grid areas in the different classes can be counted and a rank of municipalities as a function of the number of grid areas in more critical classes can be performed.

4.3 Remediation actions priority

The combination of environmental vulnerability and pollutant loads classification produces a regional classification based on priority. This classification can take into account overall sources (integrated methodology) or area/line sources and separately point sources (area and point methodology).

An example of a priority map is reported in Fig. 6 while an example of classification criteria is reported in Table 3 for integrated methodology.

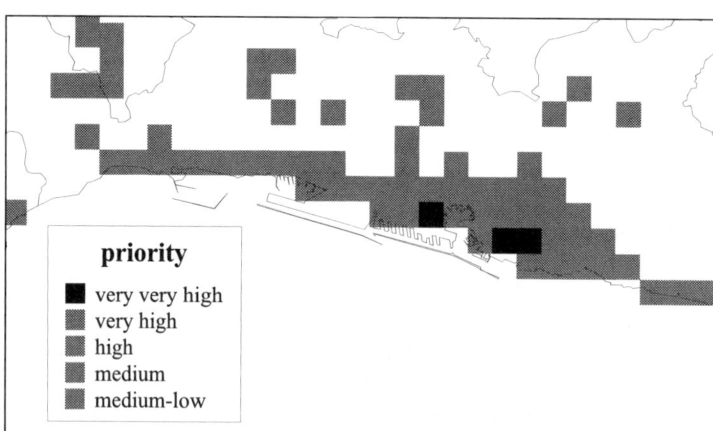

Figure 6: Liguria (Genova area) remediation action priority classification.

Table 3: Classification criteria in Umbria case study.

Priority	Sensibility	Area and line pollutant load	Point sources load
High	very high urban	average high	very high, high, average or low
		average	high or average
		average low	very high
Average–high	very high urban	highways	Low
		average high	Zero
		average	Low
		average low	average or low
	high–vegetation no biotopes	low	High
Average	very high urban	highways	Zero
		average	Zero
	high–vegetation no biotopes	average low	Average
Average–low	very high – vegetation/biotopes	average low	Zero
		low	Zero
	very high urban	average low	Zero
	high–biotopes no vegetation	highways	Zero
	high–vegetation no biotopes	average low	low or zero
		highways	Zero
		low	Low
		average	Zero
		average low	Zero
	Average	low	very high or high
		average low	High
		highways	Average
Very low	Average	low	Zero
Low		all the others	

In the area and point methodology great attention was devoted at the problem of dispersion of pollutants emitted from high sources. Classification of territory in terms of priority in remediation actions was performed only with area and line sources while priority in point sources was stated separately.

In the integrated methodology the contribution of all the sources was taken into account at the same time. In this approach great attention was devoted to the local impact of sources and to the origins of pollution. Consequently classification of territory was performed in terms of where the emissions are generated.

5 Mortality data analysis

A preliminary study of the correlation between air pollution and mortality is conducted in the special case of Friuli Venezia–Giulia region. This region has a peculiar epidemiological situation that presents higher mortality for cancer and other pathologies of the respiratory system when compared with the rest of Italy.

These differences are only partially explained with differences in age structure of the population.

In the study a statistical analysis of municipality data is performed using correlation analysis, factor analysis and cluster analysis to determine if a statistically significant relation exists between air pollutant emissions and mortality. The correlation analysis results generally show low correlation coefficients, with slightly higher values for respiratory disease and respiratory system cancer. Factor analysis takes into account the main five pollutants from area, line and point sources and extracts three factors named: “urban pollution”, “large coastal industrial plants” and “transport”. Using these factors five municipality clusters are identified and named: “port areas”, “power plants”, “large cities”, “highways” and “low pollution”. Each cluster is then characterized by mortality data to verify differences between groups.

The results show higher mortality for respiratory disease and respiratory system cancer in the cluster with lower air pollution! The conclusion is that, at least at municipality level, it is not possible to correlate emissions and mortality data. Other studies are necessary with reference to age structure of population and other "effect" indicators such as hospitalization.

6 Emissions projections

Within the realization of the air quality management plan, the projections of emissions of air pollutants are carried out 0. In Fig. 7 the data flow for future emissions evaluation is reported.

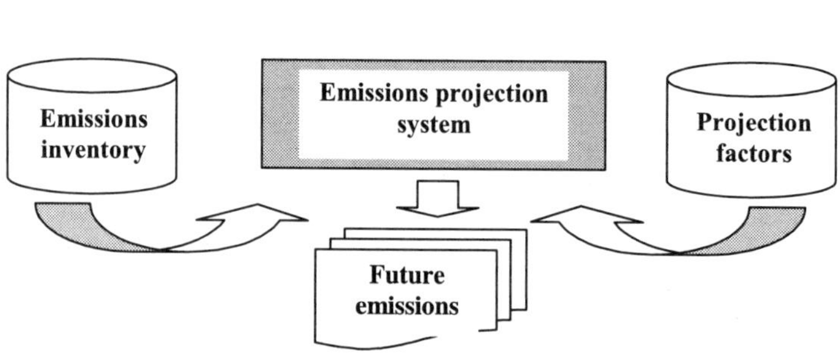

Figure 7: Emissions projection data flow.

6.1 Emissions projection methodology

Future area and line emissions are evaluated as:

$$E_{ijk} = E_{ij0}\, a_{ik}\, f_{ijk}$$

where: i, activity; j, pollutant; k, future year; E_{ij0}, emission at base year; a_{ik}, projection parameter for activity indicator; f_{ijk}, projection parameter for emission factor.

For point sources:

$$E_{ijnmk} = E_{ijnm0}\ a_{ik}\ f_{ijk}\ p_{jmnk}$$

where: i, activity; j, pollutant; k, future year; n, point source; m; point source unit; E_{ijnm0}, emissions at base year, a_{ik}, projection parameter for activity indicator; f_{ijk}, projection parameter for emission factor; p_{jmnk}, projection parameter for the specific unit of a single facility, where applicable.

6.2 Emissions projection drivers

Projection parameters for the emission factor are based on "technology drivers", projection parameters for the activity indicator are based on "activity drivers" and projection parameters for the specific unit of a single facility are based on "point source drivers". The drivers are equal to the unit in the base year.

Activity drivers A^{α} are generally defined from socio–economic variables values, S^{α}_{0} at base year and S^{α}_{k} at future year (k), as;

$$A^{\alpha} = S^{\alpha}_{k} / S^{\alpha}_{0}$$

Typical examples of socio–economic variables are: population, gross domestic product, agricultural production, waste production, number of vehicles (cars, motorcycles, off–road vehicles), number of takeoff/landing movements in the airports, tons of freight transported by road and sea, etc.

Technology drivers T^{β} are defined from parameters that affect emissions, V^{β}_{0} at base year and V^{β}_{k} at future year (k), as:

$$T^{\beta} = V^{\beta}_{k} / V^{\beta}_{0}$$

Typical examples of parameters that affect emissions are: sulfur content of fuels, new regulation limits on emissions from point sources (for power plant stacks, industrial stacks, etc.) and mobile sources (for cars, trucks, motorcycles, off–road vehicles, etc.), solvent content of products (paints, inks, glues, etc.), modification of technologies (application of best available technologies), modification of practices (application of best available practices), etc.

Point source drivers P^{γ} can be used for modelling modifications on facilities units and are defined from emissions V^{γ}_{0} before modification and emissions V^{γ}_{k} after modification as:

$$P^{\gamma} = V^{\gamma}_{k} / V^{\gamma}_{0}$$

In such a way various activities can be modelled: introduction of specific abatement systems (such as desulphuration, denitrification, particulate abatement, incineration of volatile organic compounds, etc.); modification in process technology; repowering or dismission of units; etc.

6.3 Scenarios definition

When many technology, activity and point source drivers are defined, the following scenarios are introduced:

- Activity scenarios, associating to selected activities, specific activity drivers: for every *i* and *k* exists a α such that $a_{ik} \equiv A^{\alpha}$;
- Technology scenarios, associating to selected activities and pollutants, specific technology drivers; for every *i* , *j* and *k* exists a β such that $f_{ijk} \equiv T^{\beta}$;
- Point sources scenarios, associating to selected units of facilities, specific point source drivers; for every *j*, *k*, *m* and *n* exists a γ such that $p_{jkmn} \equiv P^{\gamma}$.

Technology and activity drivers can be shared by several activities. For example the activity driver "population" can be used to project "domestic paint consumptions" and "residential fuel consumptions"; the technology driver "distillate oil sulfur content" can be used to project sulfur oxides emission factor for "domestic distillate oil consumption" and "ports–distillate oil ships".

The combination of activity, technology and point sources scenarios produce emissions scenarios for air pollution planning.

6.4 Projection of emissions without emissions reduction measures

Within the realization of the air quality management plan, the projections of emissions of air pollutants for 5–15 years in the absence of interventions (“do nothing” scenario) are carried out. Fig. 8 reports “do nothing” projections coming from the Liguria plan preparatory study.

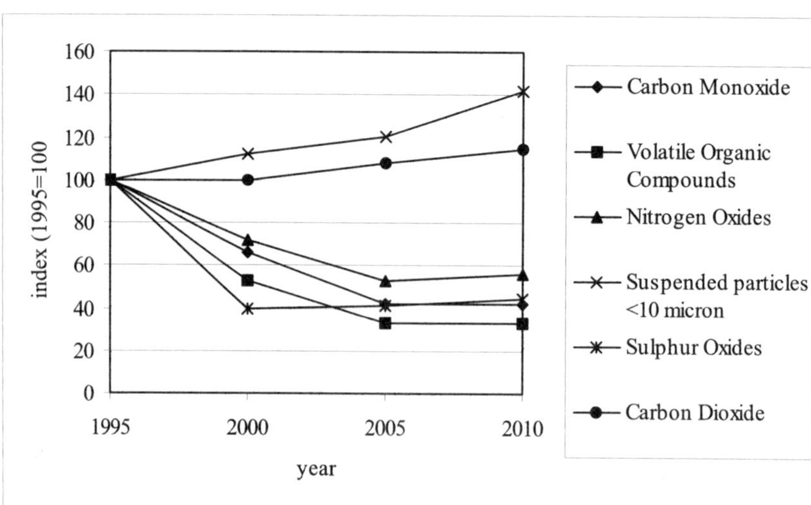

Figure 8: Example of emissions projection in do nothing scenario (Liguria Region total traffic emissions).

Evaluation use projection of population, energy demand, relevant industrial production, vehicle park, mileage and consumption, social and economical variables connected to the activity of domestic type and evaluating the different technologies of emission reduction. The different aspects of regional planning (transport, energy, etc.), are analyzed in order to obtain information about the projections of activity indicators. The trend is common to all the regions and presents decreasing NO_x, SO_x, CO, VOC emissions and increasing PM_{10} and CO_2 emissions.

7 Definition of air quality objectives

From the point of view of air quality objectives are defined for each pollutant:

- The maximum desirable level as the long–term goal of air quality that provides a basis for a policy to protect particular areas, and spurs continuing improvement in control technology;
- The maximum acceptable level to provide adequate protection against adverse effects on human health, vegetation and animals;
- The maximum tolerable level as the limit beyond which, due to a diminishing margin of safety, appropriate action is required, without delay, to protect the health of the general population.

8 Strategies and scenarios for emission control

In the following, a review of the most important measures proposed in the different plans preparatory study is reported.

The plan preparatory study scenarios contain the following measures regarding fixed area sources:

- Energy saving and solar energy use in industry and tertiary combustion;
- Heat recovery for district heating in selected power plant areas;
- Incentive for combustion of wood and wood wastes in district–heating;
- Prohibition on the use of fuel oil in tertiary combustion;
- Progressive control of open burning of agricultural wastes;
- Reduction of methane emissions from stock–farming;
- Reduction of VOC emissions from paint application;
- Incentive to use low–solvent domestic products;
- Forest fire prevention;
- Reduction of VOC emissions from maritime terminals;

regarding transport:

- Reducing private road transport by increasing local and regional rail transport, "car pooling", "car pricing", "car sharing", increasing electric or hybrid (electric and methane) public urban transport, extending cycle–path, extending "traffic–closed" areas, optimize car–public transport interchange;

- Reducing road goods transport by increasing train transport and optimize ship–truck and rail–truck interchanges;
- Car maintenance;

regarding main and minor point sources:

- Conversion to gas of plants burning fuel oil, localized in areas connected to methane pipelines;
- Switch to very low content of sulfur oil (0.3%) for plants localized in vulnerable zones and not connected to methane pipelines;
- Denitrification and desulfuration of emissions of main point sources;

regarding monitoring of air quality:

- Reducing carbon monoxide and sulfur oxides points of measure;
- Reorganize ozone networks;
- Integrating the existing meteorological networks (for atmospheric stability determination);
- Planning bio–monitoring networks.

The plan preparatory studies provide for information campaigns for energy saving, reduce private transport, increase use of bicycles, efficient management of large combustion plants, communicate the plan to population. Furthermore, some administrative measure–improvements, are defined in the plan preparatory studies.

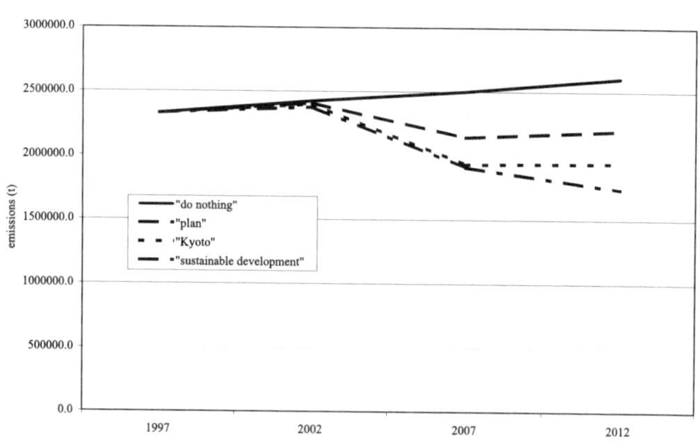

Figure 9: Example of carbon dioxide emissions in different scenarios (Bolzano Region).

9 Evaluation of emissions reduction strategies

Starting from the "do nothing" scenario it is possible to build alternative scenarios using measures introduced in the preceding section. Alternative economic and energy scenarios (activity) and technological scenarios (emission factors) are supplied. The joining of the two scenarios supplies the emission scenarios for the plan. An example of scenarios representation is reported in Fig. 9.

10 Air quality evaluation in different scenarios

Within Air Quality Management Plans preparation, a fundamental role is played by simulation models to estimate air pollution in a given area.

10.1 Long–term models

The Industrial Source Complex Long Term (ISCLT) dispersion model 0 developed by the United States Environmental Protection Agency (US EPA) is used in actual situations, for comparison with air quality network data, and in future situations, with and without plan measures, to evaluate the effect of the actions 0. In Fig. 10 an example of model application is reported.

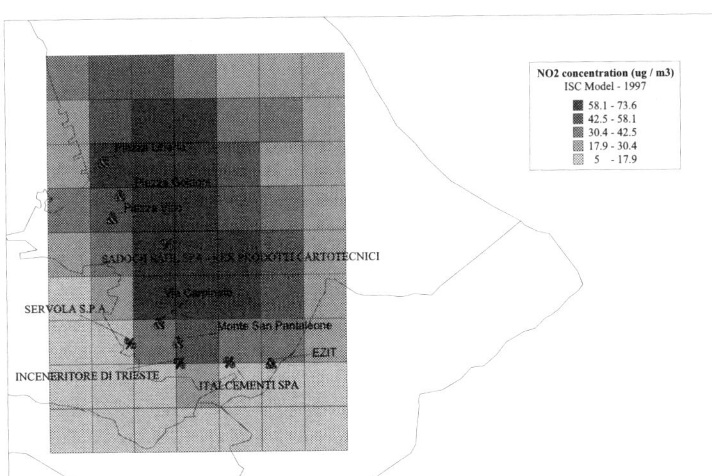

Figure 10: Example of ISCLT model computed Nitrogen oxides concentration (Friuli–Venezia Giulia Region).

The model receives as input the different sources of atmospheric pollution inside the areas:

- Point sources, modelled giving on input to the model, besides the emissions, the physical characteristics of the emissions themselves (height of the release, temperature of gases, speed of gases, inside diameter of the stack);
- Minor point sources, modelled as 50 m × 50 m squared sources and height 10 m;
- High area sources (area sources of the following CORINAIR groups: non–industrial combustion plants, combustion in manufacturing industry, production processes, solvent and other product use, other sources and sinks), modelled as area source on a 1 km × 1 km grid with height 10 m;
- Low area sources (area sources of the following CORINAIR groups: extraction and distribution of fossil fuels and geothermal energy, road transport, other mobile sources and machinery, waste treatment and disposal, agriculture), modelled as area source on a 1 km × 1 km grid with height 0.5 m.

The meteorological data (for every class of stability of Pasquill–Gilford: frequencies of the distribution of the classes of speed of the wind for sectors of the wind, median of the speeds of the wind, average temperatures) are elaborated from the data of the meteorological stations.

The results of the model are valued on 1 km × 1 km squared grid. Besides, the concentrations have been valued in an area of ray 20 metres around the air quality monitoring station specifying the height of the receptors on the level of the ground.

To evaluate the accuracy of the model the following statistical indicators 0 can be used:

- Normalized mean standard error $NMSE=\langle (C_s-C_m)^2\rangle/\langle C_s\rangle\langle C_m\rangle$
- Fractional bias $FB=2\,(\langle C_s\rangle-\langle C_m\rangle)/(\langle C_s\rangle+\langle C_m\rangle)$

where C_s is the model concentration and C_m the air quality monitoring station concentration. In Table 4 computed indexes are reported for selected pollutants and networks.

Table 4: Model performance indexes.

Index	Trento network 1995		Livorno network 1995		Test	
	SO_2	NO_2	SO_2	NO_2	Perfect	Good
NMSE	0.1	0.2	0.6	0.7	0	<1.0
FB	–0.2	–0.2	0.5	0.5	0	$0.5<FB<0.5$

In all the areas examined, the comparison, with statistical indexes, between the data calculated by the model and the values detected by the stations shows that the model can, in the long term, reconstruct the concentrations of air pollutants and can be efficiently used in air quality planning. In Table 5 an example of application of the model in Trento region air quality planning is reported.

Table 5: Example of model application.

Monitoring station	Measured data µg/m³ 2000		Model computed data µg/m³ 2000		2015 do nothing		2015 plan preparatory study	
	SO_2	NO_2	SO_2	NO_2	SO_2	NO_2	SO_2	NO_2
Trento – Gardolo	7	49	3	44	2	20	2	18
Trento – Porta Nuova	4	60	6	44	5	24	4	20
Trento – S. Chiara	6	41	6	43	5	25	4	21
Trento – Via Veneto	7	45	6	46	5	25	4	21
Grumo S. Michele	5	44	3	57	2	26	2	22
Borgo Valsugana	4	33	3	22	3	17	3	17
Rovereto – Posta	6	40	11	43	10	23	9	22
Rovereto – Benacense	5	50	12	45	11	24	9	24
Riva del Garda	7	42	12	49	11	34	10	31

10.2 Photochemical models

In several regions the main problem consists in photochemical pollution. This is the presence in troposphere of pollutants like ozone, also defined as secondary pollutants because of long, complex and not linear chemical reactions involving primary pollutants (and particularly nitrogen oxides and volatile organic compounds) from which they become catalyzed by solar radiation.

To reduce the photochemical pollution it is necessary to understand how the variation of the concentrations of secondary and oxidant pollutants, and particularly ozone, depends on the variation of primary pollutant concentrations. Moreover, it is important to take into account the effects that some meteorological factors as temperature, humidity and air pressure, wind and atmospheric stability have on photochemical smog. In fact, especially transport processes, that occur over distances of a few hundred kilometres too, can continue to enrich air masses of ozone precursory species, with possible higher values of ozone in leeward rural areas, rather than in urban or industrial areas.

To study photochemical pollution the California Air Resource Board CALGRID 0 and CALMET 0 models can be used. The meteorological model CALMET rebuilds three-dimensional wind and temperature fields and other parameters, such as stability and mixing height, descriptive of the boundary layer. The photochemical model CALGRID considers the chemical evolution of 54 species in 129 reactions 0, simulating air masses dynamical processes (transport and turbulent diffusion) due to meteorological phenomena.

Within the framework of the Liguria Region Air Quality Management Plan preparatory study, CALMET and CALGRID models are used to simulate photochemical pollution for the evaluation of the actual situation and the change in pollution in the presence of plan measures (Fig. 11) 0.

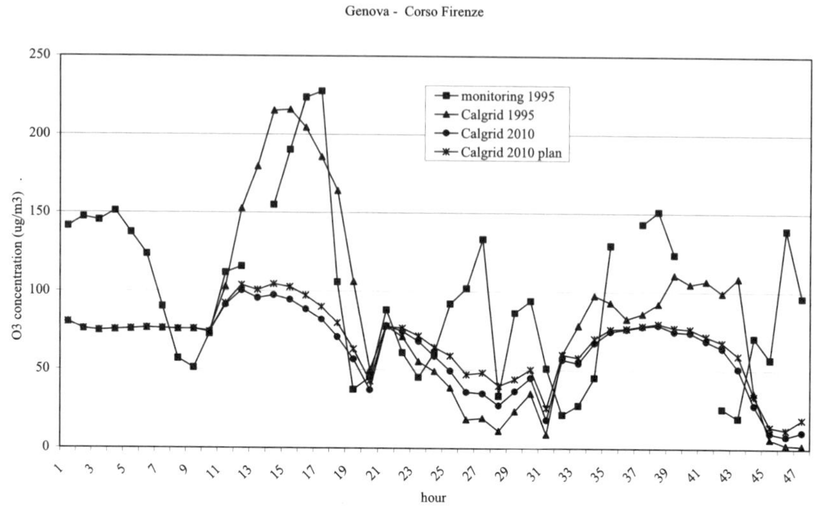

Figure 11: Example of CALGRID model computed ozone concentration (Liguria region).

The results show quite a good agreement with experimental data with the precaution that:

- An episode of intense photochemical smog is treated;
- Model–calculated values are compared with concentration values relative to monitoring stations situated in the centre of the city; these sites are influenced by mobile source emissions and sometimes show very irregular trends of measured concentrations;
- Simulated area is relatively small and so boundary conditions influence simulation results.

The comparison between data simulated at 1995 and at 2010 shows that, even if there is an uncertainty implicit in the methodology used, the change in emissions should prevent, in a similar meteorological situation, the peaks of photochemical smog. Nevertheless, the results indicate how, in the future scenarios, there is no substantial change of ozone background concentrations. The comparison between data simulated at 2010 with and without interventions shows, by virtue of the further reductions of nitrogen oxides emissions and volatile organic compounds, there is a slight increase of ozone concentrations.

The study has demonstrated that the CALGRID model is a powerful tool for air quality planning through its application in predicting photochemical pollution.

10.3 Traffic models

In the framework of the regional air quality management planning, the CALINE4 model 0, integrated with SETS emissions model, is applied. Average emission factors of generic vehicles are produced by SETS model, while the others, geometric (road type, high and coordinates), meteorological (wind speed and direction, mixed zone dimension, mixing high and stability class) and chemical (photolysis constant and bottom concentration) data are directly inserted in dispersion module.

The model is applied to some case studies [16–17] to evaluate:

- The compliance of actual and future air quality standards;
- Specific measures on traffic (for example the effects of an inversion of the traffic flow in road with gradient as in Fig. 12);
- Changes in the photochemical mechanism as a result of NO_x emissions reduction.

The results of comparison with air quality network data are good, as shown in Table 6, where the values assumed by the different indexes in the base case of 1995 were reported.

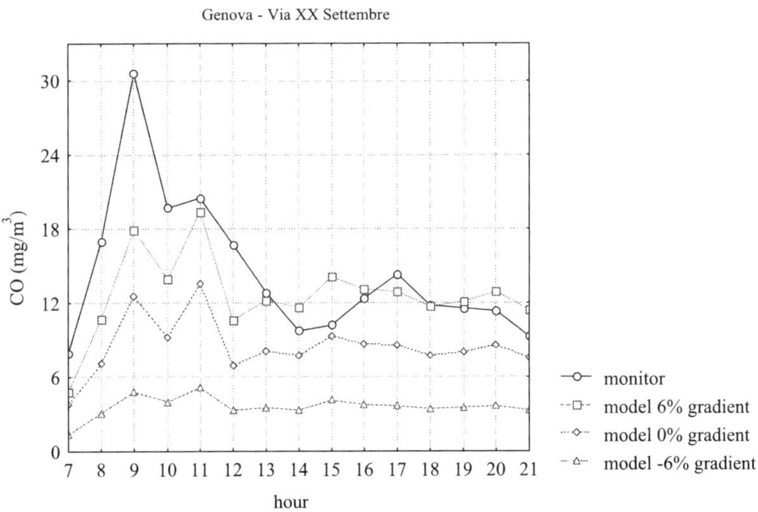

Figure 12: Example of CALINE model computed carbon monoxide concentration at different gradients (Liguria region).

Table 6: Model performance indexes.

	Carbon monoxide	Nitrogen dioxide	Test	
			Perfect	Good
NMSE	0.11	0.03	0	<1.0
FB	– 0.13	–0.04	0	0.5<FB<0.5

The case studies show that the model can be usefully used to evaluate actual and future air quality in an urban *canyon* street with gradient and that the critical parameters for concentrations evaluation are the wind intensity, the cold-start emissions for carbon monoxide (also in the central district of urban areas) and the photochemical mechanism parameters for nitrogen oxides.

11 Air quality management planning information system

The complex nature of air quality management planning involves a complex architecture of information systems to support their realization (Fig. 13).

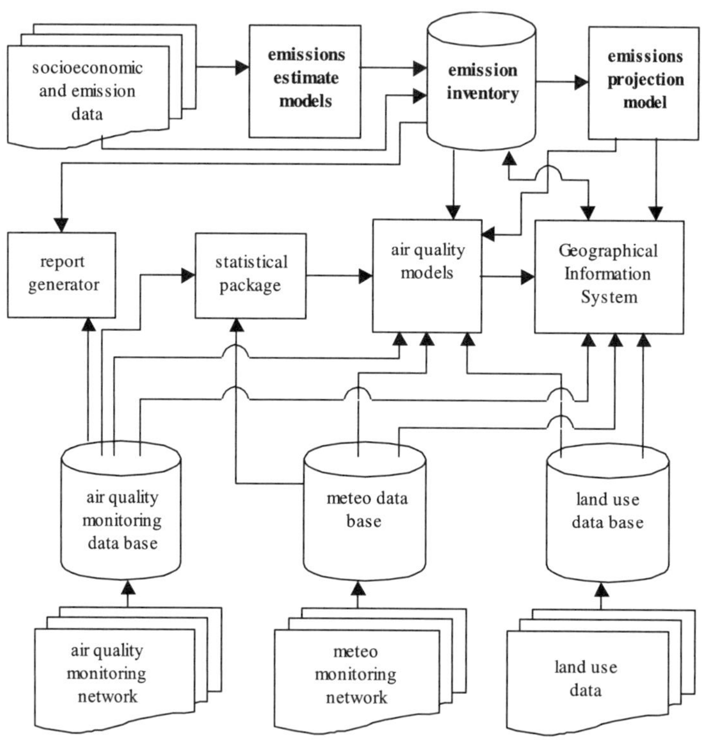

Figure 13: Structure of information system for air quality management planning.

The information system has been implemented on a network of Pentium computers with large extended RAM running Windows NT, Oracle or Access, Arc View or MapInfo, and Statistica.

The following tools are contained in the information system:

- *APEX*, an air pollutant emissions computer system originally developed in the Windows environment with the object–oriented ORACLE CARD tool [54–56] and with Visual Basic language, and available with an ORACLE database [57–58]; the system contains an emission factors data base and tools and data to estimate grid and municipal emissions from more aggregated data 0, the system uses Arc View or MapInfo for the thematic maps;

- Software models 0, developed in the Windows environment with object–oriented Visual Basic language and ACCESS database, interfaced with the *APEX* system in a unique environment: *SETS*, for emissions estimates from road traffic; *AIR AIR*, for emissions estimates from airports; *AIR FOREST*, for emissions estimates from forest; *AIR SHIPS*, for emissions estimates from ports and navigation lines *AIR FIRE*, for emissions estimates from forest fires;
- PREM 0, a software model for emissions projections and to evaluate emission control measures: the system contains a socioeconomic projection factors database, a technological projection factors database and some tools to project emissions; *PREM* is developed in the Windows environment with the object–oriented Visual Basic language, is available with an ORACLE database and is integrated with the *APEX* system;
- QUAR, a computer system for air quality data management and analysis; *QUAR* is developed in a Windows environment with object–oriented Visual Basic language, is available with an ORACLE database and use Statistica software to analyze data and ArcView or MapInfo for thematic map;
- SETS/D, a computer interface to SETS and CALINE4 models to evaluate air pollution dispersion near a road;
- AIR_ISC, a computer interface to ISC Long Term and ISC Short term integrating data coming from APEX and QUAR computer systems;
- AIR_GRID, a computer interface to CALMET and CALGRID models integrating data coming from APEX and QUAR computer systems.

12 Conclusions

The regional air quality planning in Italy has received a powerful impulse in recent years and has produced many important studies and proposals.

Today, after the conclusion of this first phase in the most advanced regions, the attention is directed at the implementation phase of the plans and at the activation of the actions need to maintain the information systems realized, with particular attention to emissions inventories. New rules of the European Community are in force today, and activities are in progress to adjust the plans.

Methodology and software used for air quality planning can also be used for other media (air, water, soil, wastes, noise). For emissions inventory, a lot of emissions factors are available for area and mobile sources, while regarding point sources, legislation was developed concerning integrated emissions registers. A new computer system was realized to build emissions inventories in different environmental media and a first Italy regional case study was realized 0.

Acknowledgement

The authors wish to thank the air quality experts of several regional administrations that have followed the activities on air quality management planning in these years and have given us many fundamental suggestions and indications. Particular thanks to M. Tava (Trento Region), M. Romanelli (Toscana Region), L. Badalato (Liguria Region), N. Lantschner and M. Guariento (Bolzano Region), P. Gubertini (Friuli Venezia–Giulia Region), M. Trinei (Umbria Region).

References

[1] Vaccaro, R. & Trozzi, C., Regional Air Quality Management Plans in Italy: balances and perspectives, *Proc. of the Air & Waste Management Association 94th Annual Conf. & Exhibition,* Air & Waste Management Association: Pittsburgh, 2001.

[2] Vaccaro, R., Trozzi, C. & Crocetti, S., Air quality management planning in Italy: Trento Alpine area case study, *Proc. of the 6th Int. Conf. on Modeling, Monitoring and Management of Air Pollution*, eds. C.A. Brebbia, C.F. Ratto & H. Power, WIT Press: Boston, pp. 343–352, 1998.

[3] Cirillo, M.C., Trozzi, C., Bocola, W. & Gaudioso, D., Emissions of air pollutants in Italy: state of art. *Proc. of the RISØ Int. Conf. on Environmental Models: emissions and consequences*, eds. J. Fenhann, H. Larsen, G.A. Mackenzie & B. Rasmussen, Elsevier, 1990.

[4] Vaccaro, R., Trozzi, C., Nicolò, L. & Gaudioso, D., Air pollutants emissions inventories at local level, *Proc. of the Worldwide Sym. on Pollution in large cities, Science and technology for planning environmental quality*, PadovaFiere, pp. 251–261, 1995.

[5] Vaccaro, R. & Trozzi, C., Emission inventories in Italy regional air quality management planning, *Proc. of the Specialty Conf. on Emission Inventory: Planning for the Future,* Air & Waste Management Association: Pittsburgh, pp. 935–946, 1997.

[6] European Commission, DGXI, *CORINAIR Technical Annexes, Volume 1 Nomenclature and software*, European Commission, DGXI, Luxembourg, EUR 12586/1EN, 1995.

[7] Trozzi, C. & Vaccaro, R., Dioxins and furans emissions estimate in Italy, *Proc. of the 10th World Clean Air Congress, Volume 1 – Emissions and Control,* eds. M. Tolvanen, P. Anttila & J. Kämäri, Helsinki, pp. 856–867, 1995.

[8] Vaccaro, R., Trozzi, C., Crocetti, S., Tabacco, M. & Romanelli, M., Air toxics emissions inventory in regional air quality management planning, *Proc. of 4th Intern. Cong. on Energy, Environment and Technological Innovation*, Università di Roma "La Sapienza", pp. 34–44, 1999.

[9] EMEP (Co–operative Programme for Monitoring and Evaluation of the Long Range Transmission of Air Pollutants in Europe), CORINAIR (The Core Inventory of Air Emissions in Europe) Task Force on Emission

Inventories, *Atmospheric Emission Inventory Guidebook, 3rd Edition*, European Environment Agency, Technical report No 30, EEA, Copenhagen 2001.

[10] United States Environmental Protection Agency, *Compilation of Air Pollutant Emission Factors: Volume I: Stationary Point and Area Sources*, United States Environmental Protection Agency AP–42, Fifth Edition, January 1995.

[11] Trozzi, C., Vaccaro, R., Di Giovandomenico, P. & Crocetti, S., SETS – Estimate of air pollutants emissions from road traffic, *Proc. of the 4th Intern. Software exhibition for environmental science and engineering*, eds. G. Guarisio & A. Rizzoli, Patron: Bologna, pp. 221–226, 1995

[12] Eggleston, H.S., Gaudioso, D., Lorissen, N., Joumard, R., Rijkeboer, R.C., Samaras, Z. & Zierock, K.–H., *Default Emission Factors from Road Traffic*; In: European Commission, DGXI, *CORINAIR Technical Annexes, Volume 2 Default Emission Factors Handbook*; European Commission, DGXI, Luxembourg, EUR 12586/2EN, pp. 129–202, 1994.

[13] European Commission, *Meet, Methodologies for calculating transport emissions and energy consumption*, European Commission, Transport Research fourth framework programme strategic research DG VII, 1999.

[14] Trozzi, C., Vaccaro, R. & Crocetti, S., Application of Corinair Baseline Methodology in Italy and in Rome Metropolitan Area, *Poster Proc. of 3rd Intern. Sym. on Transport and Air Pollution*, ed. R. Joumard, INRETS: Arcueil, France, pp. 212–217, 1994

[15] Trozzi, C., Vaccaro, R. & Crocetti, S., Speed frequency distribution in air pollutants' emissions estimate from road traffic, *Science of Total Environment*, **189/190**, pp. 181–185, 1996.

[16] Trozzi, C., Vaccaro, R., & Crocetti, S., Emission and dispersion in roads with gradients: a case study, *Poster Proc. of 4th Intern. Sym. on Transport and Air Pollution*, ed. R. Joumard, INRETS: Arcueil, France, pp. 231–236, 1997.

[17] Trozzi, C. & Colaiezzi, M., Actual and future air pollutant emissions and dispersion in urban roads with gradient: Genoa XX Settembre road case study, *Proc. of 9th Intern. Sym. on Transport and Air Pollution, Vol. 2, Poster Com.* ed. R. Joumard, INRETS: Arcueil, France, pp. 381–388, 2000.

[18] Trozzi, C., Vaccaro, R. & Crocetti, S., Actual and forecasted road traffic emissions in Air Quality Management Plans, *Proc. of Conf. on Externalities in the Urban Transport: Assessing and Reducing the Impacts*, S. Donato Milanese, Italy 26–27 October 2000

[19] Trozzi, C., Vaccaro, R., Trobbiani, R., Di Giovandomenico, P. & Piscitello, E., Air Ships: Computer model for air pollutant emissions estimates in port and in navigation, *Proc. of the Intern. Workshop on Harbour, Maritime & Industrial Logistics Modeling and Simulation*, eds. A.G. Bruzzone, Y.A. Merkuriev & R. Mosca, Society for Computer Simulation International, pp. 243–248, 1999.

[20] Trozzi, C., Vaccaro, R. & Nicolò, L., Air Pollutants Emissions Estimate from Maritime Traffic in Harbours: The Italian Harbours of Venice and Piombino, *Science of Total Environment*, **169**, pp. 257–263, 1995.

[21] Trozzi, C. & Vaccaro, R., Ships transport (part C), *Meet, Methodologies for calculating transport emissions and energy consumption*; European Commission, Transport Research fourth framework programme strategic research DG VII, 1999.

[22] Trozzi, C. & Vaccaro, R., Methodologies for estimating air pollutant emissions from ships, *Proc. of 22nd CIMAC Intern. Congress on Combustion Engines, Vol. 3,* CIMAC: Frankfurt/Germany, pp. 775–782, 1998

[23] Vaccaro, R.; Trozzi, C. & Crocetti, S., Air pollutant emissions from ships: high Tyrrhenian Sea ports case study, *Proc. of 1st Inter. Conf. On Maritime Engineering and Ports*, eds. G. Sciutto & C.A. Brebbia, WIT press: Southampton & Boston, pp. 243–254

[24] Trozzi, C. & Vaccaro, R., Actual and future air pollutant emissions from ships, *Proc. of the 8th Intern. Sym. on Transport and Air Pollution including Cost 319 Final Conf.*, ed. P.J. Sturm, Technical University of Graz, Report of Institute of Internal Combustion Engines and Thermodynamics, **76**, pp. 403–416, 1999.

[25] Trozzi, C., Vaccaro, R. & De Lauretis, R., Air pollutant emissions estimate from global ship traffic in port and in cruise: methodology and case study, *Proc. of the 11th Intern. Sym. on Transport and Air Pollution, Vol. I*, eds. P.J. Sturm & S. Minarik, Verlag der Technischen Universitat Graz, VKM–THD Mitteilungen, **81–I**, pp. 195–202, 1999

[26] Trozzi, C., Vaccaro, R., De Lauretis, R. & Romano, D., Air pollutant emissions estimate from global air traffic in airport and in cruise: methodology and case study, *Proc. of the 11th Intern. Sym. on Transport and Air Pollution, Vol. II Poster presentations*, eds. P.J. Sturm & S. Minarik, Verlag der Technischen Universitat Graz, VKM–THD Mitteilungen, **81–II**, pp. 129–136, 1999

[27] Trozzi, C., Vaccaro, R., Nicolò, L., Trobbiani, R. & Martinelli, A., AIR AIR – Estimate of air pollutants emissions from airport, *Proc. of the 4th Intern. Software exhibition for environmental science and engineering*, eds. G. Guarisio & A. Rizzoli, Patron: Bologna, pp. 209–214, 1995.

[28] Trozzi, C. & Vaccaro, R., Air Air & Noise air pollutants emissions and noise from airports, *Proc. of the 6th Intern. Conf. on Development and Application of Computer Techniques to Environmental Studies*, eds. P. Zanetti & C.A. Brebbia, Computation Mechanics Publ.: Southampton & Boston, pp. 59–68, 1996.

[29] Trozzi, C., Barbizzi, S. & Trobbiani, R., Airports Air pollutants emissions: Italy air quality management plans case studies, *Proc. of 4th Intern. Cong. on Energy, Environment and Technological Innovation*, Università di Roma "La Sapienza", pp. 27–32, 1999.

[30] Trozzi, C. & Vaccaro, R., Estimate of volatile organic compounds emissions from vegetation in regional air quality management plans, *Proc.*

of the 7th Int. Conf. on Modeling, Monitoring and Management of Air Pollution, eds. C.A. Brebbia, M. Jacobson & H. Power, WIT Press: Southampton & Boston, pp. 425–434, 1999
[31] Benjamin, M.T., Sudol, M., Bloch, L. & Winer, A.M., Low–emitting urban forests: a taxonomic methodology for assigning isoprene and monoterpene emission rates. *Atmospheric Environment*, **30**, pp. 1437–1452, 1996.
[32] Benjamin, M.T. & Winer, A.M., Estimating the ozone–forming potential of urban trees and shrubs. *Atmospheric Environment*, **32**, pp. 53–68, 1998.
[33] Guenther, A.B., Zimmerman, P.R., Harley, P.C., Monson, R.K. & Fall, R., Isoprene and monoterpene emission rate variability: model evaluations and sensitivity analyses. *Journal of Geophysical Research*, **12**, pp. 609–617, 1993.
[34] Trozzi, C., Vaccaro, R. & Piscitello, E., Emissions estimate from forest fires: methodology, software and European case studies, *Proceeding of the Inter. Conf. on Emission Inventories – Partnering for the Future,* United States Environmental Protection Agency: USA, 2002
[35] Gaudioso, D., Trozzi, C., Vaccaro, R. & Cirillo, M.C., Sensitivity Analysis of evaporative emissions from gasoline passenger cars, *Science of Total Environment*, **146/147**, pp. 325–332, 1994.
[36] Gaudioso, D., Cirillo, M.C., Trozzi, C. & Vaccaro, R., Uncertainty of NMVOC emission estimates from vegetation. *Proceeding of the International Conference and Course on The Emission Inventory – Perception and Reality* Air & Waste Management Association: Pittsburgh, pp. 318–326, 1993
[37] Council of the European Union Council Directive 96/62/EC of 27 September 1996 on ambient air quality assessment and management, *Official Journal*, **L 296**, pp. 0055–0063, 1996.
[38] Council of the European Union, Council Directive 1999/30/EC of 22 April 1999 relating to limit values for sulfur dioxide, nitrogen dioxide and oxides of nitrogen, particulate matter and lead in ambient air, *Official Journal* **L 163**, pp. 0041–0060, 1999.
[39] Council of the European Union, Directive 2000/69/EC of the European Parliament and of the Council of 16 November 2000 relating to limit values for benzene and carbon monoxide in ambient air, *Official Journal* **L 313**, pp. 0012–0021, 2000.
[40] Council of the European Union, Directive 2002/3/EC of the European Parliament and of the Council of 12 February 2002 relating to ozone in ambient air, *Official Journal*, L 067, pp. 0014–0030, 2002.
[41] Trozzi, C., Vaccaro, R. & Crocetti, S., Air quality index and its use in Italy's management plans. *Science of Total Environment*, **235**, pp. 387–389, 1999.
[42] United States Environmental Protection Agency *Measuring Air Quality, The Pollutant Standards Index*; United States Environmental Protection Agency, EPA 451/K–94–001, 1994.
[43] Trozzi, C. & Vaccaro, R., Multivariate statistical analysis for priority area localization in air quality management plans, *Proc. of the 9th Int. Conf. on*

Modeling, Monitoring and Management of Air Pollution, eds. G. Latini & C.A. Brebbia, WIT Press: Southampton & Boston, pp. 295–304, 2000

[44] MapInfo Corporation, *MapInfo Professional User Manual*, Troy, New York, 2000

[45] Trozzi, C., Crocetti, S., Vaccaro, R. & Piscitello, E., Emissions projections in air quality management plans, *Proc. of the 10th Int. Conf. on Modeling, Monitoring and Management of Air Pollution*, eds. C.A. Brebbia & J.F. Martin–Duque, WIT Press: Southampton & Boston, pp. 291–300, 2002

[46] Environmental Protection Agency (EPA) *User's guide for the Industrial Source Complex (ISC3) dispersion models. Vol. 1: User instructions. Vol. II: Description of model algorithms* EPA–454/B–95–003a–September 1995

[47] Trozzi, C., Grotti, P. & Vaccaro, L., Use of long–term gaussian dispersion model in Italy air quality management plans, *Proc. of 4th Intern. Cong. on Energy, Environment and Technological Innovation*, Università di Roma "La Sapienza", pp. 27–32, 1999.

[48] Zannetti, P., Air pollution modeling. Theories, Computational Methods and Available Software Computational Mechanics Publ.: Southampton & Boston, 1990

[49] Yamartino, R.J., Scire, J.S., Hanna, S.R., Carmichael, G.R. & Chang, Y.S., The CALGRID mesoscale photochemical grid model – I. Model Formulation, *Atmospheric Environment*, **26A**, pp. 1493–1512, 1992.

[50] Scire, J.S., Robe, F.R., Fernau, M.E. & Yamartino, R.J., *A User's Guide for the CALMET Meteorological Model (Version 5)*, Earth Tech, Inc. Concord, MA, 2000.

[51] Carter, P.L.W., *Documentation of the SAPRC–99 chemical mechanism for voc reactivity assessment*, Final Report to California Air Resources Board Contract No. 92–329, and (in part) 95–308, May 8, 2000.

[52] Trozzi, C. & Vaccaro, L., Photochemical models in Air Quality Management Plan of a complex coastal area, *Proc. of the 8th Int. Conf. on Modeling, Monitoring and Management of Air Pollution*, eds. J.W.S. Longhurst, C.A. Brebbia & H. Power, WIT Press: Southampton & Boston, pp. 169–78, 2000

[53] State of California, Department of Transportation, Division of New Technology and Research *CALINE4 – A dispersion model for predicting air pollution concentrations near roadway*. Final Report, November 1984, Revised June 1989

[54] Trozzi, C. & Vaccaro, R., Air pollutants emissions inventory and Geographic Information System, *Proc. of 4th European Conf. and Exhib. on Geographical Information Systems*, eds. J. Harts, H.F.L. Ottens & H.J. Scholten, EGIS Foundation: Utrecht/Amsterdam, pp. 47–56, 1993.

[55] Trozzi, C., Vaccaro R.; Nicolò L. Air Pollutants Emissions Inventories and Time in Geographical Information Systems *Proc. of 5th European Conf. and Exhib. on Geographical Information Systems*, eds. J. Harts, H.F.L. Ottens & H.J. Scholten, EGIS Foundation: Utrecht/Amsterdam, pp. 746–755, 1994.

[56] Trozzi, C. & Vaccaro, R., Information system for air pollutants emissions on local level *IFIP Transaction B–16. Computer support for environmental impact assessment*, Elsevier Science B.V.: Amsterdam, eds. G. Guarisio & B. Page, pp. 75–85, 1994.

[57] Trozzi, C., Vaccaro, R., Niccolò, L., Trobbiani, R., Valentini, L. & Di Giovandomenico, P., APEX – Air pollutants emissions inventory computer system, *Proc. of the 4th Intern. Software exhibition for environmental science and engineering*, eds. G. Guarisio & A. Rizzoli, Patron: Bologna, pp. 209–214, 1995

[58] Trozzi, C., Vaccaro, R., Trobbiani, R., Di Giovandomenico, P. & Piscitello, E., Emission inventory software tools for air quality management plans *Proceeding of a specialty conf. on The Emission Inventory: Living in a Global Environment,* Air & Waste Management Association: Pittsburgh, pp. 856–867, 1998.

[59] Trozzi, C., Vaccaro, R., Trobbiani, R. & Di Giovandomenico, P., Geographical Information Systems for air quality management planning *Proc of the 1st International Conference on Geographical Information Systems in the next millennium*, eds. P. Pascolo & C.A. Brebbia, Computation Mechanics Publ.: Southampton & Boston, pp. 23–32, 1998

[60] Trozzi, C., Piscitello, E., Giammarino, S. & Vaccaro, R., Advanced Pollutants Emissions Computer System (APEX 4.0), *Proc. of the 9th Intern. Conf. on Development and Application of Computer Techniques to Environmental Studies*, eds. C.A. Brebbia & P. Zanetti, WIT Press: Southampton & Boston, pp. 59–68, 1996.

Chapter 11

Air quality management – the Sussex experience

A. Deacon
Sussex Air Quality Steering Group, United Kingdom

Abstract

The Sussex Air Quality Steering Group has been in existence since before 1995, bringing together agencies responsible for the monitoring and management of air quality across two counties in the south of England. This chapter describes the operation and work of the Group, focusing in particular in work carried out to fulfil statutory air quality management obligations in the UK.

After its formation, the Group produced a joint emissions inventory, which later provided much of the data needed in the early stages of the air quality management review and assessment. The Group also appointed a Project Development Officer to provide in house expertise and operate a number of centrally managed tools, including an expanding air quality monitoring network.

As a result of the review and assessment process, no air quality management areas were declared in East or West Sussex. There were, however, a number of areas identified as being currently in exceedence of specified air quality objectives and concerns about pollutants (notably ozone) not covered by the air quality management regime.

These concerns led the Group to pursue the development of a non-statutory air quality strategy, pulling together a comprehensive picture of air quality and attempting to integrate air quality considerations into a number of local authority functions. Work leading to the development of this strategy and the future plans for the Group is also detailed.

1 Introduction

Historically, visits to the Sussex seaside were made to breathe the ozone, believed at the time to have important health-improving properties. Since those times, there has been much development of the coastal towns and cities, and improvements to the road network serving these destinations. New towns (such as Crawley) have developed and the airport at Gatwick has grown beyond all recognition. At the same time, domestic smoke pollution from coal burning has declined greatly, countered by a rise in vehicle pollution at the same time. What does all of this mean for the picture of air quality in Sussex? What has been done to monitor and manage the impacts of changing air quality? This chapter describes some of the processes and practices that have been put in place to try and deal with such issues.

All British local authorities have been engaged in a statutory process of air quality management since 1997, with the introduction of Regulations on air quality [1]. Many are working collaboratively within county-authority-based groupings, but the Sussex Air Quality Steering Group has formalised this process and has provided centralised resources to support air quality work.

Local authorities in Sussex have been working together on air quality issues for many years, since before the introduction of legislation and guidance requiring them to do so. This history of joint working has enabled the Group to compile an improved picture of the current and likely future state of the air environment, building on long-standing relationships both within and between the different agencies. Such a joint project is seldom achieved without some difficulties and tensions, and the chapter will also examine the operation of such a body and some of the difficulties encountered.

The guidance issued to local authorities at the time [2] recognised that not all local authorities would need to declare air quality management areas, but that they all had a role to play in reducing the levels of air pollutants. The development of non-statutory air quality strategies was encouraged, even in those areas declared as Air Quality Management Areas. The guidance further noted that:

> *"The Government and the National Assembly for Wales believe that air quality strategies can help local authorities to pursue their air quality policies in an integrated way".*

It was intended that these air quality strategies would not be pursued by local authorities in isolation, but that the support of local industry, community groups and businesses would be integrated.

The guidance also contained advice on work within local authorities (through the involvement of different departments/functions) and for the availability of clear and accessible information for the public. The process of review and assessment was seen as leading to greater regional co-operation:

> *"Periodic review and assessment of air quality should lead local authorities to take a strategic approach to cross-boundary issues".*

Within Sussex, as will be shown later in this chapter, the presence of the SAQSG led to a more consistent and strategic approach to the review and assessment process.

1.1 Composition of the Group

The composition of the Group has been fairly consistent since its inception, with representatives from local government, the Environment Agency, the Health Authorities and the universities. The Group is currently made up of:

- District / Unitary Local Authorities (12)
- County Councils (2)
- The Environment Agency
- Health Authorities (2)
- Universities (2)

It is interesting to note that the Group covers two county council areas. As well as enabling more effective joint working, this has allowed the Group to draw in more resources and provide an improved overview of air quality across Sussex. The Environment Agency is involved both at a regional (Southern) and local (Sussex Area Team) level. Again, this allows for the provision of strategic level resources (such as advice form the regional centre on dispersion modelling) with local-level knowledge from process inspectors.

The majority of local authority Group members are environmental health officers, with the County Councils being represented by transport and land use planners and environment team members. Regional Health Authorities across England and Wales are currently undergoing re-organisation, but they are currently represented on the Group by the Consultants in Communicable Disease Control (CCDCs). CCDCs have experience in a wide range of public health issues of which air quality (and indeed all forms of pollution) is just one.

All partners except the universities contribute financially towards the running of the Group and the purchase and operation of a number of centrally provided tools (such as air quality monitoring network software and dispersion models, described in more detail below). The Group collects contributions annually and one local authority manages the central fund.

There is one Sussex local authority (Mid-Sussex District Council in West Sussex) that has not been part of the Group for a number of years. Its officers have had some liaison with the Group, but have principally carried out their air quality duties alone. SAQSG has actively sought their participation in the Group, to complete the picture across Sussex, and they are currently considering their position.

1.2 Aims and objectives

The current aims of the Group are to:

- Improve air quality in Sussex by supporting local authorities;
- To assist in implementing the National Air Quality Strategy;
- To encourage and facilitate an Air Quality Strategy for Sussex

These aims are to be achieved through;

- Collaborative and co-ordinated effort amongst all organisations
- Collecting, interpreting and disseminating information on air quality in Sussex.

These aims and objectives are laid out in a three year business plan for the Group that was finalised in 2001 (described in further detail below). The aims recognise the importance of the National Air Quality Strategy [3], but also allow for some flexibility in the consideration of issues outside of the National Strategy remit, where it considers these appropriate. As identified above, central government has promoted the development of local air quality strategies in centrally issued guidance, and the Group is keen to adapt this approach to the particular circumstances in Sussex.

The delivery mechanisms for this work are the tools and data which the Group has available and their future development. Co-ordination and collaboration are seen as key, to avoid the duplication of work and to provide a consistent framework between authorities to allow a better strategic overview to develop.

2 The formation of the Group

Air quality monitoring had been carried out for a number of years prior to the formation of the Sussex Air Quality Steering Group (SAQSG). Historically this has consisted of measurements of smoke and sulfur dioxide and lead, carried out using well-established techniques. Records on the national air quality archive [4], show that at the end of the 1960s There were 12 smoke and sulfur dioxide monitoring sites across East and West Sussex.

With the establishment of the national nitrogen dioxide diffusion tube survey in 1993 [4] the focus of monitoring efforts nationally had switched away from industrial pollution to examine road traffic emissions.

In parallel with these developments, a sub-group of the local Chartered Institute of Environmental Health Sussex Branch Pollution Working Group had begun to meet specifically to discuss air quality issues. Much of the impetus for this group had come from developments at central government level and the promise of new legislation that would give new powers to local authorities and introduce the concept of local air quality management into the UK.

This working group commissioned a study to determine the costs involved in establishing an automated air pollution monitoring network for Sussex. At that time, all requests for additional funding were considered by the Chief Environmental Health Officers, and at the time, they decided not to proceed with the establishment of a network.

It was at about this time that the SAQSG was officially formed. The Group had a clear set of aims and objectives and worked to a tight remit based around the implications of the Environment Act 1995 and supporting legislation.

After officially forming, one of the first tasks undertaken by the group was to report on air quality across Sussex to set a baseline picture for the constituent agencies. The document that arose "Air Quality in Sussex and Options for the Future" covered a range of subjects, from sources of air pollutants and their effects on health, to the concentrations of various pollutants in the atmosphere at a number of monitoring sites.

3 Emissions inventory

The real focus for the Group was its first joint project, the production of an emissions inventory for East and West Sussex. The Group commissioned consultants CES to prepare an atmospheric emissions inventory covering East and West Sussex, and detailing sources of a wide range of pollutants. This inventory contained details of emissions to air from:

- Road Traffic (major roads)
- Regulated Industrial Processes (Part A and B operations)
- Domestic Properties (gas usage)
- Commercial Premises (gas usage)
- Air and Sea ports

For the following pollutants:

- Oxides of Nitrogen
- Sulfur Dioxide
- Carbon Monoxide
- Particulate Matter (Total, PM10 and PM2.5)
- Volatile Organic Compounds (including benzene and 1,3-butadiene)

The project was the first UK use of the CES UK Regional Air Pollution Inventory Database (UKRAPID) [5]. Data to populate the database was a mix of local-generated information (such as that covering regulated industrial processes and traffic flows) with additional national datasets (standard road vehicle emission factors and area gas usage data). The system was coded in Borland Paradox and provided a simple text-driven query screen with which to access the data (see Fig. 1). The resulting inventory was supplied to each individual member of the Group to allow them to make local emission calculations and to permit updating of the information within the central UKRAPID database.

The inventory facilitated an investigation of the source attribution of air pollution in Sussex (see Fig. 2). The inventory highlighted the prominence of the motor vehicle as a source of air pollution in Sussex, and also the importance of operations at Gatwick airport, shown as "Other" in Fig. 2.

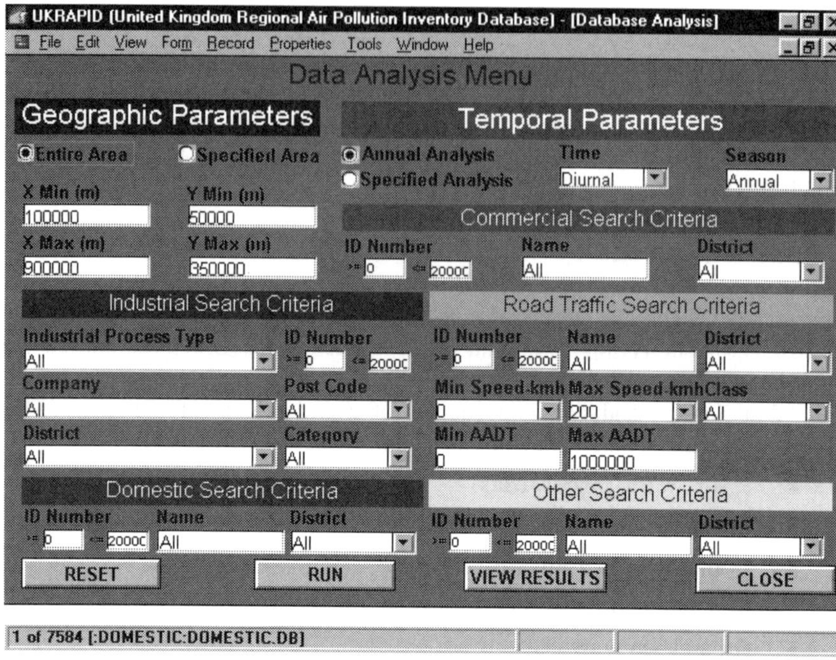

Figure 1: UKRAPID emissions inventory data query screen.

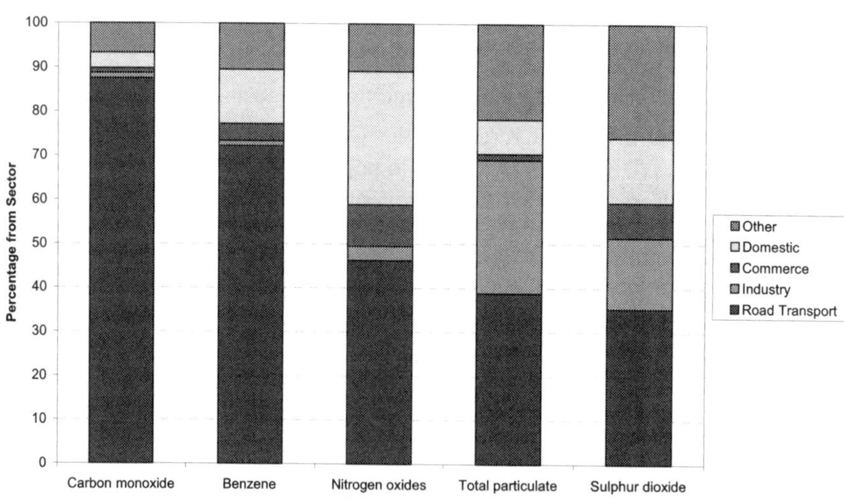

Figure 2: Percentage contributions of different sectors to air pollutant emissions in Sussex.

Subsequently, the UKRAPID system was linked to a Geographical Information System (GIS), the ESRI ArcView package. This permitted more detailed spatial querying of the data and display of the data in map format (see Fig. 3).

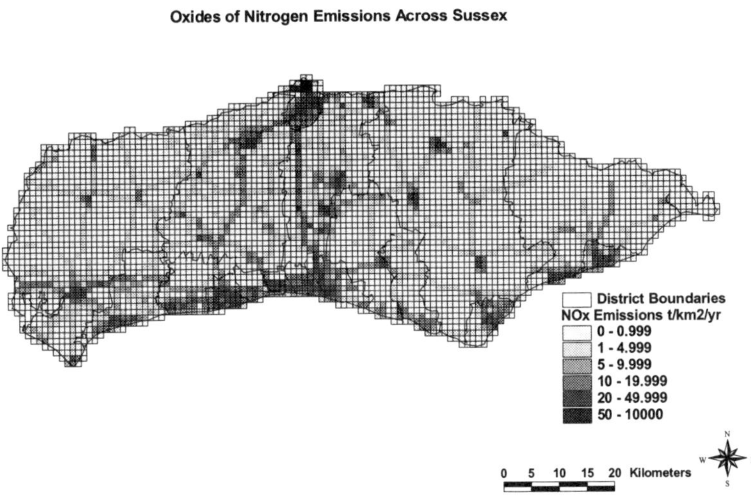

Figure 3: Map showing emissions of NO_x across Sussex in kilometre squares.

The map highlights town centres (such as Crawley, Brighton and Horsham) and major transport links, such as the A23 and A27 as sources of nitrogen dioxide, and this is borne out by the measurements made there.

4 Project Development Officer

Following on from the emissions inventory, the next step was to employ a full time member of staff to support the work of the group. The Group successfully appointed a Project Development Officer in early 1998. The post was designed to fulfil three linked roles:

- To assist local authorities in carrying out their duties under the Environment Act 1995 and in delivering completed air quality reviews and assessments;
- To link the UKRAPID system to a GIS and to keep the emissions inventory information up to date;
- To increase knowledge of air quality issues across Sussex and to raise the profile of the Group

Upon starting, the Project Development Officer began almost immediately with the local air quality management process

5 Local air quality management in Sussex

The Environment Act 1995 [6] introduced the legislative framework for local air quality management in the United Kingdom. It required the Secretary of State to produce a National Air Quality Strategy. This was originally produced in 1997 and established the Government's agenda for air quality management in the future. The Strategy was revised in 2000 (The Air Quality Strategy for England, Scotland, Wales and Northern Ireland) [3]. The Strategy set out clear objectives for improvements in air quality in the United Kingdom and stated how far national policies and measures would go towards meeting these. Alongside the Strategy, a set of Air Quality Regulations [7] came into force, requiring local authorities to review and assess air quality in their own areas, and to take action where the assessments showed that prescribed objectives for each pollutant would not be met.

The review and assessment of air quality forms the first stage of the air quality management process and has two principal objectives:

- To identify local areas where the air quality objectives will not be met
- To ensure that air quality considerations are integrated into local authorities' decision making processes, such as land use planning and traffic management

The UKRAPID emissions inventory was able to provide much of the information needed in the first screening stage of local air quality reviews. This included details of all regulated industrial processes within local authority areas, details of traffic flows on major and minor roads. In addition, some centrally available information was acquired and processed using GIS (on background pollutant concentrations) as well as the results of monitoring programmes (particularly for NO_x, one of the pollutants of most concern).

All Sussex local authorities were required to proceed to a second stage review and assessment. At this stage, screening models were utilised to screen out those areas not likely to be of concern and to identify those areas where further, more detailed modelling would be required. The Group made use of the centrally available Design Manual for Roads and Bridges (DMRB) and Guidance on Stationary Sources (Environment Agency GSS) models where possible [8]. The emissions inventory had shown that traffic flows across East and West Sussex were not particularly high, with a maximum of approximately 68,000 vehicles per day based on an annual average daily traffic flow (AADT) on the M23 close to Crawley and Gatwick Airport. Other main roads in the two counties include the A27, A259, A21, A264 and A24.

There were very few Environment Agency Part A authorisations in East and West Sussex (the largest types of industrial process). In addition to these, a number of part B processes were also screened using the same tools. These included roadstone coating operations, metal smelters, brickworks and scrapyards.

Some authorities chose to carry out advanced dispersion modelling during Stage 2, while others opted to move to a separate Stage 3 report if the screening models indicated any likelihood of exceedence. The last completed stage of review and assessment is not therefore a suitable indicator of pollution burden, as other considerations were brought into play.

For the final third stage reviews, advanced dispersion modelling was carried out centrally using Trinity Consultants Breeze ROADS and Breeze AERMOD models [9]. These were run using meteorological data also provided by Trinity and terrain data from standard Ordnance Survey products where appropriate. For road traffic sources, this modelling was carried out solely for oxides of nitrogen and PM_{10}, whilst for industrial processes, Breeze AERMOD modelling was run for sources of sulfur dioxide, oxides of nitrogen, volatile organic matter and PM_{10}.

The outcome of the process was that no Air Quality Management Areas were declared in either East or West Sussex. The outcomes of the review and assessment process for all local authorities were subject to external audit and review. With the clarification of some matters, the external assessors agreed with the determined outcomes.

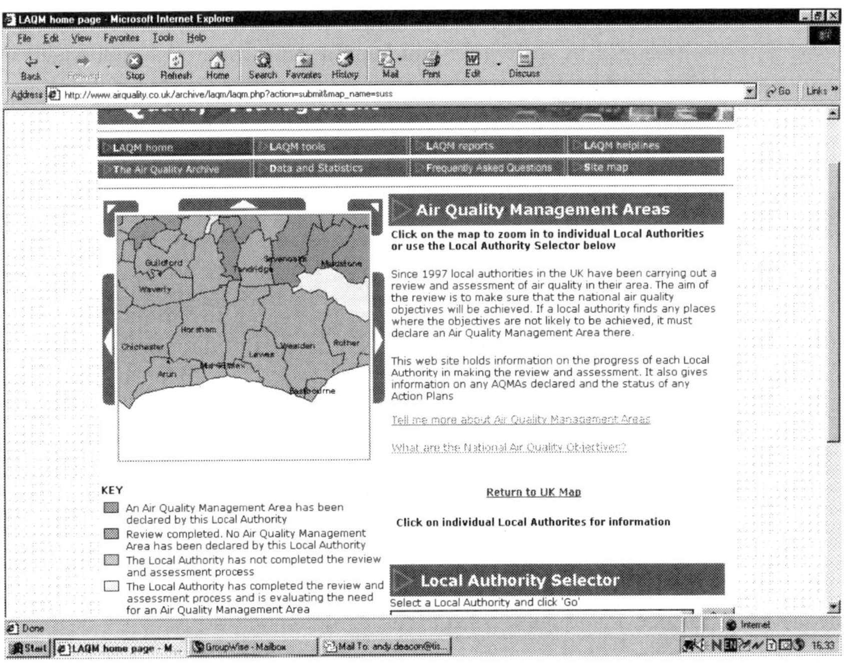

Figure 4: Air Quality Management Outcomes in Sussex (taken from AEA Technology, NETCEN website).

These conclusions seem to broadly fit with the pattern across the South East region. Parts of Surrey (M3 and Heathrow and Gatwick) and Kent (M2 and Thames Gateway) have declared Air Quality Management Areas and there is one AQMA in Hampshire, also on the M3.

It is worth noting that there are a number of areas in Sussex close to exceeding the objectives for PM_{10} and NO_2. Nationally, the review and assessment exercise is to be repeated before the end of 2003. SAQSG are keen to ensure that those areas that are close to the objectives do not become exceedence areas

One of the options recommended by central government for both those areas declaring AQMAs and those who do not declare is the preparation of a local air quality strategy. It is the intention of the Sussex local authorities to publish such a strategy. The strategy would bring together a number of elements having a bearing on local air quality, but which fall outside the remit of the local air quality management regime. These might include ozone pollution, examining the implications of waste management and disposal and better integration between air quality management and the land use planning regimes. This is discussed in further detail below.

The Sussex Air Quality Strategy needs to ensure that:

- The available information is exploited to its utmost and future effort is directed to best effect;
- The emerging Sussex air quality strategy is not only owned and influenced by all who might benefit from it but also embedded in their activities;
- Needs for future work are rigorously defined and relative priorities established.

6 Air quality monitoring in Sussex

Air quality monitoring activities in Sussex have traditionally been focussed on the monitoring of nitrogen dioxide, in most cases using diffusion tube samplers. This monitoring has been carried out since 1993 as part of the national nitrogen dioxide diffusion tube survey. In addition to this, a number of local authorities have operated automated stations for considerable periods of time. Notable amongst these is Brighton and Hove City Council, which operate sites in both Brighton and Hove as part of the national air quality monitoring network (reporting data to CEEFAX and other central information sources). East Sussex is also home to the Lullington Heath rural monitoring station, which has recorded the highest levels of ozone in the country in recent years. Ozone did not form part of the Local Air Quality Management regime, but these results have been of significance and are discussed in further detail below.

Kerbside (within 1 m of a major road) NO_2 results for 2000 across Sussex are shown in Fig. 5. This shows areas currently in excess of Government objectives for 2005. Modelling carried out as part of the review and assessment process has, however, predicted declines in the coming years to bring these areas into compliance. All intermediate and background sites in Sussex are already in compliance with the objectives.

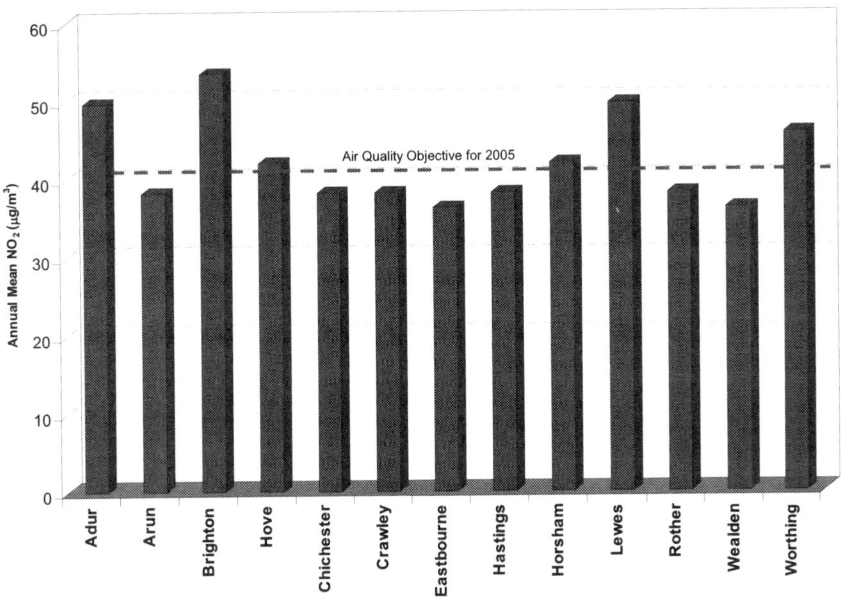

Figure 5: Annual mean kerbside NO_2 across Sussex (2000).

Local authorities in Sussex recognised the need for quality control of this data. The Group carried out work on NO_2 tube quality control early in this programme of measurements and participate in the ongoing annual interlaboratory and field trials for these measurements. The Group also made use of the London air quality manual [10], covering automated equipment at the time of its release.

In parallel with the local air quality management process, a number of local authorities were installing additional air quality monitoring equipment. In 1998, the Group was also successful in winning EC INTERREG II financial support for two projects. One, Air Rives Manche Sensibilisation, sought to improve the number of monitoring stations for the pollutants ozone and PM_{10} in East Sussex and partner regions in France (Somme and Seine-Maritime), as well as translating key documents from English to French and vice versa and raising the profile of air quality as an issue and encouraging everyone to do their bit to improve air quality.

The second, Air Rives Manche Santé, compared patterns of drug prescribing for respiratory conditions on both sides of the Channel as well as quantifying hospital admissions and mortality. This was done for both all-cause mortality and that associated specifically with respiratory conditions. This work provided the first local estimates of the impacts of air pollution on health. The research estimated that air pollution due to ozone was likely to cause 360 hospital admissions in East Sussex, Brighton and Hove Health Authority each year as a result of respiratory disease, and is a factor in 340 deaths a year. PM_{10} was estimated as likely to cause 120 respiratory disease hospital admissions in East Sussex, Brighton and Hove Health Authority each year, and was a factor in 125 deaths a year. These preliminary estimates will be refined as more air quality monitoring information becomes available.

The Air Rives Manche Sensibilisation project provided the impetus for the formation of a Sussex-wide air quality monitoring network for the first time. This was brought about through the purchase and installation of appropriate remote communications software, which enabled each monitoring station to be dialled up and data collected on a daily basis.

The project also funded the purchase and installation of three new PM_{10} monitoring sites in Lewes, Eastbourne and Rother in East Sussex and one new ozone monitoring site in Rother. This was followed by additional monitoring stations being installed in Chichester (NO_x and PM_{10}) in West Sussex. In addition to the fixed point monitoring sites, East and West Sussex County Councils and Horsham District Council operate mobile monitoring facilities. Three authorities (Adur DC, Hastings BC and Brighton and Hove) operate portable GRIMM dust monitors.

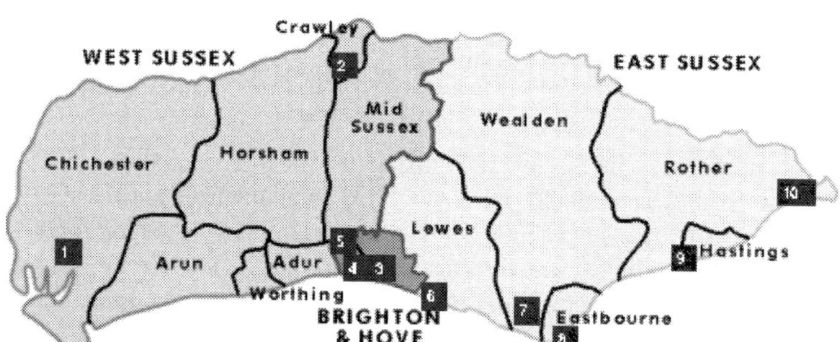

Pollutant	1	2	3	4	5	6	7	8	9	10
Site Type	I	UB	R	R	RUR	R	RUR	UB	R	RUR
Benzene										
1,3-Butadiene										
Carbon Monoxide			✓	✓						
Lead										
Nitrogen Dioxide	✓		✓	✓			✓			
Sulphur Dioxide							✓			
Ozone			✓	✓	✓		✓			✓
Particles (PM_{10})	✓	✓				✓		✓	✓	

Figure 6: Automated air pollution monitoring network in Sussex (March 2002).

At the beginning of 2002, the Group began publishing daily updates of their air quality monitoring data on the internet (www.sussex-air.net). The location of the automated air pollution monitoring sites in Sussex is shown in Fig. 6.

Alongside the automated monitoring, the Group has continued with the passive monitoring of air quality using diffusion tubes. Most authorities operate additional sites to those contributing data to the national nitrogen dioxide diffusion tube survey, with additional tubes provided by East Sussex County Council, who are not part of this survey.

A summary of observations from Sussex air quality monitoring activities is shown in Table 1.

Table 1: Observations from air quality monitoring in Sussex.

Pollutant	**Exceedences/comments from monitoring**
Benzene	Roadside and industrial monitoring (using diffusion tubes) in a number of areas shows no exceedences.
1,3-Butadiene	No monitoring of this pollutant, as no major point sources present.
Carbon monoxide	Roadside monitoring in Brighton and Hove shows no exceedences
Lead	No monitoring currently. Historically, levels dropped below standards with the introduction of unleaded fuel.
Nitrogen dioxide	Kerbside diffusion tube and automated monitoring shows exceedences currently at the roadside in a number of built up areas (e.g. Hastings, Eastbourne, Worthing, Chichester, Lewes, Crawley, Horsham). Levels expected to decline with improving vehicle and fuel quality.
Sulfur dioxide	Continuous monitoring in rural location shows no exceedences. Very few and diffuse sources of sulfur dioxide across Sussex.
Particles (PM_{10})	Monitoring in a number of locations at roadside and background sites. Levels in 2001/2002 in excess of objectives in Hastings/Bexhill.

Of particular concern are the high PM_{10} concentrations observed in Hastings, East Sussex. Results from June 2001 (the start of monitoring in Hastings) to the end of February 2002 are shown in Fig. 7. These show that levels of PM_{10} are much higher at the Hastings roadside site than at the roadside site in Lewes or the other PM_{10} monitoring site, a background station in Eastbourne.

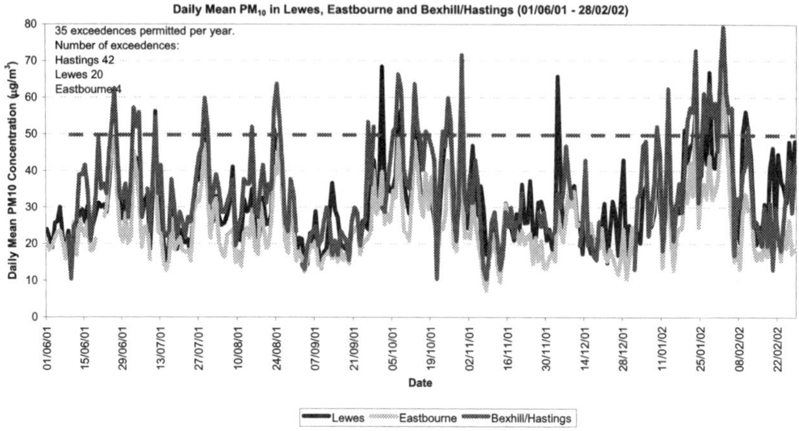

Figure 7: Daily mean PM_{10} levels in East Sussex June 2001–February 2002.

6.1 Headline air quality indicator

Air quality was chosen by the Government as one of its headline sustainable development indicators [11], the aim being to ensure that air quality continues to improve in the long term. Monitoring from the National Automatic Urban and Rural Network (AURN) is used to show progress towards meeting sustainable development objectives. The indicator includes ozone, particles (PM_{10}), sulfur dioxide, nitrogen dioxide and carbon monoxide readings. Table 2 takes an average of values from 1998 and 1999 to avoid the large inter-annual variability in ozone concentrations that are caused by changes in weather patterns from year to year and shows the number of days with moderate or higher pollution in the south-east.

Table 2: Number of days of moderate or higher pollution in the south east (1998/99)

Site	Type of Site	Number of days with moderate or higher air pollution
Southampton Centre	Urban centre	22
Reading 1	Urban b/grd	55
Harwell	Rural	40
Lullington Heath	Rural	65
Rochester	Rural	55

The rural Lullington Heath (in East Sussex) shows the highest air pollutant levels. This is due to the high ozone concentrations measured at the site, which is in the area of the country that is worst affected by ozone pollution.

A chart showing the number of days on which ozone levels were moderate or higher for Lullington Heath over the past 12 years is shown in Fig. 8.

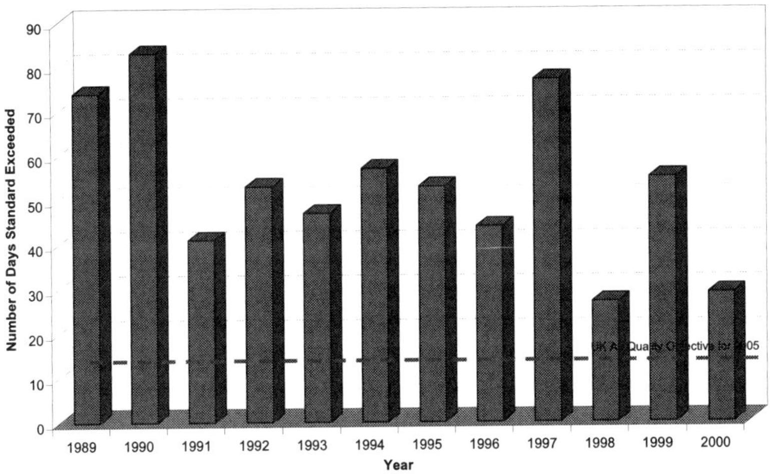

Figure 8: Ozone exceedence days at Lullington Heath 1989–2000.

This chart emphasises the interannual variability in ozone concentrations and shows that Government targets for the year 2005 are unlikely to be met in Sussex.

The UK government is currently making arrangements for national implementation of the EC Air Quality Daughter Directive covering ozone [12] and for the National Emission Ceilings Directive (NECD) [13] that is seeking to limit total national emissions of a number of precursors of ozone and other pollutants. Effective action to control the precursors of ozone will be needed if further progress is to be made towards meeting these limits.

7 Information dissemination

In order to raise the profile of the Group and to increase public awareness of air quality issues, the SAQSG has a continuing role in the dissemination of air quality information to all interested parties, and a number of campaigns have been pursued to this end.

Every year (usually during the national Don't Choke Britain campaign), local authorities officers participate in roadside "stop and test" sessions to check on-road exhaust emissions for cars. These sessions are organised through the auspices of the CIEH Sussex Branch Pollution Working Group and are conducted in conjunction with the Police and the Vehicle Inspectorate. In 1998 1352 vehicles were tested, with 64 deferred prohibition notices being served (meaning that drivers have to take an MoT emissions retest in a garage after having their car retuned and then present the results to the police). As an awareness-raising measure, all drivers received the results of their test together with advice on reducing pollution from their vehicles. Several authorities have also run additional voluntary roadside emission testing days.

Since early 2001 the Group has run a website www.sussex-air.net giving information about the Group and its activities as well as further information about monitored air pollutant levels and the results of the review and assessment exercise. From March 2002 the site has carried regularly updated air pollutant monitoring information for each of the automated monitoring sites.

The Group also makes air pollution forecasts for Sussex available through its website. These are intended to provide information for those individuals who may have a medical condition influenced by air quality and to influence travel choices. The forecast site contains links to public transport information to allow drivers to alter their transport choices where sufficient warning is given. A sample forecast is shown in Fig. 9.

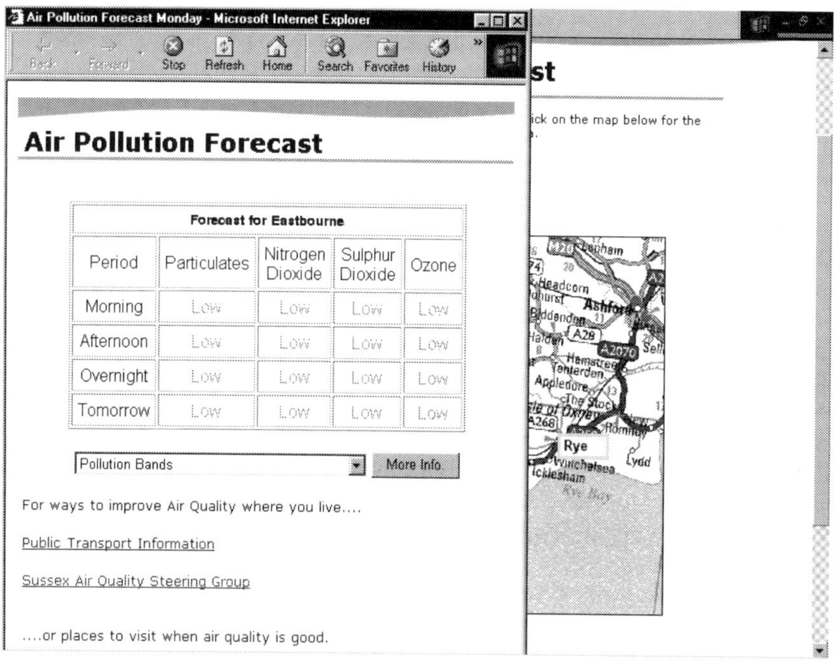

Figure 9: Sample air pollution forecast for Sussex using the Met Office system.

The Group has participated in the national Travelwise scheme (aiming to raise awareness about the pollution impacts of transport) across East and West Sussex and produced posters during 2001. Through participation in the EC INTERREG II Air Rives-Manche Sensibilisation project, a mobile display on air quality in Sussex was produced and is available for use at public events. Promotional literature, both about the project and about air quality and its effects on health has also been produced.

As an awareness-raising and pollution-reduction measure, 'Cut Engine - Cut Pollution' signs have been erected at various sites in West Sussex. A sign in Chichester is shown in Fig. 10. In a pilot study, traffic counts, crossing gate sate, numbers of switched off vehicles and pollution information were all measured over a period of several months before and after signs were erected. The study, between March and November 1998 found that 7.6% of car drivers turned engines off before the signs were erected: 17% did so 6 months after the signs were erected. Carbon monoxide levels fell by up to 17% and NO_2 levels by 15% during the morning peak hours at one crossing.

Figure 10: Cut engine, cut pollution sign at a railway crossing in Chichester.

8 Planning and air quality

In order to protect those areas where air quality is currently good and work to improve those where it is not, planning has a major role to play in air quality management. This applies both to considerations on individual developments and to the development of structure and associated Local Plans. Planning does not offer a "quick-fix" to air quality problems, but can act over a number of years to change the existing situation. Recognising this, SAQSG members have sought to ensure that policies on air quality are in place in the relevant planning instruments.

There is regional planning guidance [14] in place under which county councils, district and unitary authorities implement a set of structure and local plans. These plans contain policies designed to protect and improve air quality, and the SAQSG has reviewed their operation and commented (where timescales have permitted) and contributed to their development. There are separate plans (implemented at county level) governing waste management, which require environmental impact assessment (including air quality assessments) for any new waste management facilities. There is also a separate planning agreement (a section 106) agreement between Crawley Borough Council (as the planning authority) and BAA Gatwick. Of particular interest are the policies contained in the Local Transport Plans.

8.1 Local Transport Plans

In preparing their Local Transport Plans the County Councils and Brighton and Hove City Council were required to examine the contribution that transport improvements could make to improved air quality.

The East Sussex plan has two main targets for air quality:

- To achieve the UK air quality strategy targets by the end of 2003 (2 years early).
- Transport-related emissions of air pollutants should be no higher in 2006 than they were in 1998

The West Sussex LTP also has a series of targets and indicators for air quality:

- To Meet National Air Quality Targets across West Sussex and implement a local air quality strategy during the life of the LTP;
- To supply district and borough councils with all necessary information for on-going air quality review and assessment;
- To provide "cut engine cut pollution" signs at up to 7 railway crossings per year;
- The number of Air Quality Management Areas;
- The percentage of bus vehicle stock running on low-emission fuel.

These challenging targets have been supplemented by additional monitoring of both air quality and traffic, with annual updates on progress being submitted to central government.

9 Benefits of collaborative working

The collaboration between all Members of SAQSG has been highly productive and has delivered a number of high profile results. Continued collaboration will make it easier to maintain the network of air quality monitoring and the emissions inventory that have been established throughout Sussex under the auspices of SAQSG. Further collaboration will also facilitate production of the 2003 round of air quality reviews and assessments and ensure a consistent approach.

Predicted ambient air quality at a number of places in the County was close to statutory air quality objectives and it is essential to keep these locations under close scrutiny in advance of the next round of statutory reviews in 2003 and to consider new areas of possible exceedence.

A single point of contact on air quality in Sussex simplifies communication with similar collaborative groups in Kent, Hampshire, Surrey, France and more widely. Such communication will become increasingly important in dealing with transboundary air pollution issues, especially on ozone and PM_{10} as identified above. The grouping together of local authorities and other agencies facilitates air quality inputs to Regional planning and sustainability initiatives under the auspices of South East England Development Agency (SEEDA), South East England Regional Assembly (SEERA) and the Government Office for the South East (GOSE) and provides air quality input into Structure, Local, Waste and Minerals and Local Transport Plans.

10 An air quality forum?

For some time, the Group has been concerned to gain wider political support for its continued operation and actions. The Group holds annual seminars with elected council representatives and senior local government officers and at the most recent of these (in March 2002), there was a widely held view that a senior officers/members forum should be established across Sussex to consider air quality within a strategic framework.

It has been suggested that this forum meet at least annually to scrutinise SAQSG work, challenge and channel it where necessary and ensure ever more strategic application of results.

11 Future work

As part of its business planning activities, the group has identified work priorities for the coming years. For the period 2003–2006 they are:

- Developing and extending the picture of air quality in Sussex obtained over recent years.
- Developing links with other groups in the South-east of England and the central coastal belt of France, which would enable better management of air pollutants that cross County boundaries.

- Exploiting the air quality/pollutant emission knowledge base by promoting people's understanding of air quality issues in Sussex relevant to their interests and values. To do this it will be necessary to identify user requirements and to make available appropriate information in formats that meets the requirements of key stakeholders in Sussex.

To realise these priorities work will be needed in three main areas:

- Network and data management for air quality monitoring in Sussex.
- Improvement of the emissions inventory and modelling capabilities.
- Public information.

12 Conclusions

The picture of air quality in Sussex that has been compiled as a consequence of work in SAQSG is a valuable asset, which could be exploited more fully to the benefit of the people of Sussex. If it is kept up to date it should allow proper consideration of air quality issues within the framework of local economic and social management and development. For this to occur it will be necessary to tailor information to need and to assess the air quality implications of management and development. Such a process would lead to a non statutory air quality strategy for the County that could seek to ensure that there is no inadvertent degradation of air quality in Sussex and possibly encompass pollutants of concern in Sussex that are not presently included in the National Air Quality Strategy, such as dioxins and heavy metals.

There is now more knowledge than ever before about air quality in Sussex. The review and assessment exercise and the increasing amount of pollution monitoring have led to increased understanding of the patterns and processes influencing air quality across Sussex. The situation at present is that there are some areas currently in excess of recommended Government objectives, but these apply to future years and it is expected that concentrations of all pollutants will decrease in the coming years as a result of national and international policy measures.

What is clear is that there are a number of areas close to the objectives for PM_{10} and nitrogen dioxide. These are also the pollutants for which the most monitoring has been carried out, so the Group will be able to most closely assess progress towards meeting the objectives for these two pollutants. What this does mean is that the air quality review and assessment documents should be considered when looking at developments that may influence traffic and air quality in the affected areas. Air quality is a material consideration in the planning process and authorities will need to carefully consider what needs to be done in individual cases to ensure that the health of the population is not threatened.

Within London, such considerations have been progressed further with the issuing of additional technical guidance from the Association of London Government [15] on air quality and land use planning matters to all planning authorities. These seek to ensure not only that proper assessments of individual developments are carried out, but also that appropriate remedial measures can be taken where needed.

It is this interface between planning, transport planning and air quality that in the longer term will deliver improvements in quality of life. Whilst traditionally, air quality assessment has been an environmental health function, environmental health officers have very few powers when it comes to implementing policies to deliver improvements in air quality. A successful inter-agency partnership has been demonstrated by the Sussex Air Quality Steering Group, and it is continued work of this nature that will be needed to deliver clean air for all in Sussex. Local Authorities also need to ensure that their own house is in order, through effective fleet management, consideration of the use of alternative fuels for vehicles and policies to promote alternatives to private car use.

The forthcoming second round of the air quality review and assessment exercise will further enhance current understanding of the processes affecting air quality in Sussex and will serve as a measure of the Group's success in delivering clean air for Sussex.

Acknowledgements

Andy Deacon would like to thank all the members of the SAQSG and their supporting organisations, without whom this work would not have been possible. In particular, large debts of thanks are owed to Tim Bartlett (Group Chairman) and Andy Bryce (Vice Chair for a number of years) for their tireless effort to keep the group working. Andy Deacon's research work is supported by the School of the Environment at the University of Brighton.

References

[1] *The Air Quality Regulations 1997*, Statutory Instrument 1997 No. 3043.

[2] DETR, *Developing Local Air Quality Action Plans and Strategies: The Main Considerations*, DETR, London, 2000.

[3] DETR, *The Air Quality Strategy for England, Scotland, Wales and Northern Ireland - Working Together for Clean Air*, Cm 4548, 2000.

[4] The National Air Quality Archive – http://www.airquality.co.uk

[5] CES Ltd., *Atmospheric Emissions Inventory for East and West Sussex, Final Report*, CES Ltd, 1997.

[6] Environment Act 1995 (c. 25).

[7] Statutory Instrument No. 928, *The Air Quality (England) Regulations 2000.*

[8] DETR, *Review and Assessment: Pollutant Specific Guidance*, LAQM.TG4 00, 2000.

[9] http://www.breeze-software.com/

[10] SEIPH, *London Air Quality Manual*, SEIPH, London, 199x.

[11] DETR, *A Better Quality of Life: A Strategy for Sustainable Development for the UK* (Cm 4345), 1999.

[12] Directive 2002/3/EC of the European Parliament and of the Council on target values for ozone.
[13] Directive 2001/81/EC of the European Parliament and of the Council on national emission ceilings for certain atmospheric pollutants.
[14] Association of London Government Transport and Environment Committee, *Air Quality Assessments for Planning applications - Technical Guidance Note*, ALG, London, 2001.
[15] Government Office for the South East, Regional Planning Guidance for the South East RPG9, GOSE, Guildford, 2001.